The Fiber Optic LAN Handbook

*By the
Engineering Staff
of
Codenoll Technology
Corporation*

The Fiber Optic LAN Handbook—THIRD EDITION

Prepared by the Engineering Staff of Codenoll Technology Corporation

Copyright © Codenoll Technology Corporation, 1990.
All rights reserved.
Codenoll Technology Corporation
1086 North Broadway
Yonkers, New York 10701 U.S.A.
914-965-6300

Part Number: 05-0050-00-0318

Price: $17.95

Published April, 1990
Second Edition in October, 1990
Third Edition in January, 1991

ISBN 0-9626933-3-2

Printed in the U.S.A.

PREFACE

The past decade has seen many major developments in communications technology. Paramount among these have been the increasing use of glass fiber—rather than copper conductors—to transmit information and the networking of computing devices. Optical fibers offer virtually unlimited bandwidth for data communication. Their low attenuation makes possible long transmission spans without repeaters. They are unaffected by electrical noise and ground loops and, being lighter, thinner and more flexible than copper, they are easier to install. Computer networking has built upon the advances in semiconductor technology which allows more and more computing power to reside in smaller, low cost machines. Networking then allows sharing of data bases and expensive peripherals, such as printers, between a large number of individual workstation computers.

At Codenoll Technology Corporation, we have had the exciting experience of working at the intersection of these two major technologies—fiber optics and local area networks (LANs). The perspective thus gained is probably unique. Most of the papers in this volume have been contributed by members of Codenoll Technology Corporation's Engineering Staff. However, to give a more fully rounded picture of fiber optic LAN technology, we have also included contributions from colleagues at Hoechst AG, Packard Electric Division of The General Motors Corporation and Southwestern Bell Telephone Company. We would particularly like to acknowledge the contributions of these individuals.

Michael H. Coden
President,
Codenoll Technology Corporation

CONTRIBUTORS

CODENOLL TECHNOLOGY CORPORATION

Peter G. Abbott
Stephen J. Anderson
John Cocomello
Michael H. Coden
Dr. Bulusu V. Dutt
Richard J. Focht
Dr. Walter B. Hatfield
Jan H. Helbers
Robert W. Kilgore
Richard H. Lefkowitz
Donald W. Maley
Ernest M. Raasch
James H. Racette
Dr. Frederick W. Scholl
Timothy E. Zack

HOECHST AG

Dr. Ulrich von Alpen

KABELWERKE REINSHAGEN GmbH

H. J. Schmitt

PACKARD ELECTRIC DIVISION, GENERAL MOTORS CORPORATION

Laura K. DiLiello
Mark A. Lynn
Dominic A. Messuri
Gregory D. Miller
Robert E. Steele

SOUTHWESTERN BELL TELEPHONE COMPANY

David E. Stein

CONTENTS

The Fiber Optic LAN Handbook

Introduction   p. *xv*

I
Overview of Fiber Optics and LAN Technology   p. 33

II
Fiber Optic Components   p. 113

III
Fiber Optic Data Links   p. 169

IV
Fiber Optic Local Area Networks   p. 177

V
Plastic Optical Fiber   p. 297

VI
Glossary of Terms   p. 343

VII
Fiber Optic Ethernet Transceivers and PC Adapter Cards   p. 413

VIII
Plastic Optical Fiber Ethernet Adapter Cards and Accessories   p. 427

IX
Fiber Optic Ethernet Hubs   p. 441

X
FDDI Adapter Cards and Concentrators   p. 457

XI
Fiber Optic Cable Assemblies   p. 463

TABLE OF CONTENTS

I Overview of Fiber Optics and LAN Technology

"Every thing you ever wanted to know about Fiber Optics and LANs, but were afraid to ask!" p. 35
  M. H. Coden

II Fiber Optic Components

High Power Light Emitting Diodes for Fiber Communication, p. 115
  F. W. Scholl, S. J. Anderson, M. H. Coden

Reliability of Components for Use in Fiber Optic LANs, p. 121
  F. W. Scholl, M. H. Coden, S. J. Anderson, B. V. Dutt, J. H. Racette, W. B. Hatfield

Fiber Optic Edge Emitting LED Package for Local Area Network Applications, p. 131
  S. J. Anderson, P. G. Abbott, F. W. Scholl

AlGaInP/GaAs Red Edge-Emitting Diode for Polymer Optical Fiber Applications, p. 141
  B. V. Dutt, J. H. Racette, S. J. Anderson, F. W. Scholl

FDDI Components for Workstation Interconnection, p. 143
  S. J. Anderson, D. V. Bulusu, J. H. Racette, F. W. Scholl, T. E. Zack, P. G. Abbott

Making Fiber Optic Terminations—Step-by-Step (Part 1), p. 151
  J. Cocomello, R. Focht

Making Fiber Optic Terminations—A Compatibility Study & Evaluation (Part 2), p. 155
  J. Cocomello, R. Focht

A Fiber Optic Connection System Designed for Automotive Applications, p. 159
  D. A. Messuri, G. D. Miller, R. E. Steele

III Fiber Optic Data Links

Designing a Fiber Optic Communications System, p. 171
  E. M. Raasch, W. B. Hatfield

IV Fiber Optic Local Area Networks

Implementation of a Fiber Optic Ethernet Local Area Network, p. 179
  M. H. Coden, F. W. Scholl

IV Fiber Optic Local Area Networks (continued)

Reliability of Fiber Optic LANs,   p. 185
M. H. Coden, F. W. Scholl, W. B. Hatfield

Passive Optical Star Systems for Fiber Optic Local Area Networks,   p. 193
F. W. Scholl, M. H. Coden

Development of Fiber Optics for Passenger Car Applications,   p. 205
R. E. Steele, H. J. Schmitt

Fiber Optic Ethernet in the Factory Automation Environment,   p. 215
W. B. Hatfield, F. W. Scholl, M. H. Coden, R. W. Kilgore, J. H. Helbers

Fiber Optic LANs for the Manufacturing Environment,   p. 221
W. B. Hatfield, M. H. Coden, F. W. Scholl

Investment Bank Trades up to Fiber Optic Ethernet,   p. 227
S. J. Anderson, F. W. Scholl

High-Rise Optical Communications Networks,   p. 237
D. E. Stein

A Few Fiber Optics Applications,   p. 243
D. W. Maley

Passive Star Based Optical Network for Automotive Applications,   p. 245
L. K. DiLiello, G. D. Miller, R. E. Steele

Introduction to CSMA/CD Network Design,   p. 259
W. B. Hatfield

A Migration Strategy to FDDI for the Fiber Optic LAN Cable Plant,   p. 279
W. B. Hatfield, M. H. Coden

V Plastic Optical Fiber

Plastic Optical Fibers Take Aim at LANs,   p. 299
F. W. Scholl

Applications of Plastic Optical Fiber to Local Area Networks,   p. 303
F. W. Scholl, M. H. Coden, S. J. Anderson, B. V. Dutt

Communicating in New Ways,   p. 309
U. von Alpen, M. H. Coden

Implementation of a Passive Star Based Fiber Optic Network for Full Vehicle Control,   p. 317
R. H. Lefkowitz, M. H. Coden, F. W. Scholl, U. von Alpen

High Speed Polymer Optical Fiber Networks,   p. 329
D. V. Bulusu, T. E. Zack, F. W. Scholl, M. H. Coden, R. E. Steele, G. D. Miller, M. A. Lynn

VI Glossary of Terms

VII Fiber Optic Ethernet Transceivers and PC Adapter Cards, and Network Management System

CodeNet AllNet Enterprise Network Manager System,　p. 414
CodeNet 8380/8382,　p. 416
CodeNet 8381,　p. 418
CodeNet 8330/8331/8332,　p. 420
CodeNet 8300/8301,　p. 422
CodeNet 8320/8321/8322,　p. 424

VIII Plastic Optical Fiber Ethernet Adapter Cards and Accessories

CodeNet 8681,　p. 428
CodeNet 8601,　p. 430
CodeNet 8621,　p. 432
CodeNet 8631,　p. 434
Plastic Optical Fiber Connector System,　p. 436
Plastic Optical Fiber Tools and Accessories,　p. 438

IX Fiber Optic Ethernet Hubs

CodeStar Optical Star Coupler,　p. 442
CodeNet 4300 MultiStar Multiport Repeater,　p. 444
CodeNet 3311 FOIRL Module,　p. 446
CodeNet 8310/8312 Passive Module,　p. 448
CodeNet 8611 POF Active Star Module,　p. 450
CodeNet 4350 Thin Coax Module,　p. 452
CodeNet 4351 Thick Coax Module,　p. 454

X Fiber Optic FDDI PC Adapter Cards and Concentrators

CodeNet 9540/9543, 9340/9343,　p. 458
CodeNet 9041,　p. 460

XI Fiber Optic Cable Assemblies

CFOC 50/125, 62.5/125, 85/125, 100/140,　p. 464
Codenoll POF–Duplex Cable Assemblies,　p. 466

INTRODUCTION

The use of fiber optics for communications purposes dates back about 20 years. Since, during that time, numerous books and articles have been written about this revolutionary technology, an obvious question might be—why yet another collection of articles on fiber optics?

The answer to this is two-fold. Fiber optics found its first commercial application in the area of telecommunications, and early articles on the subject were written from this perspective. It is only within the past few years that the focus has changed from telecom to datacom applications.

The second reason follows from the first. Codenoll Technology has been actively engaged in the development and manufacture of fiber optic data communications, and particularly Local Area Network (LAN), products for the past ten years. During this time, we have acquired a unique corporate insight into the requirements of fiber optic data communications, both present and future. This insight and experience have been distilled into the articles which comprise this collection.

Codenoll Technology Corporation is a recognized leader in the field of fiber optics. The Company was founded in 1980 for the purpose of developing fiber optic components, subsystems and systems using Gallium Arsenide and Indium Phosphide compound semiconductor technology. To accomplish this, Codenoll built production facilities from wafer fabrication, thin film and hybrid manufacture, to the production of complete systems and networks. The many proprietary products and processes Codenoll has developed have positioned the Company uniquely as the only publicly-held, vertically-integrated manufacturer of fiber optic data communications equipment in the world.

Codenoll introduced the first commercial Fiber Optic Ethernet in September of 1982. Subsequently, Codenoll has developed standards-based fiber optic LANs conforming to the requirements of IEEE 802.3 (CSMA/CD), IEEE 802.4 (Token Bus) and FDDI (Token Ring). With more than 40,000 nodes functioning in networks located throughout the world, Codenoll is today by far the leading manufacturer of fiber optic LANs.

The papers included in this collection are based upon hundreds of man-years of engineering experience and cover topics which this experience has shown to be crucial to the implementation of high performance, high reliability fiber optic LANs. These include the design of high efficiency LED emitters for fiber optic applications, fiber optic LAN design and reliability issues. Also included are applications oriented articles which demonstrate the broad versatility of fiber optic LANs. Finally, there is a series of articles covering Plastic Optical Fiber (POF), a very important topic related to the future evolution of fiber optic LAN technology.

Frederick W. Scholl and Walter B. Hatfield, Editors

Corporate Background
Codenoll Technology Corporation
(914) 965-6300
(914) 965-9811 FAX

November, 1990

Company Profile

Codenoll Technology Corporation was founded in 1980 by Michael H. Coden and Dr. Frederick W. Scholl for the purpose of designing, developing, manufacturing and marketing high-performance fiber optics products and technology. Codenoll has pioneered and continues to lead the development and commercialization of fiber optic and optoelectronic technologies for use in Local Area Networks (LANs), Metropolitan Area Networks (MANs) and Wide Area Networks (WANs).

Today, Codenoll Technology Corporation is the leading provider of fiber optic LAN products for both the government and commercial markets, with thousands of network installations based on Codenoll products worldwide.

Codenoll's LAN products are marketed under the CodeNet product name. Codenoll is a public company, and is traded on the NASDAQ National Market System under the symbol CODN.

Corporate Charter

Codenoll's business is to provide LAN users with high performance, high quality fiber optic interfaces and *associated system (hardware and software) components* at increasingly affordable costs through the development of new fiber optics products based on international standards such as Ethernet, FDDI, SNMP, Token-Ring, MAN (802.6) and MAP.

Market Position

Sales of Codenoll products have increased from $432,000 in 1982, when the company's products entered the end-user market, to $8.5 million in 1989.

Codenoll has been instrumental in the increasing availability and widespread acceptance of fiber optics technology for LANs,-a market that is currently growing at a rate of more than 41% per year (AAGR).

Independent market research indicates that the fiber optic LAN business will have quadrupled during the time period between 1988 and 1992-1993.--In fact, the use of fiber optics in networks is growing faster
than the networking market as a whole.

Although other LAN product vendors have now entered the fiber optic LAN market, Codenoll continues to be the only independent, vertically-integrated supplier of fiber optic products for LANs.

Codenoll has been, and continues to be, an active participant in the development of international

Copyright Codenoll Technology Corporation, 1990
Page 1

standards such as Ethernet (IEEE 802.3/ISO 8802-3), MAP (IEEE 802.4), Token-Ring (IEEE 802.5), Fiber Distributed Data Interface/FDDI (ANSI-X3T9.5) and Metropolitan Area Network (IEEE 802.6).

As the market for fiber optics technology matures, Codenoll intends to continue to lead the network industry in making fiber optics readily available and more easily affordable for all varieties of LANs. Likewise, the company intends to lead the development of new network products that support FDDI.

Operations

The company currently conducts its sales and marketing efforts from its corporate headquarters at 1086 North Broadway Street in Yonkers, New York, and from regional locations throughout the U.S. Regional offices are located in New York, Washington, D.C. and Mountain View, Calif. International sales are handled from the headquarters facility in Yonkers, New York.

Distribution

Codenoll products are marketed through computer/network systems integrators, value-added resellers (VARs), original equipment manufacturers (OEMs), distributors and, sometimes, direct sales. Pioneer Standard and Pioneer Technologies with offices in more than twenty major cities are Codenoll's domestic distributor. The company's international distribution includes active marketing partners throughout Europe, Scandinavia, the Middle East, Asia, and the Pacific Rim, besides its business in North America. Companies such as Groupe Bull, NYNEX, Southwestern Bell, PacBell, 3Com, DEC, Bosch, Infotec, Contel, Johnson Controls, GTE, and TRW sell Codenoll systems and components to their customers.

Codenoll systems are used to connect office networks, campus buildings, financial trading floors, factory automation systems, to backbones and existing network segments, and mission-critical networks in high-security environments.

Uses of Codenoll networks include business, education, government and military organizations throughout the world.

Major Accounts

Of the thousands of Codenoll networks currently in use, primary clients include Boeing, DuPont, Electricite de France, Rods, General Electric, Goldman Sachs, Hoechst AG, Interpublic Group of Advertising Agencies, Lockheed, Motorola, New York Stock Exchange, Paine Webber, Lintas Worldwide, Schlumberger, Shearson Lehman Hutton, Sun Microsystems, the Universities of Chicago, Minnesota, Ohio, Strasbourg and Harvard University, the U.S. Department of Defense, and Volvo.

Primary Products

Since 1980, Codenoll has developed fiber optic products in three areas: 1) Fiber optic LAN products for use with personal computers, minicomputers, mainframes, communications servers and gateways; 2) Point-to-point fiber optic transmission equipment for commercial and military data communications and telecommunications; and 3) Fiber optic transmitters, receivers, LEDs, detectors, and

Copyright Codenoll Technology Corporation, 1990
Page 2

Non-Linear Optical (NOL) polymer and polymer (plastic) optical fiber devices.

The company's CodeNet® Fiber Optic Ethernet products are available in three versions: Standard Commercial, Military Ruggedized and fully Tempest Qualified to NACSIM 5100A standards. The exceptional security characteristics of Fiber Optic Networks are used both for secured government and military applications as well as banking, financial and engineering applications.

Fiber Optic LAN Products

The CodeNet LAN products are designed to provide cost-effective implementations of popular LAN standards on fiber optic cable instead of, or in combination with, coaxial or twisted pair copper wire.

The advantages of CodeNet fiber optics include greater reliability, support for longer distances, immunity to electric interference and the ability to upgrade security, speed and the number of network users. Codenoll manufacturers both active and passive fiber optic Ethernet hubs, in addition to dual and single ring interface and concentrator based FDDI products. CodeNet products allow for the seamless integration of mixed systems, ensuring optimum performance and cost.

CodeNet Ethernet LAN Products

Fiber Optic Ethernet Transceivers

These devices connect computers and other network equipment to the fiber optic Ethernet network. They are compatible with all applicable Ethernet, IEEE 802.3 and ISO 8802.3 standards for minicomputers, mainframe computers, communications gateways, personal computers, network servers, bridges and repeaters for structured star cabled and point to point applications.

Ethernet Network Interface Cards

These NICs connect any computer with an ISA bus, such as the IBM PC/AT, or an EISA bus computer, directly to a CodeNet fiber optic Ethernet network. They provide high throughput performance at a significantly lower price. The EISA cards employ BUS Mastering architecture to capitalize on the EISA computers' exceptionally high rate of throughput.

Active Hub/MultiMedia Ethernet Repeater

These systems, jointly developed by Codenoll and 3Com, provide a multimedia active star capability as well as performing all repeater functions. CodeNet repeaters are used to build large networks without any hardware or software modifications.

CodeStar-LightBus® Passive LAN Couplers

These are completely passive, optical network segment hubs or multistation access units. Without electrical or electronic components, they connect up to 32 computers, repeaters, bridges, gateways, or network segments.

CodeNet FDDI LAN Products

Copyright Codenoll Technology Corporation, 1990
Page 3

FDDI Interface Cards

CodeNet ISA/FDDI network interfaces allow computers with an Industry Standard bus to attach to a 100 Mbps Fiber Distributed Data Interface network as Class A or Class B active stations. The CodeNet FDDI interfaces can be configured for single ring (CodeNet-9540) and dual ring redundant counter-rotating ring (CodeNet-9543) attachments.

The American National Standards Institute (ANSI) developed the FDDI standard for very high speed token passing ring networks: ANSI X3T9.5, Fiber Distributed Data Interface (FDDI). The standard specifies substantial performance and configuration improvements over current LANs: 100 Mbps data rate; up to two kilometers between stations and up to 500 physical connections per ring.

These EISA/FDDI network interfaces connect Extended Industry Standard Architecture (EISA) computers to a 100 Mbps Fiber Distributed Data Interface networks as Class A or Class B active stations. These boards capitalize on the 33 MB bus transfer rate of the EISA architecture to yield the highest possible system throughput. EISA users can take full advantage of the inherent speed and power of their 386 and 486 EISA computers and the high 100Mbps capacity of FDDI. Each board contains a large packet buffer and an onboard co-processor with its own memory, which permits the LAN drivers to run concurrently with the host CPU(s). The CodeNet FDDI interfaces can be configured for single ring (CodeNet-9500) and dual ring redundant counter-rotating ring (CodeNet-9503) attachments.

All CodeNet FDDI network interfaces conform to the approved ANSI standards, including the data transfer rate of 100Mbps. They allow a standard ISA (IBM PC/AT or compatible) or Extended-ISA bus network server or personal computer to function on a backbone, in a workgroup or as a high performance workstation.

FDDI Concentrator

The first of the Company's FDDI system products is a modular CodeNet FDDI Concentrator which connects Class B FDDI stations in a star configuration to a FDDI concentrator that may be connected directly to the dual ring or can be connected in a star configuration to another concentrator. The concentrator is assembled on site using standard CodeNet FDDI interface boards, the CodeNet Concentrator software and an industry standard PC as the base.

FDDI Bridge

Codenoll has also announced the development of a modular FDDI bridge. This bridge would provide transparent network connections between FDDI, Ethernet, Token Ring, T1 WAN, and 802.6/SMDS WAN network segments.

Corporate Timeline

1981: Codenoll introduced its first fiber optic link products for sale
to OEMs that manufacture data communications, telecommunications, military and government equipment.

Copyright Codenoll Technology Corporation, 1990
Page 4

1982-1983: Codenoll introduced the Company's patented high power Gallium Arsenide Light Emitting Diode (LED) and invented the CodeNet fiber optic Ethernet LAN.

1984: Codenoll expanded its CodeLink fiber optic product line and introduced the CodeLink 2000 point-to-point communications equipment.

1985: Codenoll introduced its Indium Phosphide high power LEDs for local telecommunications, military, faster FDDI applications and high speed data communications applications and introduced the CodeLink-2000 M. The CodeLink-2000 M is a modular set of high speed digital fiber optic modem products capable of point-to-point transmission of a very wide range of digital telecommunications and computer data over distances exceeding 30 kilometers.

1986: The company announced alliances with major companies in the LAN market, including 3Com, Sytek (Hughes LAN Systems), Computrol and Novell.

Through a joint development with 3Com, Codenoll developed a fiber optic version of 3Com's EtherLink personal computer adapter card. Together with Sytek (Hughes LAN Systems), Codenoll developed a fiber optic version of the IBM PC Network that is compatible with all IBM NETBIOS and Token-Ring personal computer software.

Together with the Computrol division of Gould Modcomp (a subsidiary of AEG, which is a subsidiary of Daimler Benz), Intel and Motorola, Codenoll developed a fiber optic version of the IEEE 802.4 MAP (Manufacturing Automation Protocol) network for factory automation.

Celanese Corporation (subsequently merged into Hoechst Celanese) and Codenoll began a technical alliance to develop optical and fiber optical products based on non-linear optical (NLO) polymers.

1987: In September, 1987, the company jointly introduced with 3Com the MultiStar/MultiConnect Multimedia Multiport Ethernet Repeater. Codenoll currently produces two products for the MultiStar/MultiConnect series.

Two other key developments were the demonstration in May, 1987, of what is thought to be the world's first electro-optical polymer device, jointly developed with Hoescht Celanese. In November, 1987, Codenoll and Hoechst AT jointly unveiled the first Ethernet LAN to operate at 10Mbps on plastic optical fiber (POF) in a star configuration using a passive optical hub and 2-3 meter fiber segments.

1988: Hoescht AG and Codenoll signed a 10-year technical cooperation agreement for the joint development of polymer (plastic) optical fiber (POF) materials, components, systems and technology. The first result was the demonstration in April, 1988, at the Hanover Fair of the first publicly demonstrated fiber optic LAN controlled automobile, developed largely by Codenoll in cooperation with Hoescht.

The company installed a 150 computer LAN at the Pentagon in late 1988, and received an award via Contel in March 1989 for the entire Office of the Secretary of Defense LAN. The company now has thousands of nodes in the Pentagon as well as installations in many government and military sites.

The company's entire CodeNet Fiber Optic Ethernet product line was Tempest certified, opening up the National Security market for the first time in 1989.

1989: Codenoll announced a number of joint agreements:

Copyright Codenoll Technology Corporation, 1990
Page 5

Bull HN, the International Computer Systems Division of Groupe Bull, and Codenoll signed an OEM agreement under which Bull now resells Codenoll products in more than 90 countries worldwide. The agreement is valued at an estimated $10 million over seven years.

Compaq Computer Corporation selected Codenoll to develop fiber optic Ethernet and FDDI network interface cards for the new line of EISA bus computers.

Contel Federal Systems was awarded a contract by the Office of the Secretary of Defense to install the Office Automation Secure Information System (OASIS) in the Pentagon. Codenoll is supplying connectivity equipment for personal computers, workstations, and servers on the entire system. This contract extends over seven years.

Eurolan S.p.A., jointly owned by Olivetti and SIRTI of the Italian IRI-STET Group, is a network supplier throughout Europe. Eurolan is now selling the Codenoll product line based on an agreement with Codenoll that is valued at more than $5 million over the next three years.

Net One Systems, a joint venture of Mitsubishi and Ungermann-Bass, is supplying Codenoll products to a Japanese network market that is estimated to be growing at a rate of 200 percent annually. Net One is also selling Codenoll products in Korea, Hong Kong, Taiwan and Singapore.

Robert Bosch GmBH in Germany is now selling the complete line of Codenoll network systems through its operations in Germany, France, Switzerland, Sweden, Norway Denmark, The Netherlands, Belgium and other countries.

Soliton Systems, K.K., one of the leading network suppliers in Japan, is now selling Codenoll LAN systems under a contract estimated at between $3 and $5 million over the next four years.

The U.S. Government. In 1988, Codenoll signed several agreements designed to strengthen its government business including an agreement to produce Quality Tempest Product (QTP) versions of each Codenoll LAN product, and Microwave Modules and Devices (MMD), which supply government markets with militarized (ruggedized) versions of Codenoll products.

Major government subcontracts won in 1989 include the Secure Management Information System (SMIS) of the U.S. Army in the Pentagon, which uses Codenoll equipment supplied through Westco Automated Systems and Sales.

1990: In early 1990, Codenoll signed a five-year agreement with NYNEX to cooperate in the development of high speed fiber optic-based products and other LAN products and technology. Work on jointly-developed products has already begun.

Codenoll announces a contract manufacturing agreement with the Packard Electric Division of General Motors to manufacture its most popular LAN components at a state-of-the art electronic assembly facility in Arizona.

SCO and Codenoll announce support for 10Mbps and 100Mbps fiber optic Ethernet and FDDI interfaces for systems using SCO's System V UNIX operating system.

Codenoll publishes *Enterprise*, a quarterly newsletter.

Western Digital and Codenoll jointly announce that Codenoll will utilize Western Digital proven Ethernet technology to manufacture compatible fiber optic OEM products.

Codenoll publishes industry's first comprehensive Fiber Optic LAN Handbook, which becomes a desktop reference standard.

At the Dallas NetWorld Show in September 1990, Codenoll demonstrated the use of Plastic Optical Fiber (POF) for LAN networks. POF, as a network media, costs less than unshielded twisted pair and is easier to install.

At Comdex, Codenoll announces Ethernet POF products and a complete connector system for POF which will ship in the first quarter 1991.

In December, Codenoll and Pioneer Standard and Pioneer Technology announced that Pioneer has added the Codenoll line of Fiber Optic LAN products. Pioneer is a national distributor of computer products and systems with offices in more than 20 major cities.

Copyright Codenoll Technology Corporation, 1990
Page 7

CODENOLL Technology Corporation
1086 North Broadway, Yonkers, NY 10701 • (914) 965-6300 • Telex 646 159

SPECIAL COMDEX RELEASE
For Release, November 13, 1990, 11:15am

CODENOLL AND GM INTRODUCE PLASTIC FIBER OPTIC ETHERNET LAN SYSTEMS, COMPONENTS AND CONNECTORS AT COMDEX

Revolutionary Plastic fiber LANs to revolutionize LANs and simplify media choices

LAS VEGAS, NV -- Codenoll Technology Corporation (NASDAQ: CODN) with the Packard Electric Division of General Motors today announced a complete line of 10 Mbps Ethernet LAN systems, components and a structured cabling system for Plastic Optical Fiber (POF). This revolutionary new LAN media satisfies workgroup LAN needs immediately, and resolves concerns about future data speeds.

Announced today was a complete line of network products to connect PCs, minicomputers, terminal servers and PCs into an Ethernet workgroup network using low cost, easy to use plastic fiber. The family of CodeNet® POF products include; integrated network interfaces for ISA, EISA and Micro Channel PC platforms; a POF Ethernet transceiver, that provides network connection for any type of device; a POF module for the Codenoll MultiStar® or 3Com MultiConnect active hub; connectors for POF; wall plates; patch panels and a variety of preassembled cables.

The plastic optical networks introduced by Codenoll are configured in a star configuration, and use the same signalling conventions as the IEEE FOIRL (Fiber Optic Inter-Repeater Link) Ethernet standard. Networks configured using the current CodeNet design can be more than 100 meter apart or 50 meters from the POF concentrator.

At the event, Codenoll and Packard Electric also unveiled a complete structured cabling and connector system for Plastic Optical Fiber. Based upon a patented design, the connector makes POF easier to use and install than any other type of LAN wiring, including unshielded twisted pair (UTP). As demonstrated at the announcement, the POF connectors can be installed in less than a minute with simple and inexpensive hand-held tools. Once installed, the POF connectors are more durable, reliable and trouble-free than any other form of LAN connector.

The connector and cabling system includes: the basic POF connector; POF wall outlets; a 19" POF patch panel, a receptacle with integrated optical LED and detector for use

GM and Codenoll Developing Plastic Optic LANs, Page 2

on network adapters and concentrators; jump cables for the connection from the user's computer and the wall outlet; and prefabricated patch cables for use in the wiring closet. The price of the connectors is directly competitive with twisted pair connectors. The duplex POF connector is retail priced at $1.95. Comparable connectors such as the FDDI duplex fiber connector lists at approximately $34.00; Token Ring data connectors for twisted pair which list for approximately $7.95; simplex glass fiber optic fiber connectors which list from $10-$15 each; and RJ45 Telco connectors which list from $1.00 to $1.95 each. This announcement follows a demonstration of the POF technology at NetWorld in September. Codenoll and Packard Electric have a jointly developed this technology over the past six years. Plastic fiber as a LAN media has been designed to obsolete both coax and twisted pair wiring. It is less expensive, easier to install and easier to test than either shielded or unshielded twisted pair, while providing the higher performance and reliability associated with fiber optics.

The increasing use of 386 and more powerful machines for graphics and image processing coupled with the increased growth of the networking of computers is creating a need to share information at high network speeds. Only fiber optic technology with its inherently high data rates addresses these needs now and in the future. The same plastic optical fiber cable used for the Ethernet products introduced today can be used to run higher performance LANs such as FDDI (100 Mbps) and the new Metropolitan Area Network (MAN), at speeds of 100 Mbps and 300 Mbps.

POF shares the inherent performance, reliability and flexibility advantages of fiber optic networks: There can be no shorts, static buildup or ground loops in a fiber optic network; light transmission is immune to Electromagnetic Interference (EMI) and can be installed near copy machines, X-ray units and elevator shafts. "Fiber networks are the only choice where security or reliability are factors," said GM and Codenoll officials.

"POF will do for FDDI what twisted pair did for Ethernet -- make it cost effective for every desktop," said Michael H. Coden, president of Codenoll. "Put in plastic fiber and CodeNet POF Ethernet today, and upgrade to POF FDDI tomorrow without changing cable. Plastic Optical Fiber networks are now the most cost effective network option for workgroups and have the capacity to transmit data at 10 Megabits per second, 100 Megabits, 300 or more, as this network technology is introduced. Only optical fiber offers that. And for longer distances, glass fiber will provide backbones, inter-building and other connections required by large enterprise-wide LANs."

Bill Collins, Packard Electric's Manager of Specialty Products, commented on the effort and design behind the POF connector. " Our goal was to make the POF connector as easy to use and install as the RJ45 telephone connector and we were successful. Anyone who can install a telephone or change an auto fuse can understand and immediately use our POF connectors. We looked at all current LAN connectors and requirements, and designed a simple, snap-in, duplex plastic connector with several unique features. First, it is practically indestructible, just like the plastic fiber it uses. The connector is polarized and can only be inserted the 'right' way into the wall outlet or into the network adapter in the computer. Once inserted, the connector locks into place, and cannot be accidentally dislodged. Also included is a shroud and molded strain relief connector to protect the fiber. Best of all, these features are available at a price competitive with any twisted pair connector."

"We expect the POF connector and cable components to become a standard," said Brian Ramsey, Codenoll's Director of Marketing. "This morning we are also making a presentation on these developments in POF Ethernet to the IEEE meeting in La Jolla, CA. We will be working with the industry to make POF an open standard," he continued. "We are committed to making the connectors, ICs and opto-electronic parts available to other vendors to accelerate the adoption of the POF medium in the LAN market." Codenoll will begin marketing the connector system to other manufacturers and distributors immediately. Samples of the connector are available for immediate shipment.

Jack Olin, Director of Advanced Engineering for Packard Electronics, indicated that "Plastic Optical Fiber has been available for several years. It is now available in all of the forms needed for a strategic building wiring system, and it is made by a number of companies such as: Mitsubishi Rayon, Optectron, Toray, and Asahi Chemical." "For the LAN applications we announced today, we use a duplex, or two fiber, package. The light is transmitted within the PMMA (the same material used in plexiglass) core of the fiber, which is 1000 microns, or 1 mm thick. The fiber is protected by a PVC or plenum-rated jacket just like twisted pair," he continued. "Plastic fiber is virtually indestructible -- you just can't break it. Even if you try to crush or pinch it, it just springs right back to its original form. Its far more flexible and durable than most twisted pair copper wire."

Codenoll said that the cost of Plastic Optical Fiber will be between $0.10 and $0.25 per foot, depending upon the jacket type and the quantity purchased. "As this type of media becomes more widely used, we expect the price to stabilize below $0.10," said Mr. Coden. "POF is easier and less risky to install than the confusing variety of twisted pair wire that is required by different types of networks."

Codenoll expects to ship POF network products in January, 1991. The retail price of the CodeNet Network Interface cards start at $495, and $395 is the retail price of the POF transceiver. The POF module for the Codenoll MultiStar active hub is priced at $495. The price of the POF connector is below one dollar in large quantity lots.

Packard Electric Division, a member of GM's automotive component group, is the world's leader in automobile power and signal distribution systems. Packard supplies complete wiring systems for over 9.5 million vehicles annually, is a supplier of systems and components to 15 of the top automotive manufacturers, and has over 9,000 total customers. Packard's worldwide capabilities include 55,000 employees, 8 engineering support centers and 124 manufacturing facilities in 20 countries on six continents.

Codenoll Technology Corporation is the leading supplier of fiber optic Ethernet computer network systems with thousands of installations worldwide. In 1989 the company announced two lines of FDDI high speed computer network products under its CodeNet trade name. Codenoll products are resold worldwide by companies such as: Bosch, Bull, C&P Telephone, Contel, Digital Equipment Corporation, Pioneer Standard Electronics, Eurolan, Southwestern Bell, 3Com and hundreds of other VARs, System Integrators and OEMs. Codenoll stock is publicly traded in the Over The Counter Market, NASDAQ: CODN.

###

Contacts:

Codenoll Technology Corporation
Brian Ramsey
914/965-6300

Packard Electric Division
Michael Hissam
216/373-2364

or

Cheryl Snapp
Snapp and Associates
801/225-7888

CodeNet and Codenoll are registered trademarks of Codenoll Technology Corporation. All other trade names are the property of other companies.

GM and Codenoll Developing Plastic Optic LANs, Page 5

Price and Availability of Announced Products

CodeNet 8631	POF ISA Ethernet NIC	$495	Q1
CodeNet 8621	POF Micro Channel NIC	$795	Q1
CodeNet 8601	POF EISA Ethernet NIC	$1295	Q1
CodeNet 8681	POF Ethernet Transceiver	$395	Q1
CodeNet 8611	POF MultiStar Concentrator Module	$495	Q1
CN CWS-POF	Duplex POF Connector w/shroud	$1.95	Q1
CN CNS-POF	Duplex POF Connector no shroud	$1.95	Q1
CN WO-POF	Duplex POF Wall Outlet	$4.95	Q1
CN CSR-POF	Cable Strain Relief Sleeve	$1.75	Q1
CN 2M-1000-POF	2 Meter Preassembled Drop Cable	$24.50	Q1
CN 5M-1000-POF	5 Meter Preassembled Drop Cable	$37.50	Q1
CN 1M-1000-POF	1 Meter Preassembled Patch Cable	$19.50	Q1
CN PP-POF	24 connector 19" Patch Panel	$175	Q1

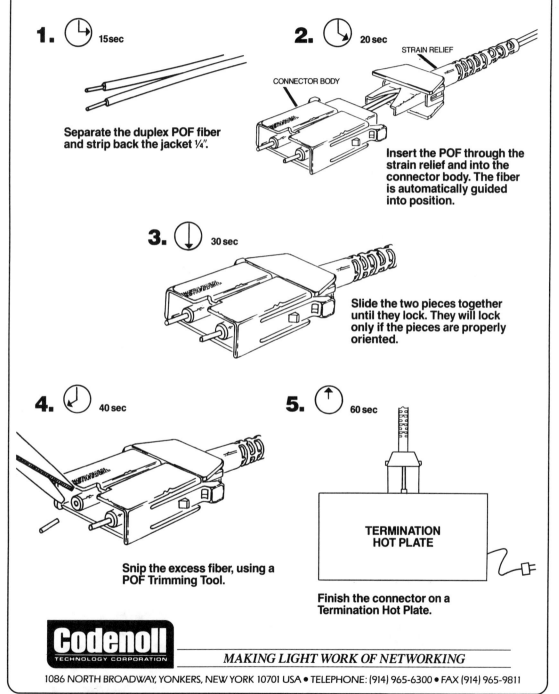

Overview of Fiber Optics and LAN Technology

"Every thing you ever wanted to know about Fiber Optics and LANs, but were afraid to ask!"   p. 35
   M. H. Coden

Everything you ever wanted to know about Fiber Optics and LANs, but were afraid to ask!

NOTES

...the primer on Fiber Optic and LAN technology.

Everything You Ever Wanted To Know About Fiber Optics and LANs, But Were Afraid To Ask!

by
Michael H. Coden
President, Codenoll Technology Corporation

Copyright 1990 Codenoll Technology Corporation

Everything you ever wanted to know about Fiber Optics and LANs, but were afraid to ask!

PCWEEK

NET ASSETS

Spreading the Good Word About High-Fiber Networks

BARRY GERBER

If you're ever at a trade show where Michael Coden, president of Codenoll Technology, is speaking, be sure to catch his act. He'll really open your eyes regarding fiber-optic networking technology.

About a month ago, I chaired a session on "High Fiber LANs" at PC Expo in New York. Coden was on the panel along with Paul Callahan, one of Digital Equipment Corp.'s FDDI experts.

The two made a wowie-zowie team, debunking myth after myth, offering a variety of network configuration options and giving the audience what one attendee called "the most useful, practical experience I've had at an industry conference in a long, long time."

As the manager of a site that's used fiber backbones for nearly five years and as PC Week's resident evangelist on building high-performance LANs, I was in hog heaven.

Aware that no one would listen to whatever else he had to say until he dealt with the cost issue, Coden showed us some numbers comparing the per-workstation cost of a twisted-pair, a shielded twisted-pair and a fiber-optic network.

Fiber beat out shielded twisted-pair and wasn't very much more expensive than plain old twisted-pair. And that was with fiber running right up to the workstation, not just serving as a backbone.

As David Strom pointed out in his column awhile back, the heaviest costs in running cable are fixed, no matter what technology you use. That's the cost of the labor to install your cable and any required conduit. The per-foot cost of the different cabling options is minor compared with those labor costs.

After polishing off the cost issue, Coden took off after the myth that fiber is hard to install. Here, I must admit, he taught me a thing or two.

First, he took a 100-foot piece of fiber cable and shined a flashlight into it. You could see the little 3mm point of light at the other end of the cable all the way across the large conference room in New York's Javits Convention Center.

Then Coden made a pretzel out of the cable. I mean, he really tied a knot in the thing.

"My god," I thought, "he's wrecking the stuff." But, guess what, we could still see the light at the other end of the cable, knot and all. Furthermore, Coden assured us, no damage had been done to the cable, though he did warn that this wasn't meant to be a demonstration of proper cable handling procedures but simply a way of reassuring his audience that you don't need kid gloves to work with fiber.

Look, No Hands

Driving the last nail in the coffin of the it's-hard-to-work-with myth, Coden actually installed a fiber-optic connector while he was in the middle of his talk. It was something to watch. It was a snap. One disbelieving member of the audience came up to the front of the room and examined Coden's connector components and tools to prove to himself there had been no sleight-of-hand.

Having convinced those of us in the conference room that we had nothing to fear from fiber, Coden then proceeded to debunk the myth that fiber is only a technology of the future.

First, he showed us regular Ethernet adapters that run on fiber rather than thick or thin copper cable. Though these don't offer any performance advantage over copper-based adapters, they're more secure against eavesdropping than the copper-based kind. This is because fiber doesn't emit electromagnetic waves that can be intercepted by nefarious governmental or industrial spies.

Next, Coden argued that 100M-bps FDDI token-passing technology is here now and reminded us that FDDI only runs on fiber.

Playing the magician, Coden pulled a fiber Ethernet adapter out of this magic briefcase he carries with him. Then he extracted FDDI adapters for a variety of PC buses. When those ran out, he pulled out active and passive fiber-optic hubs.

DEC's Paul Callahan followed Coden and firmly restated his company's plans to introduce its first FDDI products by the end of this year.

Time to put some fiber in your LAN's diet? You can get a neat 200-page book on fiber from Codenoll. Call them at (914) 965-7300. Tell them you want "The Fiber Optic LAN Handbook." It's free. ∎

Reprinted from PC WEEK July 16, 1990
Copyright © 1990 Ziff Communications Company

Copyright 1990 Codenoll Technology Corporation

PREFACE

The following paper is based on an edited transcript of a speech given by Michael Coden in June 1990.

The speech was one of two given in a session called "High Fiber LANs" chaired by Mr. Barry Gerber, then columnist of PC Week and currently Senior Editor at Network Computing, and a respected member of the faculty at UCLA.

In his column in PC Week on July 16, 1990, Mr. Gerber quotes a member of the audience: "what one attendee called 'the most useful, practical experience I've had at an industry conference in a long, long time.'"

Mr. Gerber's article is reprinted, in its entirety, on the previous page, with the permission of PC Week. We hope some day you will be able to attend one of Michael Coden's seminars. In the meantime however, we hope this paper can give you some insight into the fascinating disciplines of Fiber Optics and Local Area Networks.

iv Everything you ever wanted to know about Fiber Optics and LANs, but were afraid to ask!

The author hopes that this document can serve to provide technical background material for a variety of purposes regarding Fiber Optics and General LAN Technology with special emphasis on Ethernet, Token Ring, FDDI, 802.6 MAN, SMDS, and SONET.

Any portion of this document may be quoted, with attribution to Michael Coden, president, Codenoll Technology Corporation or to Codenoll Technology Corporation. All of the Tables, Figures and Photographs used herein are available in reproducible format. Many are available in color. Simply request the material you wish from:
       1. Mr. Brian Ramsey, Director of Marketing, or
       2. Mr. Robert Neilley, Customer Service Manager, at
           Codenoll Technology Corporation
           1086 North Broadway
           Yonkers, NY 10701 USA
           Telephone: 1-914-965-6300
           Fax Server: 1-914-965-9811

> **Everything You Ever Wanted To Know About Fiber Optics and LANs, But Were Afraid To Ask!**
> by Michael H. Coden
> President, Codenoll Technology Corporation

TABLE OF CONTENTS

PREFACE iii
TABLE OF CONTENTS v
ACKNOWLEDGEMENT vi
DEDICATION vi
PROLOGUE - Fiber Optics Does Not Cost More! 1
INTRODUCTION & ABSTRACT 5
Chapter 1 - FUNDAMENTALS OF FIBER OPTIC TECHNOLOGY 7
Chapter 2 - THE FOUR MYTHS OF FIBER OPTICS 13
    Myth 1: Fiber Optics is a Future Technology 13
    Myth 2: Fiber Optics, Incompatible with my Equipment 14
        The IEEE, ANSI & ISO Standards - brief overview 16
    Myth 3: Fiber Optics is difficult to install 18
    Myth 4: Fiber Optics is too expensive 21
        Twisted Pair FDDI? 23
Chapter 3 - FIBER OPTIC CABLING, the UNIVERSAL SOLUTION 25
Chapter 4 - FUNDAMENTALS OF NETWORKING TECHNOLOGY 27
    Alchemy -- Copper to Glass, Electrons to Photons 30
    Tapping the Light Fantastic - Couplers and Hubs 31
    Tokens and Collisions 33
    Conclusion - Fundamental Theorem of LANs 39
Chapter 5 - CONFIGURING FIBER OPTIC LANs 41
Chapter 6 - UPGRADE MIGRATION PATHS 45
Chapter 7 - MIGRATION TO FDDI AT 100 Mbps 47
    The FDDI Concentrator 47
Chapter 8 - REVIEW OF FDDI PRINCIPLES 49
Chapter 9 - BRIDGING BETWEEN DIFFERENT NETWORKS 55
Appendix A - The IEEE, ANSI and ISO Standards - Detail 61
    The IEEE 802.3 Fiber Optic LAN Standards 64
Appendix B - TABLE OF FIGURES 69

ACKNOWLEDGEMENT

I would like to express my thanks and deep appreciation for all the tireless help and thoughtful suggestions that I received from Codenoll's employees, our strategic partners, our customers and our board of directors. These people have created a new generation of information communication that will revolutionize the world as we know it. Hopefully this paper will explain what we have accomplished, why we are so excited about the potentials, and how you may personally benefit from this easy to use, high performance technology today.

DEDICATION

This paper is dedicated to the employees of Codenoll, whose teamwork, creativity and dedication keep turning technological dreams into reality.

PROLOGUE - Fiber Optics Does Not Cost More!

Most people have a number of incorrect preconceived notions about fiber optics. The most important of these is:

Fiber optic LAN technology is much too expensive for you to use in your own LAN environment.

Until we dispose of this **misperception**, you will not seriously consider using Fiber Optic technology in your own computer LAN. The general perception, throughout the world, is that fiber optics is an interesting curiosity. "Oh, it's wonderful technology, but its practical uses will first come in the year 2000." Furthermore, "it's much too expensive for me to use now." Nothing could be further from the truth.

So first, we look at the real retail prices (Shown in Figure 1), taken out of price lists for personal computer local area network (LAN) products:

1. unshielded twisted pair cable (UTP), using adapter cards for AT or ISA (Industry Standard Architecture) class computers,

2 Everything you ever wanted to know about Fiber Optics and LANs, but were afraid to ask!

2. shielded twisted pair cable (STP), using the same adapter cards for AT/ISA computers, and

3. fiber optic Ethernet using adapter cards for AT/ISA computers, and based principally on the 10Base-F draft standard which is currently in letter ballot by the IEEE 802.3 standards committee.

We see that fiber optic cable costs a little more than unshielded twisted pair (UTP), but a whole lot less than shielded twisted pair (STP). Connectors for fiber optic cable have come down to less than $7.00 each. Labor is about the same whether you are pulling "platinum wire", fiber optic cable, UTP, STP or coax. It takes about 10 to 15 minutes to put on a fiber optic connector, in the field, which is a lot less time it takes to put on a DB25 type RS232 or DB15 type Ethernet transceiver cable connector.

Ethernet Retail Price Comparison

	Unshielded Twisted Pair	Shielded Twisted Pair	CodeNet Fiber Optic
ISA Adapter Card	$499	$499	$595
Cable to Hub (50m)	50	338	100
Connectors & Patch Cables	3	53	56
Labor ($45/hr)	30	40	45
Hub ($ per port)	365	365	115
Total $ per port	$947	$1295	$911

Figure 1: Comparison of twisted pair to fiber optic Ethernet costs.

Please notice that the hub costs are significantly different. In the case of twisted pair LANs, the hubs are substantially more expensive, than in the case of fiber optic LANs. In Fiber Optic Ethernet, unlike the twisted pair cases, one can use passive as well as active hubs. In the case of FDDI (Fiber-optic Distributed Data Interface) the 100 Mbps LAN standard which we will be discussing in the second part of this paper, we can actually do without hubs, although we get some

...the primer on Fiber Optic and LAN technology.
PROLOGUE - Fiber Optics Does Not Cost More!

important advantages by using special FDDI hubs called "concentrators".

The bottom line is:

1. that it is no more expensive in 1990 to install a fiber optic Ethernet to the desktop than it is to install a twisted pair Ethernet to the desktop,

2. that fiber optic Ethernet is a lot less expensive than standard Ethernet yellow coax cable, for use in backbone and enterprise LANs, and only a slightly more expensive than thin Ethernet (which is commonly called "Cheapernet").

The real major savings and important difference is:

> **If you install fiber optic cable today, then you can upgrade to the coming networking standards such as FDDI at 100 Mbps and the new IEEE 802.6 dual 150 Mbps bus standard.**

The IEEE 802.6 networking standard, which we will discuss in the second part of this paper, and about which you will be hearing much more in 1990 and 1991, is the Metropolitan Area Network (MAN) standard that is being promoted very heavily by the Regional Bell Operating Companies (RBOCs). IEEE 802.6 will be available in several configurations with dual 1.5 Mbps to dual 150 Mbps pairs of busses which result in 3 Mbps to 300 Mbps of throughput.

4 Everything you ever wanted to know about Fiber Optics and LANs, but were afraid to ask!

The real major savings and important difference is:

> **If you install fiber optic cable today, then you can upgrade to the coming networking standards such as FDDI at 100 Mbps and the new IEEE 802.6 dual 150 Mbps bus standard.**

INTRODUCTION & ABSTRACT

In this paper, we will only have time to give a very brief overview of the state of the art of Fiber Optic LAN technology from PC LANs at 10 Mbps and 100 Mbps, that are available today, to MANs at 300 Mbps that will be available shortly.

At Codenoll we have the benefit of participating in a series of joint ventures with companies that cover a broad range of current software and hardware issues. These relationships are summarized and diagrammed in Figure 2.

We work together with Microsoft, Novell, Banyan and SCO, making FDDI products which plug into personal computers and run the Microsoft LAN Manager, SCO UNIX Version 5, Banyan Vines and Novell NetWare. Together, we are running 100 Mbps out to the desktop now. We jointly make Ethernet products for backbones as well as the desktop with Western Digital, 3Com, Novell and Compaq Computer. Together, we are stretching the limits of distributed processing.

6 Everything you ever wanted to know about Fiber Optics and LANs, but were afraid to ask!

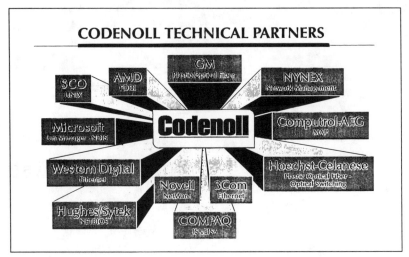

Figure 2: Diagram of Codenoll's joint development relationships.

Together with General Motors Corporation, Hoechst Celanese and Mitsubishi, we are developing a new generation of technology based not on glass fiber, but plastic optical fiber (POF). We are working together with Hoechst Celanese and Mitsubishi who are developing plastic optical fiber and the General Motors Corporation who is developing low cost precision plastic optical connectors. Together with these companies, we have been jointly developing both intra-building LAN cabling systems as well as a plastic fiber optic LAN systems to replace the complex copper wiring harnesses in automobiles.

Together with NYNEX, we are developing complete LAN-MAN-WAN network management technology, the IEEE 802.6 MAN standard, and flexible bridge/routers between wide area (WAN), metropolitan area (MAN) and multiple local area (LAN) networks.

This paper is essentially a summary of the work done by all of these companies together with Codenoll, in fiber optics.

Chapter 1 - FUNDAMENTALS OF FIBER OPTIC TECHNOLOGY

The basics of fiber optics are quite simple. They are explained in great detail in the Fiber Optic Lan Handbook. What follows here is a brief, understandable and comprehensive summary.

Fiber Optic Cable looks like and handles like any other cable. The fiber optic cable I normally use actually looks and feels like 18-awg lamp wire or "zip-cord". I usually show 10 meters, or 110 feet of fiber optic cable. It's very light and flexible. The way it works is simple to demonstrate:
> you shine light, from the overhead projector, or a flashlight in one end of the fiber optic cable, and you see, with your own eyes, that it comes out the other end.

To send information via fiber optic cable, we simply convert the electronic signals from a computer to optical signals. In other words, the electronic binary on-off becomes an optical binary on-off. The on-off of the light goes into one end of the fiber, through the fiber, and comes out as on-off light at the other end as in Figure 3. Once you've seen it, you really understand how simple it is, and believe it works.

8 Everything you ever wanted to know about Fiber Optics and LANs, but were afraid to ask!

But then the next question we get is: "Isn't a glass fiber cable obviously very fragile?" "It must break very easily." "I'm sure I could not do anything with it." "It obviously has to be handled with 'kid gloves' by a Phd."

At that point, I tie a double knot in the fiber optic cable and then shine the same flashlight or overhead projector into one end. The light still comes out the other end as before. Basically modern fiber optic cable can be handled without any special considerations. Fiber optic cable should be handled in the same way, and with the same "roughness", you would handle any good quality data grade coax or twisted pair cable.

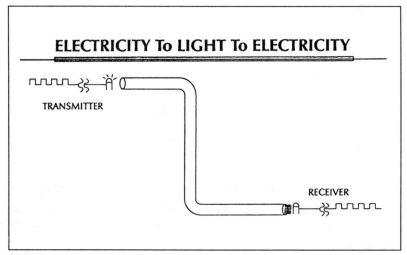

Figure 3: Transmission of data by converting electronic binary signals to light.

What's inside the cable can be seen in Figure 4. It's simply a glass fiber covered by a plastic buffer coating and surrounded by some kevlar fibers. The kevlar fibers give the cable a tremendous amount of strength. Kevlar is the same material used to make bullet-proof vests. The glass-plastic-kevlar ensemble is surrounded by a final protective sheath of plastic or Teflon.

...the primer on Fiber Optic and LAN technology.
Chapter 1 - FUNDAMENTALS OF FIBER OPTIC TECHNOLOGY

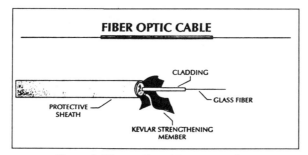

Figure 4: Fiber optic cable construction.

The "secret" of how fiber optics works is shown in Figure 5. The thin glass fiber actually consists of two parts: an inner glass cylindrical core, and an outer concentric glass cladding.

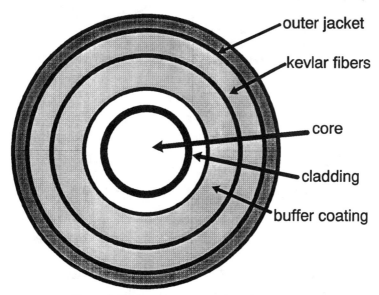

Figure 5: Horizontal cross section of fiber cable.

Figure 6 shows the longitudinal cross section of an optical fiber. The outer cladding of the glass is reflective, while the inner core of the glass is very transparent. So the light goes through the transparent core, staying in the core by bouncing off the reflective cladding.

The cladding is like a cylindrical mirror that is around the core. Thus, the

10 Everything you ever wanted to know about Fiber Optics and LANs, but were afraid to ask!

light stays in the fiber core as if it were a "light pipe", the same way water stays in the hollow core of a metal clad pipe.

It is interesting to note just how transparent the glass is in the core of an optical fiber. It is composed of pure silicon dioxide (SiO_2). Consider the picture window you have at home in your living room, it is 1/8th inch plate glass. You could replace it with a window made of fiber optic core glass that was three miles thick, and you would get the same bright image coming through the 3 mile thick window that you currently do with the 1/8th inch plate glass window.

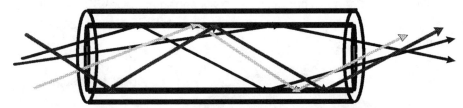

Figure 6: Longitudinal cross section of fiber showing rays of light.

Fundamentally that's all you really need to know about fiber optics. It's that simple. However, if you go to cocktail parties where people might talk to you about single-mode versus multi-mode technology, read the next paragraphs as well.

The concepts of single-mode and multi-mode are really quite straightforward and are illustrated in Figure 7. It is primarily a question of how large is the core diameter. In multi-mode fiber the core diameter ranges from 50 to 100 microns (from 1/500-th of an inch to 1/250-th of an inch). In single-mode fiber, the core diameter is a mere 7 to 9 microns (which is about 1/3000th of an inch). The new concept here is concerned with the fact that in the "wider" or "larger core" multi-mode fiber, different rays of light bounce along the fiber at different angles as they travel through the core (Figure 7). Therefore they actually travel different total distances as they go from one end of a long fiber optic cable to the other.

...the primer on Fiber Optic and LAN technology.

Chapter 1 - FUNDAMENTALS OF FIBER OPTIC TECHNOLOGY

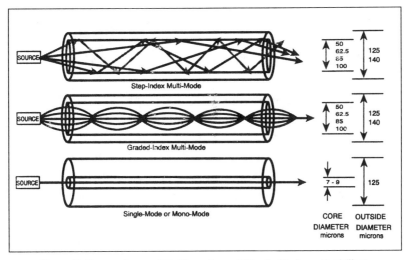

Figure 7: Comparison of Multi-mode and Single-Mode optical fiber.

Since some light rays travel longer distances and some travel shorter distances, while the speed of light is a constant, some of the rays will arrive at the other end of the cable later than others. Therefore, a square pulse of light power that goes in one end of a long fiber may come out the other end exhibiting a little pulse spreading or "dispersion", because some of the light rays (optical power) get to the other end sooner and some of the light rays (optical power) get there later.

In single-mode we transmit optical information with a single ray of light ("single-mode" is high-tech jargon for single-ray). For single-mode applications we typically use a laser to generate the optical signals, although we can do it with an LED. In single-mode systems, when we put a square wave in one end of the fiber, we still get a square wave out the other end, even over very long distances (tens of kilometers) and at very high data rates (hundreds of Mbps). However, the only times you really have to consider this effect is if your networks have data rates in excess of 150 Mbps and your distances are in excess of several kilometers.

In other words, if you are dealing with an average building or campus, you don't have to worry about single-mode versus multi-mode. You can easily use either for almost any application, present or future. If, on the other hand, you're dealing with wiring up the cities across a country, as the telephone companies are,

12 Everything you ever wanted to know about Fiber Optics and LANs, but were afraid to ask!

then you do worry about it, and you use single-mode fiber for those long distances (typically 30 miles - 50 km - between repeaters) and very high data rate capacities (typically 565 Mbps) then you should probably use single-mode fiber cable.

A typical LED is shown in Figure 8. LEDs are tiny semiconductor chips about 1/100-th of an inch on each side. The LED is usually mounted on some sort of a header and then the header goes into a little fiber optic connector. The optical fiber then comes up to the edge of LED chip. The optical signal, an optical version of the electronic current flowing through the LED, is emitted from the LED into the fiber. That is all we do to get optical signals into the fiber.

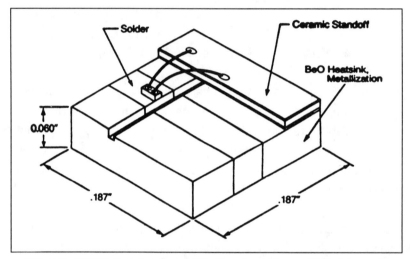

Figure 8: A 1/100-th inch square LED chip mounted on a 1/4 inch ceramic header with two wire bonds.

The optical signal is created by sending current through the small wires shown in Figure 8. As the computer turns the electronic current on and off, or the electronic voltage on and off, the optical (light) output of the LED goes on and off. The LED is a transducer converting electronic binary code into optical binary code.

Chapter 2 - THE FOUR MYTHS OF FIBER OPTICS

Now, with a background in fiber optic technology, we can begin a discussion of fiber optic LANs. We start with what we call the Four Myths of Fiber Optics.

Myth 1: Fiber Optics is a Future Technology

The first myth of fiber optics is: that it is only a future technology. It is pretty hard for me to believe that fiber optic LANs are a future technology when Codenoll alone has over 50,000 nodes installed at customers ranging from the United States Air Force, to Goldman Sachs investment banking firm, to Harvard University, the Pentagon, Shearson American Express and Campbell Ewald advertising. The applications range from Ronzoni Macaroni Inc. for an accounting office in Brooklyn, NY, to the Joint Chiefs of Staff in the Pentagon. The reasons these diverse users chose fiber optics are typically the same:
1. that fiber optic networks are no more expensive than copper networks,
2. that fiber optic LANs are considerably more reliable than copper

LANs, and
3. that fiber optic LANs allow you to expand the number of users and distances much more easily than copper LANs.
4. that fiber optic LANs and cabling can be upgraded to higher speed higher performance LAN protocols, data rates and technology using the same cable.

That fiber optic LANs automatically meet most Tempest electronic security criteria certainly makes them desirable in the government and military. But, security is also an issue for engineering and banking companies who are worried about industrial sabotage as well. There is no way to remotely spy or eavesdrop on a fiber optic local area network. And it is almost impossible to tap into a fiber optic LAN without permission or being detected.

Myth 2: Fiber Optics, Incompatible with my Equipment

The second myth is that: fiber optic LAN equipment is incompatible with existing computer technology. Nothing could be further from the truth. A wide variety of fiber optic LAN products on the market today, from a number of manufacturers, are designed to work with equipment that meets IEEE, ANSI and ISO standards. The primary LAN standards are:
- IEEE 802.3, Ethernet,
- IEEE 802.4, MAP (Manufacturing Automation Protocol),
- IEEE 802.5, Token Ring,
- IEEE 802.6, DQDB (Dual Queue Dual Bus),
    MAN (Metropolitan Area Network), and
- ANSI-X3T9.5 FDDI (Fiber Distributed Data Interface),

and the ISO (International Standards Organization) versions of these standards.

Many people ask how the standards organizations work. They are a rather complex hierarchy of organizations, that are motivated by politics and profit much more than by technology. I have included a rather complete description of the international standards making process and the current status of Fiber Optic LAN standards as of September 1990 as Appendix A, on page 61. A brief summary of Appendix A is included below for completeness.

Chapter 2 - THE FOUR MYTHS OF FIBER OPTICS

Fiber optic products that meet the IEEE, ANSI and ISO standards are **transparent to the hardware, and the software, that you currently have**. In other words, you can connect fiber optic LAN equipment to your existing computer systems as easily as you can connect copper based network equipment.

Moreover, there is a lot of equipment on the market today that allows you to connect fiber optic LAN equipment to existing copper based LANs and expand them using fiber optics. This conversion equipment is readily available and very inexpensive.

It is not necessary to think that, to go over to a fiber optic LAN, you might have to pull out any of the copper already installed. If you have copper LAN segments already installed, and they work, you don't need to touch them. You can simply connect your copper LANs with a fiber optic backbone, or expand your copper LANs using fiber optics from now on forward. If the copper LAN is installed, and it is not working, maybe you really need to replace it with fiber optics.

Typical examples of the kinds of equipment you can connect using fiber optic LAN products are shown in Figure 9. They include personal computers (AT/ISA type), MicroChannel computers (P/S-2 type), all types of work stations, terminals, mini computers, repeaters, bridges, gateways, etc. There is virtually no type of computer equipment on the market today that is not compatible with fiber optic equipment already available for it.

16 Everything you ever wanted to know about Fiber Optics and LANs, but were afraid to ask!

FIBER OPTIC LAN STANDARDS

USA		Data Rate	ISO Committee
Committee	Subcommittee		
IEEE 802.3	10BASE-F	10Mbps	8802-3
IEEE 802.4	802.4H	10Mbps	8002-4
IEEE 802.5	802.5J	4 & 16 Mbps	8802-5
IEEE 802.6	SMDS	46 Mbps	8802-6
ANSI-X3T9.5	FDDI	100 Mbps	9384-1,2,3

ANSI = American National Standards Institute
FDDI = Fiber (Optic) Distributed Data Interface
SMDS = Switched Multimegabit Data Service
ISO = International Standards Organization
IEEE = Institute of Electrical and Electronics Engineers

Figure 9: Examples of equipment for which standardized fiber optic LAN equipment is readily available.

The IEEE, ANSI & ISO Standards - brief overview

ISO, ANSI and the IEEE either have developed or are developing a number of emerging fiber optic standards.

As shown in Figure 10, IEEE 802.3 has created a subcommittee called 10Base-F (not-surprisingly similar to the name 10Base-T for Twisted pair). 10Base-F is a full 10 Mbps LAN system standard that is fully compatible with the other Ethernet standards, including the point-to-point FOIRL (Fiber Optic Inter-Repeater Link) standard, which already exists in the latest version of Chapter 9 of the 802.3 standard. The full 10Base-F standards effort and FOIRL have been endorsed by ISO. The IEEE 802.3 10Base-F standard consist of 3 parts:
1. 10Base-FP, Passive Hub segments,
2. 10Base-FA, Active Hub segments, and
3. 10Base-FM, Media specifications: cable & connectors.

...the primer on Fiber Optic and LAN technology.
Chapter 2 - THE FOUR MYTHS OF FIBER OPTICS

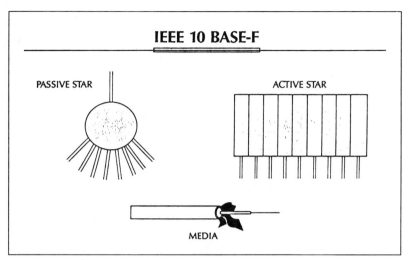

Figure 10: IEEE 802.3 10Base-F: Passive Hubs, Active Hubs, Cable and Connectors.

The entire 10Base-F standard is currently very near completion. All three parts of 10Base-F are in letter ballot by the entire IEEE 802.3 committee. At the time of this writing, the letter ballots are being reviewed and 10Base-F is expected to become a standard either at the November, 1990 meeting of the IEEE or the following meeting in March, 1991.

IEEE 802.4, the MAP Manufacturing Automation Protocol token passing bus standard committee has already approved its fiber optic standard called 802.4H. The IEEE 802.5 token ring standard committee is working on a draft fiber optic standard called 802.5J.

IEEE 802.6 is the new Metropolitan Area Network (MAN). It is the first data rate independent network protocol. It is envisioned to work over a range of data rates from 1.5 Mbps to 150 Mbps and higher on each of its two busses. The protocol, called "DQDB" for Dual Queue Dual Bus actually has a total throughput of twice the base data rate because of the use of dual busses. The first IEEE 802.6 MANs will be introduced by the telephone companies late in 1990 and 1991 together with higher level software called SMDS which stands for Switched Multi-Megabit Data Service.

18 Everything you ever wanted to know about Fiber Optics and LANs, but were afraid to ask!

FDDI is the ANSI Standard LAN which operates at 100 Mbps and has been endorsed by virtually every computer company. The FDDI standard consists of 5 parts. Four of these parts, covering the hardware aspects of FDDI have officially been approved as standards. The fifth part of the standard, called Station Management (SMT), which can be implemented entirely in software, has been approved by the working group and is soon expected to become a standard as well.

Myth 3: Fiber Optics is difficult to install

The third myth of fiber optics is: that it is difficult to install. There is only one way I have ever found to prove that this myth is not true, and that is to put fiber optic connector, on a cable, right in front of the audience. No pictures have ever fully convinced someone how easy this is to do, because the mythology about how difficult fiber optics is to use, is so strong. However, since this is a written paper, I'll try my best, and describe what I normally do at a presentation.

First, I take out my tools and parts and put them on top of the overhead projector. The connector parts (all four of them) come in a little plastic bag. They are: 1) a crimp ring, 2) a little plastic insert, 3) the connector, and 4) a piece of heat shrink tubing as shown in Figure 11. As you'll see, I usually forget to put the heat shrink tubing on before jumping ahead and stripping the cable. Quite frankly, I find remembering to put the heat shrink tubing on first, is the biggest problem with fiber optic connector installation.

...the primer on Fiber Optic and LAN technology.

Chapter 2 - THE FOUR MYTHS OF FIBER OPTICS

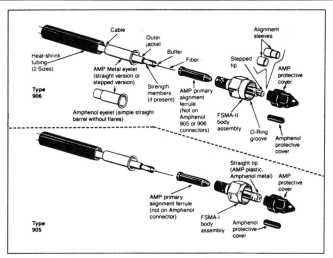

Figure 11: Fiber Optic Connector Parts and Tools.

Please put the heat shrink tubing on first. (Don't be like me and forget to put on the heat shrink tubing until after I've got most of the connector done). The second step is then to put on the crimp ring which holds the kevlar fibers mechanically to the connector body (I usually forget to put this on until after I've stripped the fiber, it's easier to do it before stripping the fiber).

Then we take this rather complicated tool I bought at Radio Shack, called a Wire Strippers, and we strip off the outer jacket of this fiber optic cable. When we look inside we find the yellow kevlar fibers that I described before, and a plastic coated glass fiber as in Figure 12. This plastic coated glass fiber is every bit as strong as any twisted pair wires, such as those in RS-232 type cable, and Ethernet twisted pair cable, a transceiver cable, or any other twisted pair type cable that you deal with. I demonstrate just how strong the fiber is by tying the thin 0.25 mm plastic coated glass fiber into a tight knot, and place it on top of the overhead projector. That projects a silhouette of the tiny optical glass fiber tied into a knot on the screen causing general amazement. You can handle the glass fiber with impunity as long as it has the plastic buffer on it.

Then I take the very same fiber, untie the knot and take another inexpensive tool, called a No-Nick Wire Strippers, which every technician has in his lab for dealing with hook-up wire, and use that to mechanically strip the plastic buffer jacket off the glass fiber. Then I put the stripped fiber on the overhead projector and one can clearly see in silhouette the glass fiber, the plastic buffer coating, the kevlar fibers and the outer jacket as in Figure 12.

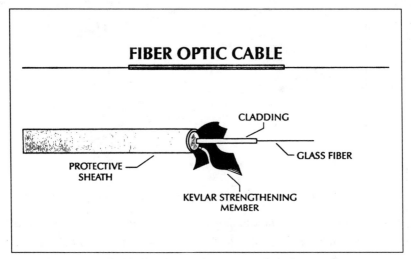

Figure 12: Fiber optic cable construction.

Now we put a little dab of epoxy in the connector, stick the little plastic insert in the connector and then insert the glass fiber into the connector. I put the assembly on the overhead projector and you actually see the glass going through the connector. (Considering that I thread the glass fiber into the connector with the light of the overhead projector shining in my eyes, it can't be that tough to do.)

Then we bring up the crimp ring and crimp the kevlar onto the connector with a standard cable crimping tool. Therefore, it's actually the kevlar which now absorbs all the mechanical stress and strain when we pull on the connector or the cable.

Next we break off the excess little piece of fiber that sticks out of the front of the connector. We are then left with small bead of epoxy residue on the tip.

Now we come to the most famous and most feared part of the fiber optic connectorization process: the polishing.

We attach the fiber/connector assembly to the 2 inch diameter polishing

We attach the fiber/connector assembly to the 2 inch diameter polishing disk. Then we take out three sheets of "lapping film". They look just like fancy sand paper. One is pink, one is green, and the other is grey. The technical names for them are: coarse, medium and fine. We place the polishing disk, with connector and fiber, onto the pink (coarse) lapping film and move the disk in figure eight movements for twenty seconds. (Yes, I actually do this on the overhead projector, the little 2 inch polishing disk makes very nice figure eight silhouettes on the screen.) Then we polish for twenty seconds on the green (medium) lapping film and finally twenty seconds on the grey (fine) lapping film. Then we take off the disk and we are done. We've put on an optical connector, in about 5 minutes, sometimes in front of an audience of 200 to 300 people.

At Codenoll we have a program with the state of New York where we hire mentally handicapped people through a program called Job Path. They come to our headquarters, we train them how to put on the optical connectors. They have a wonderful time putting on these connectors all day long. I must say, it is a very satisfying program for me. These are people who throughout their lives have been thrown out of one job after another, people never trusted them. Then we give them a job that they can actually do. For the first month they are very suspicious of us, "How long will it be until we throw them out?" Then, when they find out that they really do have a job their whole personality changes. They become very happy people. To be able to work eight hours a day and get paid every two weeks like everybody else and have a job and a home to go to, they feel part of society. It's really quite exciting.

We also trained real cable installers in many places throughout the world. In some cases that is more difficult.

Experienced fiber optic cable installers are readily available throughout the world. In fact there are over a thousand members of Local Three of the International Brotherhood of Electrical Workers currently trained and certified to put on fiber optic connectors on the island of Manhattan in the City of New York.

Myth 4: Fiber Optics is too expensive

22 Everything you ever wanted to know about Fiber Optics and LANs, but were afraid to ask!

Finally, we get into the myth: that fiber optics is too expensive. The answer is epitomized in the first Figure in this paper, Figure 1, which is repeated here for your convenience as Figure 13.

Duplex (two fiber) fiber optic cable in 1990, runs anywhere from $0.28 per foot to $0.75 per foot depending on the quantity and quality of the cable. Data grade unshielded twisted pair is about $0.14 a foot, and data grade shielded twisted pair starts at $0.62 a foot.

ISA/ETHERNET RETAIL PRICE COMPARISON

	Unshielded Twisted Pair	Shielded Twisted Pair	CodeNet Fiber Optic
ISA Network Board	$499	$499	$595
Cable to HUB (50m)	50	338	100
Connectors & Patch Cables	3	53	56
Labor ($45/hr)	30	40	45
Hub ($ Per Port)	365	365	115
Total $ Per Port	$947	$1295	$911

Figure 13: Comparison of LAN costs - ISA.

While Figure 13 above shows that for standard type personal computers a fiber optic LAN is either no more expensive, or less expensive than a twisted pair LAN; Figure 14, below shows that as the performance of the network increases, for example with Bus Master Extended-ISA machines, the cost of fiber optic LANs is considerably lower than twisted pair.

EISA/ETHERNET RETAIL PRICE COMPARISON

	Unshielded Twisted Pair	Shielded Twisted Pair	CodeNet Fiber Optic
EISA Network Board	$1150	$1150	$ 995
Cable to HUB (50m)	50	338	100
Connectors & Patch Cables	3	53	56
Labor ($45/hr)	30	40	45
Hub ($ Per Port)	365	365	115
Total $ Per Port	$1597	$1946	$1311

Figure 14: Comparison of LAN costs - EISA.

Twisted Pair FDDI?

Just last month I asked Belden Cable to quote on all three different types of cable, to compare the costs of doing a test application -- an apples to apples comparison. Unshielded twisted pair was $.14 a foot, fiber optic cable was $.40 a foot, and shielded twisted pair was $.62 a foot.

In fact, I did this to help a reporter. His question concerned claims by two companies that in theory they could run 100 Mbps FDDI on shielded twisted pair and that would provide substantial cost savings for users. The reporter wanted to know what the cost savings were, since the companies making the announcement didn't want to give out any figures. Well, with fiber optic cable costing $0.40 a foot and shielded twisted pair costing $0.62 a foot, neither one of us could quite see exactly where the cost savings could come from.

Chapter 3 - FIBER OPTIC CABLING, the UNIVERSAL SOLUTION

Our goal at Codenoll was to develop a total cabling solution that would allow a facility to be cabled one time and allow the network technology to change by only changing the end node equipment -- i.e. never needing to change a cable.

The computer industry has been messing up users lives for decades now. In Figure 15 I have shown some of the many cable changes that the computer industry has imposed upon you as users. Starting out with five wire RS-232 cable through various impedances of coax, twin-ax and going up to unshielded and shielded twisted pair. Every time the industry came out with a "better" technology you had to install a new cable.

My favorite story on this subject is that of a major computer company's development programming group, where they had 50 programmers in a very large open office type environment. Every time they changed the computer equipment they threw the new cable over the standard hung, acoustic tile, ceiling. One day, about six years ago, the ceiling came down on top of the fifty programmers.

In another example, we have a customer facility in California where they have determined that to run shielded twisted pair cables they would have to do $300,000 worth of construction to strengthen the structure of the building, because

26 Everything you ever wanted to know about Fiber Optics and LANs, but were afraid to ask!

it won't carry the weight of the shielded twisted pair copper cable.

Our goal at Codenoll was to come up with a universal cabling solution that would handle all of today's networking protocols and problems as well as tomorrow's. After examining the chart in Figure 15, we could see that all of the future standards, as well as some of the present standards, will run on fiber optic cable. In many cases, because of the data rate, only on fiber optic cable.

AVOID OBSOLESCENCE

Network	Data Rate	Original Cable
ASCII	19.2 KBPS	TWISTED PAIR
IBM SYSTEM 36/38	1.0 MBPS	TWIN-AX
PC NETWORK	2.0 MBPS	75 OHM COAX
IBM 3278/9	2.35 MBPS	92 OHM COAX
TOKEN RING	4.0 MBPS	DUAL SHIELDED TWISTED PAIR
IEEE 802.4 MAP(GM)	5/10 MBPS	FIBER OPTIC OR 75 OHM COAX
IEEE 802.3 ETHERNET	10 MBPS	FIBER OPTIC/COAX/TWISTED PAIR
ADVANCED TOKEN RING	16 MBPS	FIBER OPTIC/TWISTED PAIR
IEEE 802.6	50 MBPS	FIBER OPTIC
ANSI - FDDI	100 MBPS	FIBER OPTIC

Figure 15: The various networking protocols, data rates and cables.

Therefore, the problem became one of determining how to run all the different types of LANs on a single fiber optic cabling system. In order to do this we had to go back and study the very fundametal principles of LANs and networking. Once we fully understood the basic principles behind networking we could solve the universal problem.

In the section that follows, I describe in rather simple and understandable terms (I believe), the result of several man years of basic research in networking. I believe both the least technical and the most technical reader will find this description enLIGHTening.

Chapter 4 - FUNDAMENTALS OF NETWORKING TECHNOLOGY

There are only three types of networking technologies. Number one is the ring (Figure 16), which has point-to-point connections between all the nodes.

Basic LAN Technology 1: The Ring

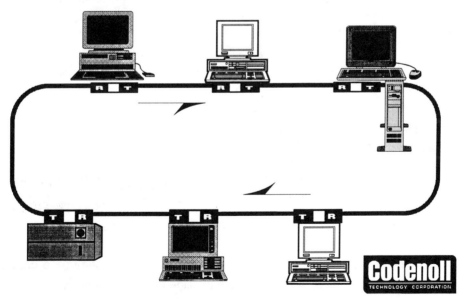

Figure 16: Basic LAN Technology Number 1 of 3: The Ring.

> [Note: We shall use the term "node" to mean any piece of equipment on a network that uses information -- e.g. personal computers, bridges, gateways, mainframes, etc., as opposed to a "hub", which only passes information through, unless it is also a node.]

In the ring LAN technology, every node is a repeater, every node has the intelligence to detect addresses, every node has the intelligence to determine whether to use the packet of information or just repeat the information and every node has the intelligence to manage the ring in case of errors.

The second major type of networking technology is the active hub, active star, switching, and multiplexing type systems that are shown in Figure 17.

Basic LAN Technology 2: The Active Hub

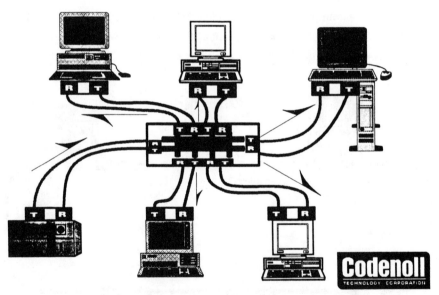

Figure 17: Basic LAN Technology Number 2 of 3: The Active Hub.

The chief characteristic of these systems is that they depend on a central hub unit that contains most, if not all, of the intelligence in the network. Although, some of the intelligence may be distributed intelligence out in the nodes, the central hub is a repeater for all the packets. The individual nodes are not repeaters as in the ring. While each node is not a repeater, and therefore a less intelligent and simpler device than in the ring, the central unit is a repeater for all signals, data, packets etc. Therefore, although this system may be configured as a bus (active hub Ethernet) as well as a switch (as in a telephone PBX), all connections between the nodes are actually double point-to-point connections between the nodes and the hub.

The third major type of network technology is the so called Linear Bus System shown in its familiar drawing in Figure 18. It is really not a linear bus system, it is a **Broadcast System**. Personally, I think the name "bus" is a misleading misnomer for these type systems which are the most commonly used in the world. Broadcast type LANs include: Arcnet, Ethernet and the new IEEE 802.6 MAN (Metropolitan Area Network).

Basic LAN Technology 3: The Bus = Broadcast

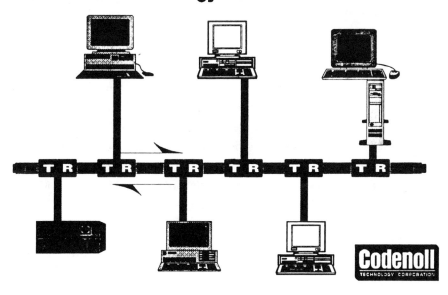

Figure 18: Basic LAN Technology Number 3 of 3: The Bus or Broadcast System.

Let's examine how a bus/broadcast network really works. If Node 1 wants to talk to Node 4, Node 1 sends its message out on the "media", a coax cable if you must. The packet travels up and down the "media" so that every node on the "media" sees the same packet. Nodes 2, 3 and 5 will ignore the packet because it does not have their addresses in the "destination address field" of the packet. Node 4 on the other hand does see its address, reads in the packet and uses the information in the message.

Well, if we got rid of the coax cable, and put radio-frequency-transceivers in place of the coaxial-cable-transceivers, the LAN functions the same. When Node 1 wants to talk to node four it will radio-broadcast the message "all over". All the radio receivers will pick up the signal but only Node 4 will recognize its address and use the packet. The other nodes ignore the packet.

The LAN functions just like our broadcast television system! Channel 2 is broadcasting its signal all over. Every TV in the area actually receives all of the signals of all the TV stations all of the time! (Sorry Mr. Lincoln.) But only those TVs whose channel selectors are set to Channel 2 will actually use the information!

So we can see that an Ethernet LAN is truly a broadcast system environment. The only function of the coax cable in the original Ethernet is to contain the electromagnetic radiation of the packet information so that the data signal does not really radiate "all over" and interfere with public frequency allocations, the telephone system, the operation of computer equipment, etc. The data rate of Ethernet, by the way, is 10 Mbps and the signals used are 10 MHz. These signals are right between the AM radio band (550 kHz to 1.6 MHz) and Channel 2 television (50 MHz), and radiate very well.

One last note: it is this quality of Ethernet, that each node <u>broadcasts</u> its signals "all over" the LAN (i.e. they are seen by every node) that gives <u>Ethernet</u> its name. In times long past, scientists and engineers thought that electromagnetic energy was transmitted through the "Ether" all around us.

Alchemy -- Copper to Glass, Electrons to Photons

In the first LAN technology, the Ring type LANs, it is fairly easy to see how we can convert the point-to-point connections in the ring from copper to fiber optics. Each node is a computer type device which uses CMOS, TTL or ECL standardized electronic circuitry inside. Since standard CMOS, TTL and ECL has been designed and optimized for use inside computer equipment, these circuits can only drive electronic signals for distances of 6 inches. Therefore, it is necessary to have a cable drive circuit and a cable receive circuit in each node to send and receive the data over the long inter-nodal ring cables. These circuits are indicated by the small boxes labeled "T" and "R" in each node in the ring Figure 16.

To convert a ring to fiber optics we need only replace the copper cable drive circuit (labeled "T") with a fiber optic transmitter and the copper cable receiver circuit (labeled "R") with a fiber optic receiver. Then we have a fiber optic ring.

In the second LAN technology, the active hub type LANs, once again, since all the connections are point-to-point connections and the computer type equipment uses CMOS, TTL or ECL signals, each cable needs a cable driver ("T") and a cable receiver ("R"). Once again we can simply replace these point-to-point copper cable driver and receiver circuits with fiber optic transmitters and receivers to make a fiber optic active hub local area network.

While we could easily see how to implement in fiber optics the point-to-point links in the Ring and Active Hub type LANs, it was not so obvious how to make a fiber optic implementation of the multi-drop (as opposed to point-to-point) Bus/Broadcast type LANs such as Ethernet. We needed an optical bus/broadcast "tap".

Tapping the Light Fantastic - Couplers and Hubs

To implement the bus/broadcast type LAN in fiber optics we developed the passive optical tap, splitter, or passive star. In this true broadcast device, we simply take a bunch of fibers and fuse them together in the center. Then, as in Figure 19, when an input signal comes in on the left side, it spreads out in the optical divider. Then one-nth of the signal goes out into each of the output fibers.

32 Everything you ever wanted to know about Fiber Optics and LANs, but were afraid to ask!

Figure 19: Operating principle of a passive optical star coupler.

Thus, as we can see in Figure 20, if a computer transmits into the passive optical hub, the signal is divided (in this case by four) and comes out to all the other computers on the network as well as back to itself. It comes back to itself so you get complete self-checking that your signal was in fact properly transmitted. This adds an even higher degree of reliability to your transmission.

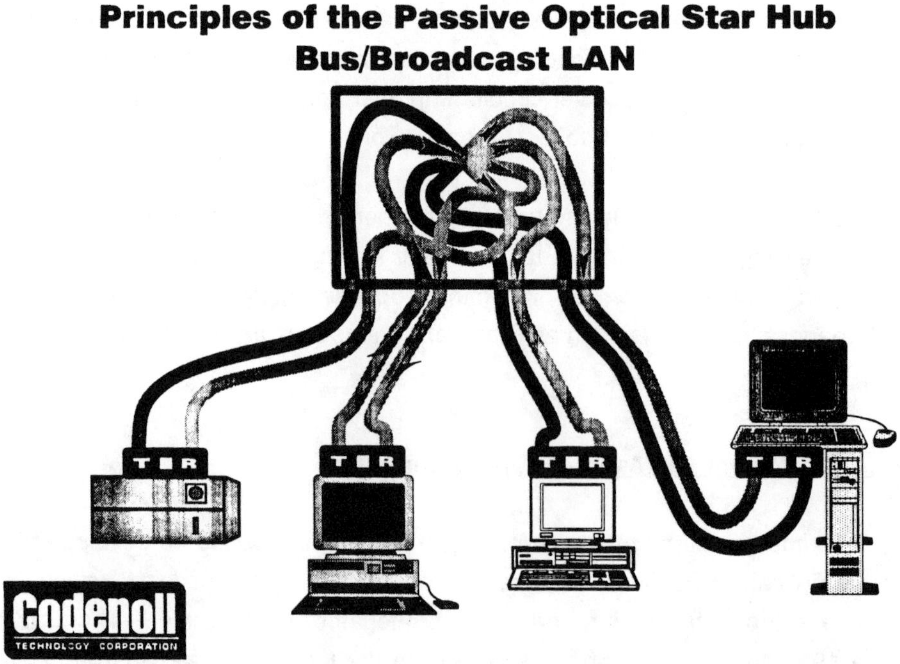

Principles of the Passive Optical Star Hub Bus/Broadcast LAN

Figure 20: Using a passive optical star to implement a Bus/Broadcast LAN.

Chapter 4 - FUNDAMENTALS OF NETWORKING TECHNOLOGY

Tokens and Collisions

Now, what about all those cocktail party terms like token passing and collision detection? Where do they fit in our theory? They are "access methods". Referring to Figure 21, we can see that access methods fall into two major categories:

1) deterministic = collision avoidance, and
2) stochastic = collision detection with contention resolution.

The most common way of providing collision avoidance, which means that no two nodes can transmit at the same time, is through token passing. Token passing ring is IEEE 802.5, and token passing bus is IEEE 802.4 (yes, a bus can be deterministic).

The stochastic systems rely on a very strong body of mathematics, called Finite Markov Theory, which shows that the probability that two nodes will actually want to transmit at the "same time", where "same time" in Ethernet means within the same 52 microsecond period, is very close to zero. Furthermore, the very few times it happens, it can be detected and resolved. The resolution occurs by having both transmitting nodes stop and wait different random period of time (actually random multiples of 52 microseconds in Ethernet). Then one of the nodes will transmit first, acquire the bus for its full transmission and the other node will patiently wait for the first node to finish (that's the Carrier Sense part of CSMA/CD = Carrier Sense Multiple Access with Collision Detection).

34 Everything you ever wanted to know about Fiber Optics and LANs, but were afraid to ask!

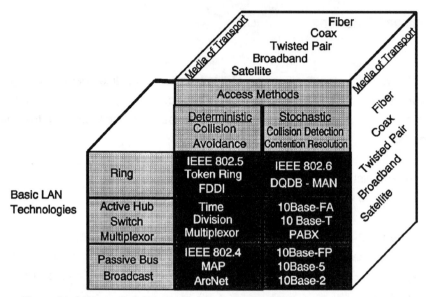

Figure 21: 3-Dimensional Matrix Model of: Basic LAN Technologies vs Access Methods vs Media of Transport.

Let's accept that ring, active-hub and passive-bus/broadcast are the three fundamental types of network technologies, with the two basic types of access methods. Then, we can show that we can cable all of the existing and future types of networks in a single hierarchical structured fiber optic Star cabling system.

Using the model in Figure 22 we can construct many layers of network architecture. At the bottom layer you have, for example, desks in a work group, with two fibers running from every desk to some central location where you might put a passive hub. Perhaps on the wall next to the secretary's desk, in the ceiling, or on a bookcase.

...the primer on Fiber Optic and LAN technology.

Chapter 4 - FUNDAMENTALS OF NETWORKING TECHNOLOGY

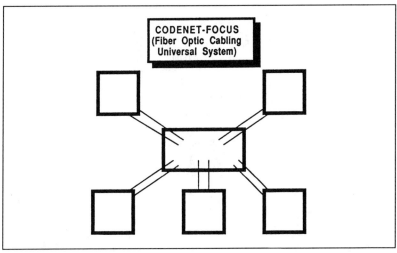

Figure 22: Hierarchical Star Fibered Tree Structured Cabling System: Basic cell.

Then, using the same architecture at the next higher level, you bring two fibers from every workgroup hub to a satellite distribution closet. At the next higher level you bring two fibers from each of the satellite distribution closets to an intermediate distribution frame using the same architecture. At the next higher level you run two fibers from the intermediate frames into a main distribution frame and then two fibers from each of the main distribution frames in different buildings into a central campus distribution location.

Then, referring to Figure 23, we see that any of these "hierarchically star cabled" levels can become a Bus/Broadcast system by putting a passive optical star like the CodeStar LightBus series at the hub or concentration point of each star cabled level.

36 Everything you ever wanted to know about Fiber Optics and LANs, but were afraid to ask!

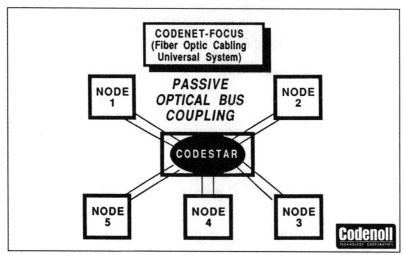

Figure 23: Hierarchical Star Fibered Tree Structured Cabling System: Bus/Broadcast LAN.

The same cable can also be used, now or later, for an active-hub/switch/multiplex system simply by replacing the passive optical hub with the appropriate active hub, such as the CodeNet MultiStar as in Figure 24.

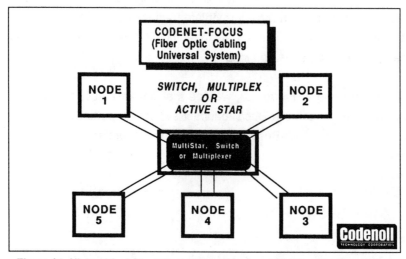

Figure 24: Hierarchical Star Fibered Tree Structured Cabling System: Active Hub.

...the primer on Fiber Optic and LAN technology.
Chapter 4 - FUNDAMENTALS OF NETWORKING TECHNOLOGY

Please note, that we don't have to change the cabling, even though we may have changed technology, since fiber optic cable is for all intents and purposes bandwidth independent.

We can even use the same cabling system for a ring type system. We simply take the transmit fibers from prior nodes and connect them, at the hub location, to the receive fibers of subsequent nodes as in Figure 25. This creates what we call a Star Wired Ring, a name which is used by IBM as the only way to implement their Token Ring systems (IEEE 802.5), and there are significant, important and expensive reasons for that design rule.

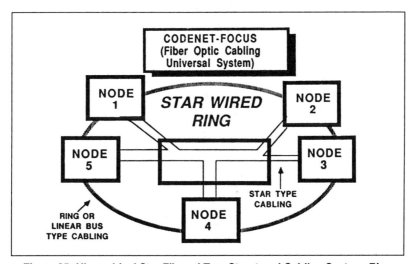

Figure 25: Hierarchical Star Fibered Tree Structured Cabling System: Ring LAN.

If you try to wire a ring system using point-to-point or circumferential ring cabling as in Figure 26, a failure is very difficult, time consuming and therefore expensive to locate, diagnose and repair. If you use radial Star Wired Rings it's considerably simpler. You go to one known place to diagnose and repair the failure, in a very short time. This makes the mean time to get the network up and running again very short.

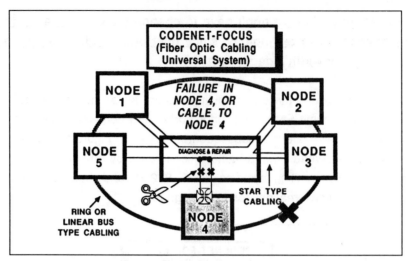

Figure 26: Maintenance of Ring Failure.

The second major reason for using star cabled rings or bus/broadcast systems is Murphy's First Law of Networks. This law states that "Wherever you run this cable the next computer you want to install will be in the wrong place". Therefore, as shown in Figure 27, you would have to run two new cables, adding considerably to the time required to install the node and insert it into the ring or bus.

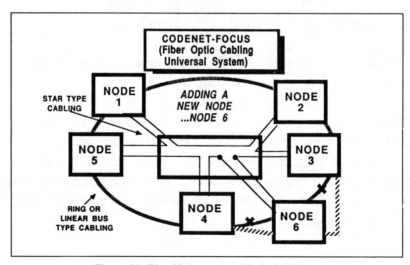

Figure 27: Ring Maintenance: Node Addition.

If you have a Star Wired Ring or Bus, you simply run one cable to the "known place", the wiring closet and you can very quickly insert the new computer into the ring or onto the bus.

Conclusion - Fundamental Theorem of LANs

> ### Michael Coden's Fundamental Theorem of LANs
>
> Thus we have proven that:
> 1. There are only three types of LAN technology:
>   - 1.1 Ring
>   - 1.2 Active Hub/Switch/Multiplexor
>   - 1.3 Passive Bus/Broadcast
> 2. There are only two types of access methods:
>   - 2.1 Collision Avoidance = Deterministic
>   - 2.2 Contention Resolution = Stochastic
> 3. All of the past, present and future networking technologies fit into this three by two matrix
> 4. All of the technologies covered by this matrix can be accommodated by a hierarchical star cabled fiber optic architecture, and therefore,
> 5. All past, present and future network technologies can be run on a single tree structured hierarchical star fibered cabling system. [QED]

Chapter 5 - CONFIGURING FIBER OPTIC LANs

With this technical background, we can start building fiber optic LANs level by level. We can start with a basic workgroup LAN consisting of a passive star and some PC adapter cards as in Figure 28.

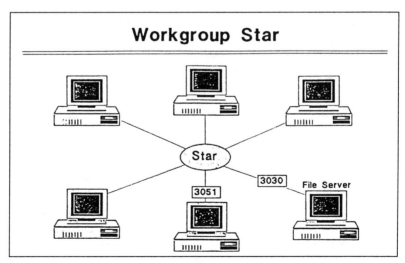

Figure 28: Basic Fiber Optic Ethernet Workgroup.

Then we can build hierarchies of stars, connected by repeaters, as in Figure 29.

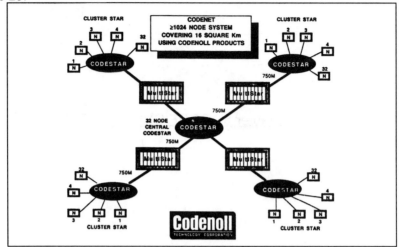

Figure 29: Hierarchical structure of Workgroup LANs.

Then by using MultiStar multiport repeaters as active hubs we can go all the way up to nine levels of stars. Where we can have as in Figure 30 passive - active - passive - active - passive - active, and so on.

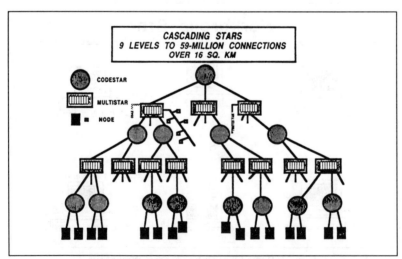

Figure 30: The full 9 level fiber optic Ethernet structured cabling system.

Chapter 5 - CONFIGURING FIBER OPTIC LANs

Figure 30 is the ultimate general purpose networking diagram. It represents the most general case of all network configurations. For convenience, in this Figure we have used circles to represent CodeStar passive optical star hubs, and rectangles to represent CodeNet MultiStar active hubs. We assume the initial configuration of our "general case" network is Ethernet, although it could just as well be IEEE 802.5 Token Ring.

Note, if we turn the diagram 90 degrees on its side, we have a major skyscraper office automation LAN, such as Southwestern Bell's 44 floor Corporate Headquarters building in St. Louis, Missouri. Eight hundred transceivers connecting 2,500 users and 2,500 mainframe computer ports. The lowest level of our Figure represents the transceivers and PC adapters on the desktops, connecting terminal servers, gateways and personal computers to passive stars in the East and West wiring closets on each floor. Active hubs connect groups of five floor wiring closets together. Then another layer of CodeStar passive stars connects all of the group wiring closets.

The hierarchy eventually comes to a main hub on the twenty-second floor where all the group wiring closets are connected together. That hub is connected by bridges to all the buildings in the St. Louis area together and by gateways through a wide area network to over 150 other buildings in Southwestern Bell's five state area. In total, Southwestern Bell has over 50,000 users attached with CodeNet Fiber Optic Ethernet.

Chapter 6 - UPGRADE MIGRATION PATHS

As the network traffic increases, network response time may slow down. This is a curious process. As users become more and more enamored of the network and what it does for them, they use the network more. This increases the traffic load, which eventually causes the network to appear slower to the users. The users however, have been spoiled by the previous fast response and insist on maintaining the previous level of response even as they use the network more and run more complex applications on it. In other words, the more successful the network is, the more it is used. The more the network is used, the more performance suffers. The more performance suffers, the more performance the users want added.

In any IEEE 802, ANSI, or ISO network, every packet is seen by all the nodes (equipment) on the network. In an Ethernet LAN, or Token Passing Bus, every packet is broadcast and seen by every node. In a Token Ring LAN, every packet is repeated by every node. Therefore, the more users, and the more each user uses the network, the more packets that are jamming a fixed bandwidth (capacity) highway.

We can increase the capacity of a LAN by dividing it into several sub-LANs connected by devices we call Filtering Bridges. A filtering bridge is a device that replaces a hub in our general purpose hierarchical star cabling system (Figure 30). Instead of repeating the packets from the sub-LAN on one side of the hub to the sub-LAN on the other sides, the filtering bridge reads every packet address and looks in a table to see where the node with that address is located. Then, if the destination of the packet is on the same side of the bridge as the source of the packet, the bridge does not repeat the packet to the other sides, reducing the amount of traffic on the other sides of the bridge. Of course if the bridge finds that the destination address is not located on the same side as the source address, it does forward the packet to the appropriate sub-LAN(s). This can greatly reduce the overall traffic load of a network and therefore greatly increase the performance of a network, since studies have shown that in an average network, 80% of all packets normally are destined for a node on the same sub-LAN and only 20% of the packets need to be repeated to the other sub-LAN(s).

We can add the filtering bridge capability in a very cost effective way if we have used the recommended CodeNet FOCUS hierarchical tree structured cabling system. Here, as in Figure 30, we can begin putting in bridges one at a time, only at the highest level of the tree structure, where we can achieve quite a large increase in performance by replacing the smallest number of hubs. The packets that need to, will travel all throughout the network. But those packets that are used locally are kept localized. So a packet that might have previously gone through the whole network, but is really only needed on the local sub-LAN, will just stay on one side of a bridge and not be passed on to the rest of the network, thereby increasing the bandwidth of the whole network.

So, one at a time, as traffic increases we can change each hub to a bridge, increasing the capacity of the network each time. We work our way down the tree as time goes on and network utilization goes up. This allows us to start with a very small number of bridges and increase the number of bridges as bridge prices come down. It's very cost effective.

Chapter 7 - MIGRATION TO FDDI AT 100 Mbps

At some point we will have such high network utilization that we have used up all the bandwidth of the Ethernet or Token Ring network. If we used the CodeNet FOCUS fiber optic hierarchical tree structured cabling system then we can easily convert any portion of the network to FDDI without changing a single cable. All we need are FDDI-to-Ethernet and FDDI-to-Token Ring Bridges. These bridges can then be put in place of any existing sub-LAN bridge or hub. On one side of the bridge/hub the star fiber cabling runs Ethernet and/or Token Ring and on the other side of the bridge/hub the star fiber cabling runs FDDI. Referring again to Figure 30, we could make the first layer of active hubs into multiport FDDI to Ethernet bridges. Then the top hub, which was originally a passive hub can be reconfigured to be a star cabled FDDI ring hub between the bridges. Or better yet we can replace the top level hub with an FDDI Concentrator hub, like the CodeNet 9041.

The FDDI Concentrator

What's a concentrator? It is an active FDDI hub, that performs for the 100 Mbps FDDI ring the equivalent star cabling functions that the MAU (Multistation

Access Unit) does in a star cabled Token Ring IEEE 802.5 LAN. These functions include "Configuration Management" which is the inserting of a ring node when it is successfully powered up, the de-inserting of a failed or powered down node, and other network management functions described below.

FDDI Concentrator: Star Wired Ring

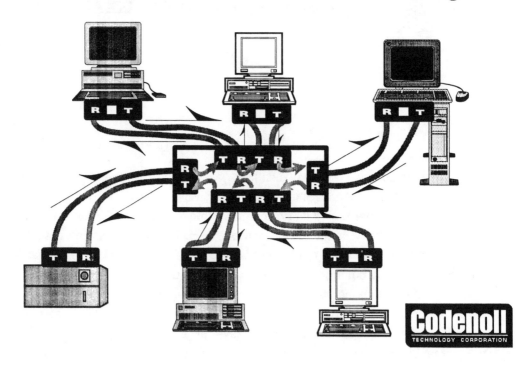

Chapter 8 - REVIEW OF FDDI PRINCIPLES

Before we go further with our example of upgrading the fiber optic structured star cabling system from Ethernet and Token Ring to FDDI it may pay to review both the basics and a little bit of the history of FDDI. This will enable us to better understand how a few (like three) building blocks can be used to construct all the various FDDI configurations and also enable bridging between FDDI and Ethernet Token Ring!

The basic building blocks of FDDI are adapter cards, hubs and cable. FDDI is a 100 Mbps token passing ring as shown in Figure 31. The main problem with token passing rings is: "what happens if you have a break in the ring?" Answer: "the whole system is down".

50   Everything you ever wanted to know about Fiber Optics and LANs, but were afraid to ask!

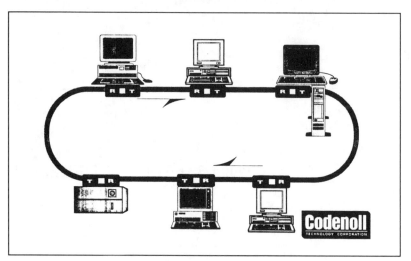

Figure 31: FDDI single ring configuration.

To solve that very serious problem the FDDI committee, in its wisdom, developed a dual ring where the signals on the two rings are traveling in opposite directions as in Figure 32. This technique is known by the simple name "Dual Counterrotating Rings".

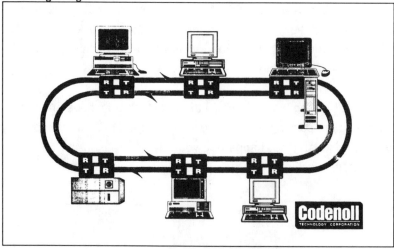

Figure 32: FDDI Dual Counterrotating Rings.

The reason for using the dual counterrotating rings is shown in Figure 33. If there is a cable fault, or a node failure any one place in the system, the nodes can be intelligent enough to do "end wraps". Thus the FDDI 100 Mbps ring will wrap around to remain a continuous ring, although about double its original length.

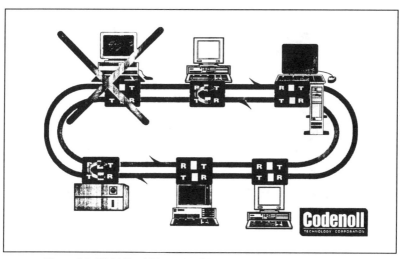

Figure 33: FDDI Dual Ring with Single Break and Wrap-Around.

What happens if the ring breaks in two places? We end up with two disconnected rings. This is not a desirable situation. In fact, this was not the original intent of the FDDI design.

If we go back several years ago to the beginning of the ANSI committee's history, when I was personally attending, the original design of FDDI consisted of a Single Link (as opposed to dual) from each node to a Concentrator hub. The concentrator is an active hub such that the FDDI ring is really connected as a star-fibered ring. The data path is: station to concentrator, back to station, to concentrator, back to station, to concentrator, and so on, as we described before. As we can see in Figure 34, with a star cabled ring you don't need the dual ring because the concentrator hub can actively electronically bypass any non-functioning areas in the network. Therefore, you save the expense and complexity of having two fiber optic interfaces (called PHYs) in each end node (workstation). Similarly, in this case the only device that needs expensive and complex configuration management intelligence is the concentrator hub. The individual nodes do not need to have the "wrap around" intelligence and expense of the dual counterrotating ring fiber optic interfaces.

52 Everything you ever wanted to know about Fiber Optics and LANs, but were afraid to ask!

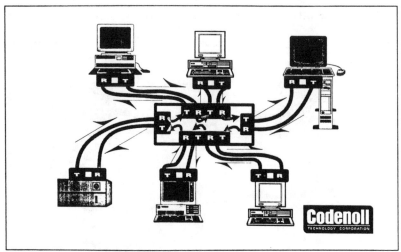

Figure 34: FDDI Concentrator, Star Wired Ring.

Well, what is a concentrator? How do we make a concentrator? The CodeNet 9041 FDDI concentrator is a personal computer with some number **of** differently configured CodeNet 9000 series, modular personal computer adapter cards, running the CodeNet Concentrator software package.

Each concentrator needs only one MAC (Media Access Controller), the module that understands FDDI packets and allows the concentrator to interpret and respond to network management packets. Therefore, at least one of the plug-in boards in a concentrator must have at least one MAC. This is the board that is usually connected to the next higher level in the structured cabling system. In much of the FDDI literature today, FDDI is shown as a "ring of trees", where the highest level of the structure (usually called the backbone) consists of a dual counterrotating ring that is interconnecting concentrators which act as star-wired-ring (or tree) hubs to the lower level concentrators and various computer equipment as in Figure 35.

...the primer on Fiber Optic and LAN technology.

Chapter 8 - REVIEW OF FDDI PRINCIPLES

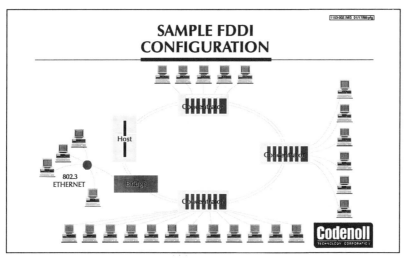

Figure 35: FDDI Standard Ring of Trees Diagram.

In the CodeNet 9000 Modular FDDI system, each plug-in adapter card consists of three parts:

1) a bus interface board (e.g. AT, ISA, EISA, MCA, VME, etc.),
2) one or two MAC daughter board modules, and
3) one or two PHY daughter board modules.

This creates a product family of six different combinations as shown in Figure 36.

The rest of the plug-in bus interface boards do not need the MAC daughter boards, only the PHY daughter board modules. All the other boards are used as the concentrator end of "single attach" connections to the workstations and are MAC-less. The fiber optic cable physical connection runs from one of the PHY modules in the concentrator to a single PHY FDDI adapter board in a personal computer or workstation or to single PHY boards in other (lower level in the tree) concentrators, and then out to individual personal computers and work stations. In this manner we can build the same tree structure that we had before.

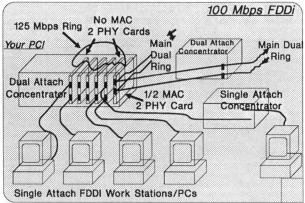

Figure 36: FDDI Concentrator, Internal operation.

As I mentioned before, the conventional drawing that you will see in the literature today is the one in Figure 37 which shows the dual ring of concentrators and bridges, with the concentrators connected to individual work stations. My opinion is that the dual ring will not achieve wide popularity. What I believe will happen is that we will find a concentrator in the center of every major FDDI installation, with star wiring to all the main units, because of the same important reasons of maintenance, quality, network downtime and network management that I described before in Chapters 4 and 5.

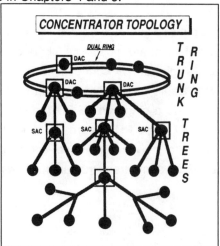

Figure 37: FDDI Ring of Concentrator Trees.

Chapter 9 - BRIDGING BETWEEN DIFFERENT NETWORKS

Any migration path from IEEE 802 networks like 10 Mbps Ethernet and 16 Mbps Token Ring, to LANs and MANs like 100 Mbps FDDI and IEEE 802.6, needs to include a device that converts data from one network format and speed to another. Such a device is called a Bridge. While most bridges today are built as specialized dedicated boxes of equipment, two new developments in technology have enabled us to modularize and standardize a multitude of such devices.

The first is the development of "Bus Master" technology in modern personal computers. This is a technology that allows a plug-in adapter card to transfer data very quickly over the internal computers data bus: 33 Million Bytes per second in the case of EISA (Extended Industry Standard Architecture and 40 Million Bytes per second in the case of MCA (MicroChannel Architecture). The second is the development of the modular network adapter cards at Codenoll with specially developed low cost hardware/software address filtering technology that distributes the bridging functions efficiently among multiple adapter cards and the host CPU or multiple CPUs.

During the last 3 years Codenoll has devoted the same kind of thorough study to making the most general purpose modular hardware and software modules as we did to developing the CodeNet-FOCUS Fiber Optic Cabling Universal Solution. The development of the general purpose hardware modularity was code named "Verdi" at Codenoll and has provided a unique ability to provide many complex functions with a simple set of interchangeable modules. The development of the modular software that allows the combining of the hardware modules to make an almost infinite variety of systems products based on personal computer platforms was code named "Mozart", and had been jointly developed by Codenoll and NYNEX Science and Technology.

Using these advances we can use a modern bus master (EISA or MCA) personal computer as the platform for a high speed multi-protocol bridge. We simply plug-in a variety of CodeNet modular network adapter cards, such as one or more FDDI adapter cards, one or more Ethernet adapter cards, one or more Token Ring adapter cards and a modular software package creating a multiport FDDI-Ethernet-Token-Ring bridge.

Once we have this new modular tool, which we call a "MultiBrouter" we can begin to upgrade our CodeNet-FOCUS hierarchical-star-fibered-tree-structured-cabling-system LAN from Ethernet and/or Token Ring to FDDI -- by only switching hubs and adapter cards! We do not have to change a single cable!

Referring to Figure 38 we see that we can again start at the top level of the structured cabling system. In the top spot we exchange the previous Ethernet or Token Ring hub for an FDDI concentrator (a personal computer with cards and Mozart software). Then we replace each of the next level hubs with CodeNet FDDI-Ethernet-Token Ring MultiBrouters (a personal computer with cards and Mozart software) and, at a very low cost, without changing a single piece of cable, we have an FDDI backbone.

...the primer on Fiber Optic and LAN technology.

Chapter 9 - BRIDGING BETWEEN DIFFERENT NETWORKS

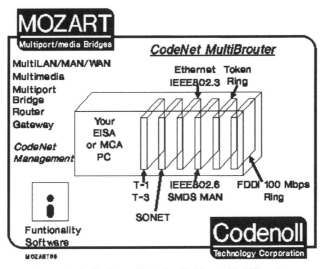

Figure 38: CodeNet Modular Multiport MultiBrouter

We can follow the same procedure for any hub area in the entire **cabling** system as different areas of the LAN require the higher speed performance. For example, in graphics intensive applications (like at our customer Campbell Ewald Advertising), we can bring FDDI right out to desktop.

For as little as $3995 retail, complete CodeNet 9000 series FDDI cards can allow you to connect a personal computer to an FDDI ring or concentrator, turn a personal computer into an FDDI bridge, or interconnect a multitude of NetWare, LAN Manager and UNIX file servers in a 100 Mbps backbone. So the reports that FDDI costs are from $11,000 to $12,500 per node are false.

There are four significant advantages to the Codenoll Mozart approach of using standard personal computer platforms and adapter cards for purposes that were previously required special custom boxes:

1. You save a significant amount of money by using standardized, mass produced personal computer platforms. You purchase the personal computer from the most economic source. Then you purchase the CodeNet cards and software as a kit from Codenoll. You do not have to pay several extra "mark-ups" on the box, fan, power supply,

mother board and CPU that are inside your bridge, router, concentrator, etc.

2. The same low cost modular parts can be used to make bridges, routers, concentrators and attach personal computer/workstations for high throughput. You buy only what you need and you can upgrade or modify your LAN simply by buying and exchanging modules -- not entire boxes. Moreover, as you modify one part of your LAN, <u>any</u> parts that you remove can be used somewhere else in your network. For example, if you start with Ethernet-Ethernet Mozart bridges and then want to upgrade to Ethernet-FDDI, remove one Ethernet card -- (And use it in a workstation!) -- and replace it with an FDDI card. After loading the Mozart software upgrade you have an Ethernet-FDDI bridge.

3. You buy only the platform with the performance you need today -- and you can upgrade the platform at any time. For example use 80286 based machines for concentrators and low throughput bridges. When your network requires higher performance you can plug the same cards and software into an i386, i486, a multiprocessor platform or an i586 when it becomes available. Not only can you upgrade the performance of the system at any time by just changing the box, but the box you are replacing is a reusable general purpose personal computer that can be put into a multitude of other applications or on a deserving desk top.

4. The "box" you have purchased can be continuously modified and expanded to additional future technologies such as: T-1, T-3, IEEE 802.6 MAN (Metropolitan Area Network), SMDS (Switched Multi-megabit Data Service), and SONET the new internationally standardized Synchronous Optical NETwork.

So we have an interesting situation where if you take modern personal computer technology, such as the Extended-ISA Bus which has a 33 megabytes per second throughput rate or the new MicroChannel Architecture which has a 40

Chapter 9 - BRIDGING BETWEEN DIFFERENT NETWORKS

megabytes per second throughput rate, and FDDI, Ethernet, Token Ring, the new IEEE 802.6 MAN, T-carrier cards and the new Sonnet, Synchronous Optical Network cards, standard telephone cards such as 19.2 kbps and 56 kbps, combine them with a software package and make bridges between any or all of these different types of networks.

The ability to cost effectively use the ISA/AT, EISA and MicroChannel platforms is based on the Codenoll Verdi modularity technology. Modular daughter boards can plug onto a MicroChannel bus-interface-mother-board just the same as they can plug onto AT/ISA type bus-interface-mother-board or the ISA bus-interface-mother-board. So with a small number of mass produced modules we have all the options that we could ever want.

There is only one thing you have to do -- put in that fiber optic cable today! Then no matter what network comes out tomorrow, they're all going to run on that fiber. No matter what the speed and performance of the new network technology because the bandwidth of fiber optics is virtually infinite. You'll never run out of bandwidth, you'll never have the wrong impedance, you'll never have a short circuit, you'll never have the wrong gauge or the wrong number of twists per foot. In other words, if you put in fiber optic LANs today, you will save money on the immediate installation, on the maintenance, and on the upgrades. With fiber optic LANs, you are set for decades to come.

Appendix A - The IEEE, ANSI and ISO Standards - Detail

Many people ask how the standards organizations work. They are a rather complex hierarchy of organizations, that are motivated by politics and profit much more than by technology. In this appendix, I have attempted to include a rather complete description of the international standards making process and the current status of Fiber Optic LAN standards as of September 1990.

ISO, the International Standards Organization, has delegated the standardization of LAN technology to the United States' standards body which is ANSI, the American National Standards Institute. ANSI has in turn delegated the lower speed LAN standards (defined as 50 Mbps and below) to the IEEE, while keeping the higher speed LAN standards in its own committees, such as FDDI. Once the IEEE develops and approves a standard it is sent to ANSI. Then ANSI reviews the IEEE standard. If ANSI approves the standard it is sent to ISO. ISO then tries to make sure that the standard will work worldwide by soliciting comments from all member countries.

Fiber optic products that meet the IEEE, ANSI and ISO standards are **transparent to the hardware, and the software, that you currently have.** In other words, you can connect fiber optic LAN equipment to your existing computer

62   Everything you ever wanted to know about Fiber Optics and LANs, but were afraid to ask!

systems as easily as you can connect copper based network equipment.

There is a lot of equipment on the market today that allows you to connect fiber optic LAN equipment to existing copper based LANs and expand them using fiber optics. This conversion equipment is readily available and very inexpensive.

It is not necessary to think that, to go over to a fiber optic LAN, you might have to pull out all the copper already installed. If the copper LAN section is already installed, and it works, you don't need to touch it. You can simply connect your copper LANs with a fiber optic backbone, or expand you copper LANs using fiber optics from now on forward. If the copper LAN is installed, and it is not working, maybe you really need to replace it with fiber optics.

Typical examples of the kinds of equipment you can connect using fiber optic LAN products are shown in Figure 39. They include personal computers (AT/ISA type), MicroChannel computers (P/S-2 type), all types of work stations, terminals, mini computers, repeaters, bridges, gateways, etc. There is virtually no type of computer equipment on the market today that is not compatible with fiber optic equipment already available for it.

TOTAL CONNECTIVITY
IEEE 802.3 Ethernet Compatible

CODENOLL FIBER OPTICS CAN CONNECT:
- PC
- PS/2
- Workstations
- Terminals
- Computers
- Repeaters
- Bridges
- Gateways
- Packet Switching
- Thick Ethernet
- Thin Ethernet
- Twisted Pair Ethernet

Figure 39: Examples of equipment for which standardized fiber optic LAN equipment is readily available.

Appendix A - The IEEE, ANSI and ISO Standards - Detail

Most importantly, ISO, ANSI and the IEEE either have developed or are developing a number of emerging fiber optic standards. As shown in Figure 40, IEEE 802.3 has created a subcommittee called 10Base-F (not-surprisingly similar to the name 10Base-T for twisted pair). 10Base-F is a 10 Mbps standard that is fully compatible with the other Ethernet standards and has been endorsed by ISO. The standard is currently in letter ballot by the entire IEEE 802 committee. At the time of this writing, the letter ballots are being reviewed and 10Base-F is expected to become a standard either at the November, 1990 meeting of the IEEE or the following meeting in March, 1991.

FIBER OPTIC LAN STANDARDS

USA			
Committee	Subcommittee	Data Rate	ISO Committee
IEEE 802.3	10BASE-F	10Mbps	8802-3
IEEE 802.4	802.4H	10Mbps	8002-4
IEEE 802.5	802.5J	4 & 16 Mbps	8802-5
IEEE 802.6	SMDS	46 Mbps	8802-6
ANSI-X3T9.5	FDDI	100 Mbps	9384-1,2,3

ANSI = American National Standards Institute
FDDI = Fiber (Optic) Distributed Data Interface
SMDS = Switched Multimegabit Data Service
ISO = International Standards Organization
IEEE = Institute of Electrical and Electronics Engineers

Figure 40: The IEEE and ANSI Fiber Optic Standards Activities.

IEEE 802.4, the MAP Manufacturing Automation Protocol token passing bus standard committee has already approved its fiber optic standard called 802.4H. The IEEE 802.5 token ring standard committee is working on a draft fiber optic standard called 802.5J.

IEEE 802.6 is the new Metropolitan Area Network (MAN). It is the first data rate independent network protocol. It is envisioned to work over a range of data rates from 1.5 Mbps to 150 Mbps and higher on each of its two busses. The protocol, called "DQDB" for Dual Queue Dual Bus actually has a total throughput of twice the base data rate because of the use of dual busses. The first IEEE 802.6 MANs will be introduced by the telephone companies late in 1990 and 1991

64 Everything you ever wanted to know about Fiber Optics and LANs, but were afraid to ask!

together with higher level software called SMDS which stands for Switched Multi-Megabit Data Service.

FDDI is the ANSI Standard LAN which operates at 100 Mbps has been endorsed by virtually every computer company. The FDDI standard consists of 5 parts. Four of these parts, covering the hardware aspects of FDDI have officially been approved as standards. The fifth part of the standard, called Station Management (SMT), which can be implemented in software, has been approved by the working group and is expected to become a standard by the time this book is published. The so-called "final" version of SMT, which is being voted on as the ANSI standard is called version 6.2. It is virtually identical and fully compatible with version 6.1, which is the version currently used by most vendors. Once version 6.2 has been fully approved, manufacturers will begin to update their SMT. You should have no problem however, running both 6.1 and 6.2 on the same FDDI LAN.

The IEEE 802.3 Fiber Optic LAN Standards

The most popular LAN standard is Ethernet, or IEEE 802.3. The IEEE 802.3 standards can be a little confusing because there are so many of them. Thick coax Ethernet is 10Base5, Thin coax Ethernet is 10Base2. Broadband Ethernet is 10Broad36, twisted pair Ethernet is 10Base-T and for fiber optic Ethernet there are three standards: FOIRL, 10Base-FP and 10Base-FA. One standard already exists, 802.3 Chapter 9 Section 9 is a point-to-point connection which is called Fiber Optic Inter Repeater Link (FOIRL). It was designed solely to connect two different Ethernets together over long distances. For example two Ethernets on two different floors, or in two different buildings. It's been a standard for several years, and I am proud to say that I participated in its writing.

...the primer on Fiber Optic and LAN technology.
Appendix A - The IEEE, ANSI and ISO Standards - Detail

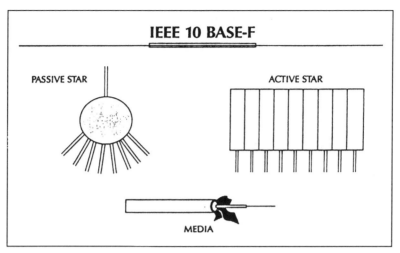

Figure 41: The Three parts of IEEE 802.3 10Base-F.

At this time, the 10Base-F Sub-committee of the Ethernet IEEE 802.3 is looking to develop a broader System Standard that is not limited to point-to-point connections between two LANs, but can be used as a multi-point LAN for interconnecting many devices using fiber optics. The Ethernet IEEE 802.3 committee has determined a need for developing two types of fiber optic LANs which are described in a three chapter draft standard (Figure 41). One chapter covers the media, which is the fiber optic cable, connectors and cabling practices.

The second chapter is on an active hub LAN system, which is basically a fiber optic version of 10Base-T. Active hubs or active stars are boxes of equipment that usually have plug-in boards each of which has fiber optic transmitters, receivers and connectors on it. The fiber optic cable is installed in a hierarchical structured star cabling system with one end of each fiber optic cable connected to a plug-in card in the active hub and the other end connected to a transceiver or a PC adapter card which also has fiber optic transmitters, receivers, and connectors. This creates a point-to-point connection between the active Star or the active Hub and an individual transceiver or PC adapter card. Then just like a twisted pair hub, the 10Base-FA hub becomes the Ethernet bus reduced to a point. The hub then is responsible through active electronic circuitry to perform the broadcast function of the original Ethernet coax cable and the collision detect function of the original Ethernet transceiver or MAU (Media Access Unit) or the PC adapter card. In this way the 10Base-FA system is exactly analogous to the 10Base-T system.

In fiber optics we have another option which does not exist in twisted pair. It is called the Passive Optical Star Hub and is the basis of Passive Fiber Optic Ethernet IEEE 802.3 10Base-FP. A typical passive optical star hub is a rack or shelf mount box with many optical connectors that make it look like a patch panel (Figure 42).

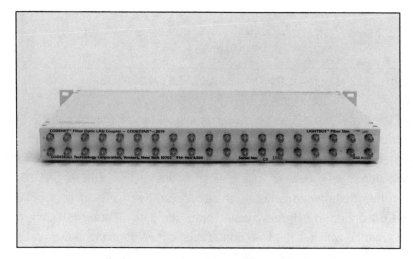

Figure 42: CodeStar passive optical star coupler

The passive optical star hub actually optically splits the signal passively. In fact there is virtually nothing in a passive optical star except some glass fussed together as in Figure 43 (it's kind of like "smoke and mirrors" without the smoke). As shown in Figure 43, when an optical signal enters one of the fibers on the right side (input fibers) it is actually divided and 1/n-th of the optical power (for an n-port passive star hub) comes out from all of the fibers on the left side (the output fibers).

...the primer on Fiber Optic and LAN technology. 67
Appendix A - The IEEE, ANSI and ISO Standards - Detail

Figure 43: Principles of passive optical star coupler.

This provides a passive optical bus that performs the same broadcast functions as the original Ethernet coax. By looking at Figure 44 we can see how the optical signal from one computer comes in an input fiber, goes through the optical divider, and is divided amongst all the output fibers which then go out to all the other computers on the LAN segment. In this manner all the Ethernet signals broadcast from any computer are received by every other computer which is the definition of a bus or broadcast type LAN such as Ethernet.

Principles of the Passive Optical Star Hub Bus/Broadcast LAN

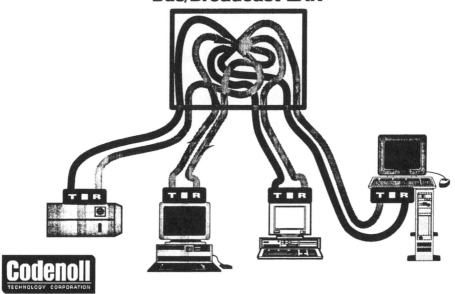

Figure 44: Bus/Broadcast system using passive optical star coupler.

68 Everything you ever wanted to know about Fiber Optics and LANs, but were afraid to ask!

Please note that the passive optical star hub is truly a passive device. There are no batteries, no power cord, and the hub needs no power supply, ventilation or cooling. Furthermore, as a completely optical or dielectric device the passive optical star hub is completely immune to all forms of electromagnetic interference. There is no noise source that can interfere with the proper operation of the LAN. And, the passive optical star hub cannot radiate any signal. Therefore, not only can it not interfere with any other equipment (such as medical instrumentation), but it is also completely secure from electronic eavesdropping and meets all requirement of the government Tempest security requirements (NACSIM-5100A).

The three different 802.3 Standards allow optimal use of fiber optics in many LAN applications. The passive star hub LANs are much less expensive than active hub LANs and therefore are often used for workgroup LANs. They are very useful in "mission critical" high reliability environments because the passive hub **cannot** go down. In other words, in a passive fiber optic Ethernet LAN there is no single point of failure as in an active hub or coax LAN. Not only are passive hubs very cost effective departmental LANs, they are perfect for backbones, where you have to connect a number of buildings and the ideal place to put the central hub is in a manhole in the middle of the parking lot. You don't really want to have an air conditioning system down in the manhole to keep active electronics from failing. At Harvard University for example, 25 buildings are interconnected using three CodeStar passive optical star hubs that are installed in the steam tunnels between the buildings.

On the active side, you can go longer distances because we are not dividing the optical signals so there is less signal attenuation.

Appendix B - TABLE OF FIGURES

Figure 1: Comparison of twisted pair to fiber optic Ethernet costs. 2
Figure 2: Diagram of Codenoll's joint development relationships..6
Figure 3: Transmission of data by converting electronic binary signals to light............................8
Figure 4: Fiber optic cable construction. 9
Figure 5: Horizontal cross section of fiber cable. 9
Figure 6: Longitudinal cross section of fiber showing rays of light. 10
Figure 7: Comparison of Multi-mode and Single-Mode optical fiber. 11
Figure 8: A 1/100-th inch square LED chip mounted on a 1/4 inch ceramic header with two wire bonds. 12
Figure 9: Examples of equipment for which standardized fiber optic LAN equipment is readily available. 16
Figure 10: IEEE 802.3 10Base-F: Passive Hubs, Active Hubs, Cable and Connectors. 17
Figure 11: Fiber Optic Connector Parts and Tools...19
Figure 12: Fiber optic cable construction..20
Figure 13: Comparison of LAN costs - ISA..22
Figure 14: Comparison of LAN costs - EISA. 23
Figure 15: The various networking protocols, data rates and cables..26
Figure 16: Basic LAN Technology Number 1 of 3: The Ring..27
Figure 17: Basic LAN Technology Number 2 of 3: The Active Hub. 28

Figure 18: Basic LAN Technology Number 3 of 3: The Bus or Broadcast System. 29
Figure 19: Operating principle of a passive optical star coupler. 32
Figure 20: Using a passive optical star to implement a Bus/Broadcast LAN. 32
Figure 21: 3-Dimensional Matrix Model of: Basic LAN Technologies vs Access Methods vs Media of Transport. 34
Figure 22: Hierarchical Star Fibered Tree Structured Cabling System: Basic cell. 35
Figure 23: Hierarchical Star Fibered Tree Structured Cabling System: Bus/Broadcast LAN........ 36
Figure 24: Hierarchical Star Fibered Tree Structured Cabling System: Active Hub. 36
Figure 25: Hierarchical Star Fibered Tree Structured Cabling System: Ring LAN. 37
Figure 26: Maintenance of Ring Failure. 38
Figure 27: Ring Maintenance: Node Addition. 38
Figure 28: Basic Fiber Optic Ethernet Workgroup. 41
Figure 29: Hierarchical structure of Workgroup LANs. 42
Figure 30: The full 9 level fiber optic Ethernet structured cabling system. 42
Figure 31: FDDI single ring configuration. 50
Figure 32: FDDI Dual Counterrotating Rings. 50
Figure 33: FDDI Dual Ring with Single Break and Wrap-Around. 51
Figure 34: FDDI Concentrator, Star Wired Ring. 52
Figure 35: FDDI Standard Ring of Trees Diagram. 53
Figure 36: FDDI Concentrator, Internal operation. 54
Figure 37: FDDI Ring of Concentrator Trees. 54
Figure 38: CodeNet Modular Multiport MultiBrouter 57
Figure 39: Examples of equipment for which standardized fiber optic LAN equipment is readily available. 62
Figure 40: The IEEE and ANSI Fiber Optic Standards Activities. 63
Figure 41: The Three parts of IEEE 802.3 10Base-F. 65
Figure 42: CodeStar passive optical star coupler 66
Figure 43: Principles of passive optical star coupler. 67
Figure 44: Bus/Broadcast system using passive optical star coupler. 67

I

III

Fiber Optic Components

High Power Light Emitting Diodes for Fiber Communication, p. 115
F. W. Scholl, S. J. Anderson, M. H. Coden

Reliability of Components for Use in Fiber Optic LANs, p. 121
F. W. Scholl, M. H. Coden, S. J. Anderson, B. V. Dutt, J. H. Racette, W. B. Hatfield

Fiber Optic Edge Emitting LED Package for Local Area Network Applications, p. 131
S. J. Anderson, P. G. Abbott, F. W. Scholl

AlGaInP/GaAs Red Edge-Emitting Diode for Polymer Optical Fiber Applications, p. 141
B. V. Dutt, J. H. Racette, S. J. Anderson, F. W. Scholl

FDDI Components for Workstation Interconnection, p. 143
S. J. Anderson, D. V. Bulusu, J. H. Racette, F. W. Scholl, T. E. Zack, P. G. Abbott

Making Fiber Optic Terminations—Step-by-Step (Part 1), p. 151
J. Cocomello, R. Focht

Making Fiber Optic Terminations—A Compatibility Study & Evaluation (Part 2), p. 155
J. Cocomello, R. Focht

A Fiber Optic Connection System Designed for Automotive Applications, p. 159
D. A. Messuri, G. D. Miller, R. E. Steele

II
114

High Power Light Emitting Diode for Fiber Communication

F. W. Scholl, S. J. Anderson, M. H. Coden
Codenoll Technology Corporation
1086 North Broadway
Yonkers, New York 10701

Abstract

In this paper we report the design and characterization of a new high power edge emitting LED. The experimental results are confined to AlGaAs devices, but the design principle is applicable to other materials systems such as InGaAsP. The coupled powers for 0.2 na, 50 micron fiber are the highest reported to date for devices without optical gain. Our new LED structure utilizes a proprietary double heterostructure emitting configuration to maximize optical output power. These devices couple very high optical flux into the core of 0.2 na, 50 micron graded index fiber. A recent sample of 20 LEDs from 6 wafers had the following CW coupled power at 125 ma: average power = 198 microwatt, minimum power = 175 microwatts, maximum power = 232 microwatts.

Rise and fall times have been measured; the results for 10–90% transition time are 5–7 nsec for undoped active layer and under 2.5 nsec for doped active layers. Reliability data has been obtained from accelerated aging conducted over the past 18 months. Extrapolated life of 2.7×10^7 hours, time to half power at 25° C ambient is obtained. In summary, the new LED we have developed will find many applications in emerging local area networks where distances, cost, temperature stability and required flux budget favor the use of GaAlAs LED emitters. Table I shows in summary form the characteristics of the devices along with the Codenoll part number.

I. Introduction

Increasing speed and speed-distance requirements (1), (2) for computer networks illustrate the need for improved optical sources and detectors. GaAlAs heterojunction light emitting diodes (LEDs) are the preferred optical source device for systems where high bandwidth—data rates in excess of 10 Mbps—and minimum link cost are both of concern. In this paper we present our results on a new edge emitting LED optimized for high bit rate systems using 50, 62.5, 85 or 100 micron core fiber. These devices have launched 175–225 microwatts of optical flux into 50 micron core, 0.2 na fiber, the highest power reported for devices without optical gain. In addition, risetimes of under 2.5 nsec have been obtained while lifetesting over 18 months has demonstrated extrapolated median time to failure (to half power) of 2.7×10^7 hours at 25° C. Our LEDs are packaged on BeO heat sinks metallized for hybrid packaging with drive electronics or other proprietary optical or electronic components.

The advantages of edge emitters over surface emitting LEDs are documented in References (3) and (4). These advantages are most significant for small core, low numerical aperture fibers. Here, the smaller cross-section of the edge emitter's emitting area and the focused far field pattern give it significant performance advantages. In spite of these advantages, limited information has been available on the performance of practical GaAlAs edge-emitting components. Earlier device designs are discussed in Refs (5)–(9). Design and performance of etched well emitters (10) and planar surface emitters (11), (12) is also discussed. A general review of optical fiber sources and detectors is available in Reference (13).

Table I. Codeled™ High Radiance Edge Emitting LEDs.

Parameter	Symbol	Units	8E001 8EP001	8E002 8EP002	8E003 8EP003	8EHS001 8EPHS001	8EHS002 8EPHS002	13E001 13EP001	13E002 13EP002
Minimum output power at $T_A = 25°C$, I_F = forward current			I_F = 100mA			I_F = 100mA		I_F = 150mA	
Into an aperture 0.5NA	P_o	µW	350	450	550	350	450		
0.3NA	P_o	µW						50	85
Coupled into fiber									
100µm 0.29 NA	P_c	µW	280	355	445	280	355	45	75
85µm 0.26 NA	P_c	µW	180	230	290	180	230	35	60
62µm 0.29 NA	P_c	µW	125	160	200	125	160	35	60
50µm 0.20 NA	P_c	µW	80	100	125	80	100	30	50
Wavelength of peak emission	λ_o	nm	810–860			810-860		1270–1330	
Typical spectral full width at half maximum	FWHM	nm	60			60		90	
Typical optical rise and fall times	t_r, t_f	ns	7			2.5		5	
Storage temperature	T_s	°C	−55 to +150						
Case operating temperature	T_c	°C	−55 to +125						

8EP series and 13EP series LEDs are packaged in 14 pin DIP package.

In Section II of this paper we describe the operating principle and design of the new LED. Section III describes device packaging. Section IV presents our experimental results, obtained from production wafers over the past 18 months. We include typical optical and electrical characteristics, fiber coupling data, and packaging and reliability information. Conclusions are presented in Section V.

II. Device Design

Our LED device design is based on the following principles which ensure that as much light as possible is extracted from the LED and launched into the fiber.

1) Light rays generated within the active layer must not get trapped there where they will be absorbed.

2) The electrical carriers injected into the active layer must be confined there to allow efficient generation of light.

3) Light rays traveling through the device must not encounter absorbing regions or interfaces on their way into the fiber core.

4) Reflections at the emitting facet must be minimized to allow the greatest amount of light to escape.

To accomplish these desired goals, a proprietary device structure was developed (U.S. Patent Pending). This structure meets the above objectives while also facilitating good device reliability and reproducible manufacturing techniques.

III. Device Packaging

The LED-fiber interface is a critical element in determining overall device performance and cost. Two design approaches are possible including direct "pigtailing" of fiber to source active area and packaging the LED in an active device mount with connectorized interface (14). The first approach offers higher coupled optical power and flexibility in choice of optical connectors. For these reasons, we have developed an LED heat sink suitable for direct fiber pigtailing. This heat sink can easily be incorporated into a variety of hybrid packages.

The overall component design is shown in Figure 1. Beryllia is used as a heat sink material because of its good machining characteristics, good thermal conductivity, and excellent thermal expansion match to GaAs. Table II compares BeO with the other heat sink materials that were considered, silicon and copper.

Table II. Comparison of heat sink materials.

Material	σth (w/°C-cm)	$\dfrac{\alpha - \alpha_{GaAs}}{\alpha_{GaAs}}$	Machinability
BeO (99.5%)[15]	2.7	−.04	Good
Copper[16]	3.8	1.5	Poor
Silicon[16]	1.5	−.62	Good

σ = thermal conductivity;
α = linear expansion coefficient.

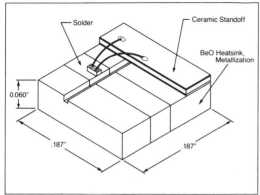

Figure 1. Physical structure on LED packaged on heat sink. Step in ceramic heat sink is provided for fiber alignment.

The heat sinks are fabricated from machined BeO blocks. Metallization is deposited over the surface of the block. An insulating standoff is then attached to supply negative bias to the LED chips. The chips are soldered to the BeO block using high temperature solder. This allows operation at case temperatures up to 200° C. The BeO heat sinks can then be epoxied or soldered into a hybrid package with standard conductive epoxy.

The step machined into the heat sink surface is used for fiber alignment. After alignment of the fiber core to the emitting area, the fiber is fixed in place on the heat sink surface. The stability of the source-fiber positioning has been demonstrated by temperature testing over the range −40° C to +85° C.

IV. Experimental Results

The experimental results reported in this Section were obtained on devices fabricated in Codenoll's semiconductor manufacturing facility. They are representative of production devices routinely obtained. All devices are burned in at 125° C for 100 hours before testing.

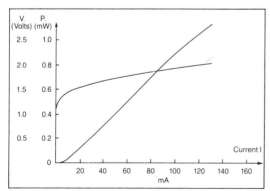

Figure 2. Typical LED current-voltage and optical power—current characteristics. Optical power measured into numerical aperture of 0.5 NA.

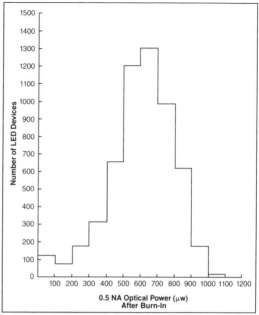

Figure 3. Statistical presentation of optical power from production devices. Sample size = 5500 devices.

A) Electrical and Optical Characteristics

Typical room temperature LED P-I and V-I characteristics are shown in Figure 2. Histogram for optical power in a 0.5 numerical aperture—our standard test condition—is shown in Figure 3. The reproducibility of device I-V characteristic determines the reproducibility of its circuit characteristics. For our devices we find (sample size = 27 wafers) device breakover voltage $V_B = 1.48$ volts ± .096 (1 σ) and series resistance $R = 4.69\Omega \pm .80$ (1σ).

Since the LEDs are designed for high fiber coupling efficiencies, near field characteristics are of great importance. Small source cross-section is desirable, limited by device heating at high current density. The near field intensity is measured with 0.4 na microscope objective and silicon vidicon camera; Figure 4 illustrates typical results at current density of 0.53 ka/cm^2 and 5.3 ka/cm^2. Increased spreading at low current densities parallel to the junction plane (w_x) results from stripe geometry current confinement and the exponential junction current-voltage characteristic. The small source size at 100 ma, $w_x = 35\mu m$, $w_y = 12\mu m$, insures high fiber coupling efficiency.

We measure coupling efficiency by comparison of power launched into fiber core to optical power emitted into 0.5 numerical aperture. Fiber ends are lensed prior to the coupling efficiency measurement. Typical coupling efficiencies observed are shown in Table III for commercially available 50, 85, and 100 micron core fiber.

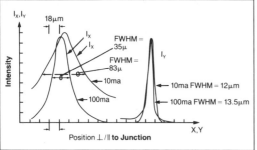

Figure 4. Near field intensity parallel (Ix) and perpendicular (Iy) to LED junction. Measurement at 10 ma shows increased spreading parallel to junction.

Table III. Typical average coupling efficiencies for 3 commercial fiber types: 50 micron, 0.2 NA; 85 micron, 0.26 NA; 100 micron, 0.29 NA. All efficiencies are normalized to power output measured into aperture with NA = 0.5. All fibers are lensed.

Wafer #	50 micron	85 micron	100 micron
1-420 A	26%	61%	91%

The combination of high coupling efficiency and high 0.5 NA power results in devices with high output power launched into a fiber pigtail. Results on 20 selected devices (Table IV) from 6 wafers are: minimum power = 175 μW, maximum power = 232 μW, average power = 193 μW, all measured at 125 ma with 50 micron core fiber. These data illustrate the uniformly high powers that can be achieved with our device structure.

Table IV. Selection of 20 high power LEDs from 6 wafers. P_{min} = 175, P_{max} = 232, $P_{average}$ = 193 μW. Power measured into 50 micron, 0.2 na fiber, at 125 ma current.

Wafer #	Sample Size	Average Power Coupled Into 50 Micron Fiber
1-393A	7	212μW
1-392A	1	201
1-349A	4	180
1-368A	5	186
1-399A	1	207
1-400A	2	210

These devices are all anti-reflection coated; optical power is coupled into spherical lensed fiber. Coupled power at 100 ma routinely is 100—140μW for 50 micron .2 NA fiber, and 315-500 μW for 100 micron 0.29 NA fiber. Over the temperature range 25° C–125° C the measured optical power drops linearly with temperature at a rate of 0.029–0.034 dB/° C.

Optical rise and fall time is extremely important for successful implementation of high speed communication links. We have fabricated two types of devices, with undoped active layer and with doped active layers. Devices are driven with a pulse generator and 47Ω series resistor. A silicon avalanche photodetector measures the optical flux. Rise and fall time of the driver and detection electronics is 1.4 ns. Typical results are summarized in Table V. The best

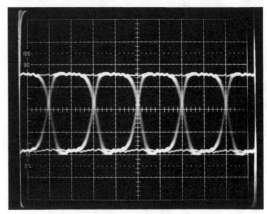

Figure 5. Eye diagram for fiber optic link using edge emitting LED, ECL driver and PIN receiver. 50 MHz Manchester coded data. 3 microwatt peak optical input power to receiver. 5 ns per division; 0.2 v per division.

Table V. Rise and fall times for LEDs with doped and undoped active layer. 47Ω driver circuit, silicon APD detector.

Wafer #	Number of Devices	10–90% Rise time	10–90% Fall time	Doping in Active layer
402	8	5.7 nsec	4.4 nsec	undoped
456	8	2.3 nsec	2.5 nsec	1×10^{18}

of the devices with doped active layer have 10–90% transition times $\tau_r \simeq \tau_f \simeq 2.0$ nanoseconds. Standard devices with $\tau = 4$–6 nanoseconds are successfully used in communication links operating up to 50 Mbps. In addition, with speed up circuitry, these devices will operate at over 100 Mbps. Figure 5 shows the eye diagram for an ECL compatible fiber optic link (Codenoll's CL-100B) using these standard LEDs. Distortion in this link (which does not retime the data) is measured by peak to peak jitter; this is less than 3 nanoseconds.

B) Device Reliability

Our devices are bonded to BeO heat sinks. The thermal expansion of beryllia matches almost perfectly to that of GaAs ($\alpha_{BeO} = 6.5 \times 10^{-6}$ C^{-1}, $\alpha_{GaAs} = 6.8 \times 10^{-6}$ C^{-1}) (17). This allows a wide operating temperature range— –55° C to +125° C—without excessive bond stress. Before any device reliability testing is conducted, each unit is burned in (100 ma) at 125° C ambient temperature for 100 hours. This removes any infant failures, but primarily eliminates the initial 2–8% typical power drop characteristic of 830 nm LEDs. After this drop, further power decrease is attributed to long term wearout mechanisms. Reliability studies for surface emitters have been presented by several authors (Refs 18–21).

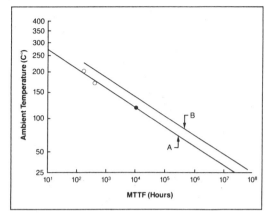

Figure 6. Accelerated life testing data at 100 ma. Curve A is a least squares fit to data obtained from all production devices summarized in Table VI and VII. Curve B is a typical curve for "superior" wafers.

Accelerated aging is conducted at 125° C, 175° C, and 200° C in laboratory ambient conditions (18). The LEDs are driven at 50, 100, and 150 ma at elevated temperatures; device light output is then measured at room temperature at periodic time intervals. Time to failure is indicated by output power reading one-half of initial value. Time to failure of devices still "good" during the course of an experiment was estimated by linear extrapolation. A random sample of standard production wafers was used as indicated in Table VI. Individual device failure times were observed to follow a

Table VI. Devices used for accelerated aging. All devices randomly selected from production lots, after burn-in at 120° C for 100 hours.

Temperature	Current	# of wafers	Sample size
120° C	100 ma	2	41
175° C	100 ma	14	144
	50 ma	1	14
	150 ma	8	28
200° C	100 ma	8	94

Table VII. Summary of reliability data.

I	100	100	100	50	150	ma
T	120	175	200	175	175	° C
MTTF	9984	391	161	750	170	hrs
σ	0.60	0.46	0.53	0.34	0.89	—

log-normal distribution. The median time to failure, MTTF, and standard deviation, σ, were calculated; the results are summarized in Table VII. The majority of our lifetesting is done at 100 ma. MTTF results for this operating current are plotted in Figure 6, where we calculate an activation energy of 0.82 ev and an extrapolated room temperature life of

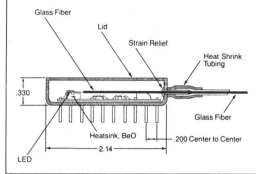

Figure 7. Cross section of hybrid integrated fiber optic transmitter. Dimensions in inches.

2.7×10^7 hours. Using the log normal distribution and our average measured $\sigma \simeq 0.5$, we calculate the arithmetic mean lifetime at 25°C to be 5.2×10^7 hours. These results are comparable or superior to results reported for GaAlAs surface emitters (19-22). The standard deviation $\sigma \simeq 0.5$ is a measure of the spread in the distribution of failure times. The narrow distribution characteristic of our manufacturing process implies a good predictability of failure times. For example, we can calculate that only 5% of our devices will have a lifetime of less than 0.41×10^7 hours under 25° C ambient conditions.

Reliability testing for higher and lower currents has also been carried out (Table VII). Our results for MTTF indicate a penalty of 0.43 when operating at higher currents (150 ma) and 1.9 increase in life when operating at lower current (50 ma). These results are normalized to the MTTF results at 100 ma. Using the activation energy measured for 100 ma operation, the corresponding extrapolated 25° C MTTF results are:

50 ma – 5.1×10^7 hours, 150 ma – 1.2×10^7 hours

C) Device Packaging

To provide a practical optoelectronic component, the fiber pigtail must be stable over a wide operating temperature range. A procedure was developed for attaching the optical fiber in place to insure maximum coupled power as well as environmental stability. The packaged LED transmitter is shown schematically in Figure 7. The LED and associated drive circuitry are encapsulated in silicone resin. This encapsulation eliminates problems from moisture condensation (23) and is utilized for devices to be operated over the full military temperature and humidity conditions. Devices assembled as in Figure 7 were subjected to MIL-STD-810C temperature testing, with operating temperature limits of −40° C and +85° C ambient. Temperature cycling between −40° C and +68° C was also carried out. The results of a series of tests on 5 transmitters are shown in Table VIII. The maximum drop in power observed is 1.7 dB. The drop in power at +85° C can be attributed largely to lower LED power output. The small drop in power (1 dB) for TX 412 at −40° C can be attributed to residual thermal expansion effects.

Table VIII. Results on pigtailed LED transmitter power output, at 25° C, at −40° C, at +85° C, and after cycling between −40° C and +68° C.

TX	Initial P, 25° C	P, −40° C	P, +85° C	After Cycling P, 25° C
409	142 μW	178 μW	112 μW	151 μW
410	70	92	64	76
406A	68	70	49	75
412	66	52	44	63
415	59	57	41	55

V. Conclusions

The results in this paper illustrate the high performance that can be obtained with 830 nm edge emitting AlGaAs LEDs. This performance is a result of device design and component packaging. Our results on manufacturing lots produced over two years indicate that these components can be used in systems operating at over 125 Mbps with optical loss between transmitter and receiver of up to 24 dB. Component packaging is compatible with both commercial systems and military systems. We believe these edge emitting devices lead to superior system performance in those local area communication systems where high data rate and operating optical gain margins are of importance.

References

1. FDDI Token Ring Physical Layer Standard, ANSI X3T9.5, Rev 6, Aug 10, 1984.
2. M. H. Coden, F. W. Scholl, Photonics Spectra, *17*, 47 (1983).
3. D. Botez, M. Ettenberg, IEEE Trans. on ED, *26*, 1230 (1979).
4. D. Marcuse, IEEE Journal of Quant. Elec., *13*, 819 (1977).
5. Y. Horikoshi, Y. Takanashi, G. Iwane, Jap. Journal of Applied Physics, *15*, 485, (1976).
6. J. P. Wittke, M. Ettenberg, H. Kressel, RCA Review, *37*, 159 (1976).
7. Y. Pang, H. Pan, Z. Cheng, P. Shen, G. Wu, Z. Xiao, IEEE Trans. on ED-30, 348 (1983).
8. H. Kressel, M. Ettenberg, Proc. of IEEE, 1360 (1975).
9. Y. Seki, Jap. Journal of Appl. Phys., *15*, 327, (1976).
10. T. P. Lee, A. G. Dentai, IEEE J. Quantum Electron., QE-14, 150 (1978).
11. R. S. Speer, B. M. Hawkins, Proc. 30th Elec. Comp. Conf. 270 (1980).
12. J. S. Escher, H. M. Berg, G. L. Lewis, C. D. Moyer, T. U. Robertson, H. A. Wey, IEEE Trans. on ED, 19, 1463 (1982).
13. D. Botez, G. J. Herskowitz, Proc. IEEE, *68*, 689, (1980).
14. H. M. Berg, D. L. Shealy, C. M. Mitchell, D. W. Stevenson, L. C. Lofgran, IEEE Trans. on Comp. Hybrids, Manuf. Tech., 6, 334 (1983).
15. Brush, Wellman, Inc., Elmore, Ohio.
16. S. K. Ghandi, *Semiconductor Power Devices*, Wiley, New York, 1977.
17. E. D. Pierron, D. L. Parker, J. B. McNeely, Journ. of Appl. Phys., *38*, 4669 (1967).
18. F. H. Reynolds, Proc. of IEEE, *62*, 212 (1974).
19. S. Yamakoshi, O. Hasegawa, H. Hamaguchi, M. Abe, T. Yamaoka, Appl. Phys. Lett., *31*, 627 (1977).
20. C. Zipfel, A. Chin, V. Keramidas, R. Saul, Proceedings of 1981 Int. Reliab. Physics Symp., 124 (1981).
21. H. O. Sorenson, Proceedings of FOC/LAN '83, 92 (1983).
22. B. Twu, H. Kung, Proceedings of SPIE, *321*, 86 (1982).
23. C. P. Wong, D. M. Rose, IEEE Trans. on CHMT, 6, 485 (1983).

"Reliability of Components for Use in Fiber Optic LANs"
Frederick W. Scholl, Michael H. Coden, Stephen Anderson,
Bulusu Dutt, James Racette, W. Bryan Hatfield

Codenoll Technology Corporation, 1086 North Broadway
Yonkers, New York 10701

Abstract

Fiber optic LAN transmission system reliability is ultimately determined by the reliability of the individual components in these systems. For LANs the optoelectronic components are predominantly GaAlAs LEDs and Si photodetectors for 830nm transmission while InGaAsP LEDs and InGaAs photodetectors are used for 1300/1500nm transmission. Component design and packaging is determined by system application and such factors as data rate, power output and fiber core size. Typical data rates for LANs vary from 2Mb - 200Mb, while fiber types of interest include single mode and 50, 62.5, 85 and 100 micron multimode fibers.

In this presentation we will first discuss the design and selection of reliable emitter and detector components for LANs. Secondly, a review of our reliability results for GaAlAs edge emitting LEDs and InGaAs PIN detectors will be presented. These data include five years of component reliability testing and quality assurance monitoring in a manufacturing environment.

Introduction

An increasing number of fiber optic local area networks (LANs) are being implemented. These systems conform to one of several industry standards such as IEEE 802.3 (Ethernet), 802.4 (MAP), 802.5 (Token Ring), or custom configurations constructed for specific application requirements. The flux budget and data rate requirements of these systems can be met by GaAlAs and InGaAsP LEDs, silicon and InGaAs photodetectors and associated electronics. The first commercially available industry standard LAN was the optical Ethernet.[1] This system is based on a passive optical star and thus has no single point of failure. Subsequent optical fiber LANs have generally been based on a passive star architecture or, for ring networks, rely on bypass switches to eliminate single points of failure.

Components for LAN systems must provide adequate performance including system margin and maintain this performance over system operating life. Codenoll has designed and developed a family of LED sources and PIN detectors that are optimized for LAN applications. Reliability in field use is the result of two efforts: (1) component design and (2) manufacturing quality control. Both aspects will be discussed below.

Codenoll's component designs have several unique features that were included to optimize reliability as well as performance. For both 830nm and 1300nm LEDs, a proprietary (Patent Pending) edge emitting diode geometry was designed. The edge emitter offers higher power output into small core fibers with lower current density. These devices are soldered directly to high conductivity BeO heat sinks for minimum thermal resistance and minimum junction temperature rise. The resulting impedance for Codenoll LED is $\theta_{jc} \approx 50°C/W$; for surface emitters $\theta_{jc} \approx 125 - 150°C$ per watt. This can translate to reduced life (MTTF = median time to failure) for surface emitters because of the proportional increase in junction temperature. Quantitative results can be predicted from the expressions:

$$MTTF = MTTF_o e^{T_o/T}$$

$$\frac{1}{MTTF_o} \frac{d}{dt}(MTTF) = -T_o/T^2 \approx -11\%/°C$$

Photodetectors used in LAN systems include low capacitance silicon devices and InGaAs devices for 1300 - 1500nm transmission. The Codenoll InGaAs devices are inverted mesa geometry construction with substrate illumination. The lowest possible capacitance and device area is obtained, consistent with requirements for reliable wire bonding and fiber alignment. This leads to a device diameter of 70 microns and maximum junction capacitance of 0.5pF (at -10V bias). Stable dark currents are obtained by use of appropriate passivating compounds.

Insuring reliable operation of devices produced in a manufacturing environment requires careful design and administration of quality controls. These controls are used to monitor wafer fabrication processes and device assembly (die and wire bond). After assembly, wafers are screened for reliability by accelerated lifetests; each chip from approved wafers is further burned-in at 135°C for 48 hours at 100mA (LEDs). Other sample tests include thermal impedance, fiber coupling efficiency, near field and far field pattern. These tests and others eliminate devices that pass both the burn-in screen and accelerated lifetest screen, but which may be expected to have reduced life.

830 NM LED

Device description

The optical emitters described in this paper are double heterostructure edge emitting LEDs with a waveguide structure optimized for coupling into multimode or single mode fibers.

The advantage of edge emitters over surface emitting LEDs are documented in Reference (2) and (3). The smaller cross-section of emitting area and the focused far field pattern give edge emitters significant performance advantages. Results for 830nm (AlGaAs/GaAs) structures are reported in this paper. The overall component design is shown in Figure 1. Typical performance of the LEDs is shown in Table 1.

Parameter	Symbol	Units	8E001 8EP001	8E002 8EP002	8E003 8EP003	8EHS001 8EPHS001	8EHS002 8EPHS002	13E001 13EP001	13E002 13EP002
Minimum output power at $T_A = 25°C$, I_f = forward current			$I_F = 100mA$			$I_F = 100mA$		$I_F = 150mA$	
Into an aperture 0.5NA	P_o	μW	350	450	550	350	450		
0.3NA	P_o	μW						50	85
Coupled into fiber									
100μm 0.29 NA	P_c	μW	280	355	445	280	355	45	75
85μm 0.26 NA	P_c	μW	180	230	290	180	230	35	60
62μm 0.29 NA	P_c	μW	125	160	200	125	100	35	60
50μm 0.20 NA	P_c	μW	80	100	125	80	100	30	50
Wavelength of peak emission	λ_o	nm	810–860			810–860		1270–1330	
Typical spectral full width at half maximum	FWHM	nm	60			60		90	
Typical optical rise and fall times	t_r, t_f	ns	7			2.5		5	
Storage temperature	T_s	°C	−55 to +150						
Case operating temperature	T_c	°C	−55 to +125						

8EP series and 13EP series LEDs are packaged in 14 pin DIP package.

Table 1. Codenoll high radiance edge emitting LEDs.

The heat sinks are fabricated from machined and metallized BeO blocks. An insulating standoff supplies negative bias to the LED chips. The chips are soldered to the BeO block using high temperature solder and the emitting facet of the LED chip is AR coated. The BeO heat sinks can then be epoxied into a hybrid package with standard conductive epoxy. The step machined into the heat sink surface is used for fixing the fiber following fiber alignment.

The thermal expansion of beryllia matches almost perfectly to that of GaAs and InP ($\alpha_{BeO} = 6.5 \times 10^{-6} \, °C^{-1}$, $\alpha_{GaAs} = 6.8 \times 10^{-6} \, °C^{-1}$, $\alpha_{InP} = 4.5 \times 10^{-6} \, °C^{-1}$,) allowing reliable operation over a wide temperature range - −55°C to +125°C.

Device reliability

Before any device reliability testing is conducted, each unit is burned in using our standard production burn-in procedure. The burn-in conditions for 830nm LEDs are 100mA at 135°C ambient for 48 hours. This removes any infant failures, but primarily eliminates the initial 2-8% typical power drop characteristic of our LEDs. After this drop, further power decrease is attributed to long term wearout mechanisms.

Accelerated aging is conducted at 125°C, 175°C, and 200°C in laboratory ambient conditions. The LEDs are driven at 50, 100, and 150 ma at the elevated temperatures; device light output is then measured at room temperature at periodic time intervals. Time to failure is indicated by output power reading one-half of initial value. Reliability testing of LEDs has been ongoing for 5 years at Codenoll. A random sample of standard production wafers was used. Individual device failure times were observed to follow a log-normal distribution.

A total of 320 devices from more than 20 wafers were lifetested. The median time to failure, MTTF, and standard deviation, σ, were calculated; the results are summarized in Table 2.

Table 2. Summary of 830 nm LEDs reliability data.

I	100	100	100	50	150	ma
T	125	175	200	175	175	°C
MTTF	9984	391	161	750	170	hrs
σ	0.60	0.46	0.53	0.34	0.89	-

The majority of our lifetesting is done at 100 ma. MTTF results for this operating current are plotted in Figure 2 where we calculate an activation energy of 0.82 ev and an extrapolated room temperature life of 2.7×10^7 hours. This is equivalent to a maximum failure rate of 3.7×10^{-8} hr^{-1} or 37 FITs. Using the log normal distribution, and an average measured $\sigma = 0.5$, we calculate that 95% of our devices will have a lifetime greater than 0.41×10^7 hours under 25°C ambient conditions.

Reliability testing for higher and lower currents has also been carried out (Table 2). Our results for MTTF indicate a penalty of 0.43 when operating at higher currents (150 ma) and 1.9 increase in life when operating at lower current (50 ma). These results are normalized to the MTTF results at 100 ma. Using the activation energy measured for 100 ma operation, the corresponding extrapolated 25°C MTTF results are:

50 ma - 5.1×10^7 hours, 150 ma - 1.2×10^7 hours

In summary, accelerated temperature testing of high radiance edge emitting LED's indicate room temperature lifetimes in excess of 10^7 hours (>1000 years) when operated continuously. In many applications, including local area networks, the emitting devices are only turned on part of the time and are run at approximately 50% duty cycle when they are on. These factors serve to further increase the expected life of the device under actual operating conditions.

1300 NM Photodetector

Device design

The structure of the back-illuminated mesa geometry photodiode currently manufactured at Codenoll is as shown in Figure 3. It consists of a diffused junction in a 3 to 4 micron thick absorbing layer of $In_{.53}Ga_{.47}As$ grown lattice-matched to a 100 - oriented InP:Sn substrate. The growth technique used is liquid epitaxy. A 70 micron diameter mesa is formed in the diffused wafer by standard photolithography. The anti-reflection coating on the substrate consists of a quarter wave film of SiO. After the device is bonded onto a ceramic carrier, a proprietary passivation procedure with a commercially available encapsulant covering the P-contact and the junction perimeter surrounding the mesa, completes the fabrication process. The device characteristics are listed in Table 3.

TABLE 3: Characteristics of InGaAs PIN Photodiodes

Junction Diameter:	70 micron
Maximum Dark Current: (-10V, 25°C)	20nA
Maximum Capacitance: (-10V, 25°C)	0.5pf
Minimum Responsivity: at 1300nm (-10V, 25°C)	0.8A/W
Minimum Breakdown Voltage:	50V

Device reliability

Since all of our devices are used in receivers for applications in local area networks, we believe that the encapsulated mesa diodes offer excellent reliability. More demanding applications such as tactical military systems or submarine cable systems will require a hermetic package for best reliability. Earlier investigators[4,5] demonstrated that the mesa devices have median lives exceeding 10^8 hours at 20° C operation in hermetic packages. Nevertheless, a comprehensive reliability evaluation of our passivated mesa devices is under way along the lines initiated by Saul et. al.[6] for nitride-passivated planar devices.

The device parameters of importance are capacitance, breakdown voltage, dark current, responsivity and response time. The capacitance is determined by the junction area, the dielectric constant and the depletion width and is not expected to change with ambient conditions. Charge accumulation, however, could cause changes. The breakdown is a function of the background doping and is attributed to the avalanche mechanism. Surface wetting for example might initiate breakdown by mechanisms other than those due to the bulk semiconductor processes. The dark current at operating temperature is primarily due to the generation-recombination mechanism and at high temperatures, the diffusion mechanism dominates. Surface contamination and wetting can cause severe leakage. Responsivity may degrade due to failure of the anti-reflection coating. Response time is a function of the capacitance and is not expected to change.

The initial phase of the reliability program for the mesa devices consisted of assessing the degradation behavior of breakdown voltage and the dark current under relatively high humidity (85%) and high temperature (85°C). Ten devices from a wafer were subjected to a 90°C, 72 hour burn-in without bias. They were then subjected to an accelerated aging test at 85% relative humidity at 85°C at -10V bias. Table 4 summarizes the data from these tests. The measurements were performed periodically at room ambient after each of the tests. One device is virtually shorted but the other nine did not show any perceptible degradation. The devices showed excellent recovery of the original characteristics, after drying at 90°C for 90 hours.

Table 4: Results of testing at 85% relative humidity at 85°C with -10V bias

Id (-10V) DARK CURRENT AT -10V BIAS, nA

DEVICE #	AFTER PASSIVATION	2hrs 85/85	24hrs 85/85	50hrs 85/85	100hrs 85/85	24hrs @ 90°C, no humidity	20hrs @ 25°C, no bias(RT)	24hrs @ 90°C, no bias
1	14	15	20	19	19	19	20	8
2	9	7	9	11	11	11	11	10
3	9	3	17	13	15	15	17	3
4	5	4	8	9	31	31	17	5
5	5	3	5	7	5	5	9	3
6	20	12	21	22	24	24	32	7
7	4	3	9	7	8	8	11	4
8	8	5	8	11	FAILED (VB 1 VOLT)		VIRTUAL SHORT	
9	5	4	23	26	16	16	90	5
10	11	5	22	40	24	24	120	4

In order to assess the long term reliability of the mesa devices, initially we tested several devices for dark current degradation at 85°C with a -10V bias. No changes were noticed in several hundred hours similar to the previous reports. At present reliability tests are in progress at 200°C and -10V bias.

Manufacturing quality screens

830 NM LED

The quality and reliability of Codenoll's high power, edge emitting LED is achieved through a combination of carefully controlled processing operations together with a series of rigorous quality screens. A simplified flow chart for the Codenoll LED manufacturing process is shown in Figure 4. Wafers of manufacturer certified orientation are inspected for surface defects and proper thickness prior to epitaxial crystal growth. Following carefully controlled growth of the epitaxial layers, the wafer surface is again inspected for crystalline defects by means of differential, interference contrast photomicrographic techniques. A small portion of each wafer is then chosen for further analysis. Using standard angle lap techniques, the thickness of the various epitaxial layers is determined. Photoluminescence measurements on the same sample provide information regarding layer composition. Proper contact metallization thickness on the processed wafer is verified using a surface profilometer.

The smoothness of the step edge of the beryllium oxide LED submount together with its curvature are critical to proper LED die placement and subsequent fiber alignment. Consequently, these parameters are carefully monitored on in-coming inspection of the unprocessed blocks. Visual inspection of the metallized block together with sample stand-off shear tests control the quality of the submount going into final device assembly.

Control of the processes by which the LED die is first die bonded and then wirebonded to the submount is maintained by conducting wirebond pull tests and die shear tests on a sample from each production lot. The uniformity of the die bond is also verified through thermal impedance measurements on production lot samples. Statistical process control techniques are utilized to maintain the process parameters within specified limits.

Prior to release of a wafer to final device assembly, an extensive series of quality related measurements are made on a sample of devices from each wafer. The device resistance, breakover voltage and optical output power are determined both before and after a 48 hour, 135 degree C burn-in during which a 100 milliamp current is maintained through the devices. Following burn-in, the wafer spectral properties are characterized in terms of wavelength of peak emission and the half power spectral width. The near field and far field emission patterns are recorded as well as the coupling of the LED optical emission to various core diameter multi-mode fibers. Finally, the expected lifetime of devices from a given wafer is determined through temperature accelerated life tests conducted at 175 degree C and 100mA device current. If a sample deviates from established limits in any of the quality assessment tests outlined above, that wafer is removed prior to production.

LED devices produced from approved wafers are subjected to a final series of quality checks prior to shipment. First, each device is burned-in under power (100mA) for 48 hours at 135 degree C. Then an automated test system determines the resistance, breakover voltage and total optical power output for each device. Again devices which fail to meet established criteria are rejected. Finally, sample fiber coupling measurements are used to sort devices according to coupled power options.

1300 NM Photodetector

The series of quality screens employed to assure reliable, high performance photodiodes differs from those employed in LED manufacture mainly in the nature of the parameters measured. As can be seen from Figure 5, quality control during the early stages of the photodiode process is identical to that for LEDs except that a wafer capacitance measurement has been added.

Wafer quality is assessed by first burning in a sample of devices from each processed wafer for 24 hours at 200 degrees C with an applied reverse voltage of 20V. Following this burn-in, breakdown voltage, dark current and responsivity are determined for each device in the sample. The sample is then subjected to an accelerated life test, the conditions of which are 85 degree C, 85 percent R.H. and 10 volts applied reverse voltage. As was the case with LEDs, failure of the wafer sample to conform to established criteria results in rejection of the wafer.

For completed devices, the final test sequence consists of the 200 degree C burn-in followed by measurement of device dark current and breakdown voltage.

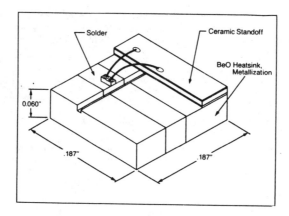

Figure 1. Physical structure of LED packaged on heat sink. Step in ceramic heat sink is provided for fiber alignment.

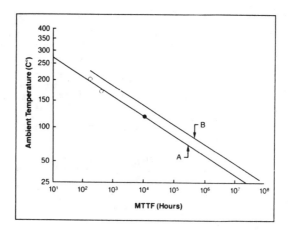

Figure 2. Accelerated life testing data at 100mA. Curve A is a least squares fit to data obtained which is summarized in Table 2.

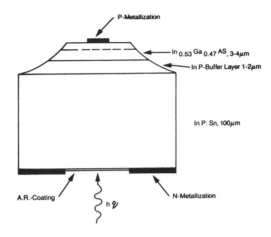

Figure 3. Cross-section of mesa photodetector.

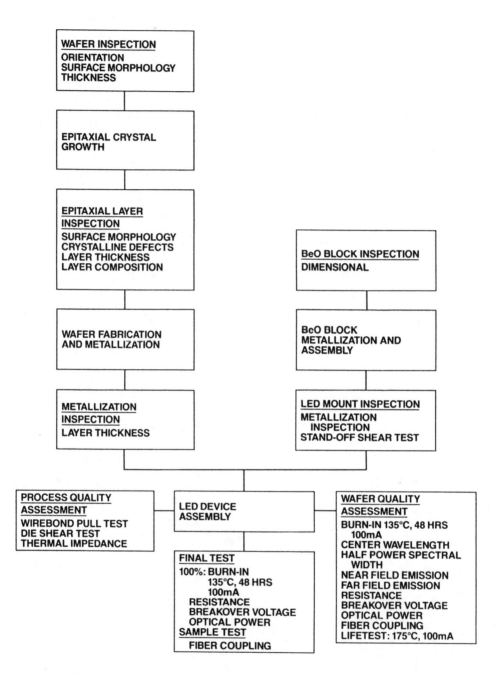

Figure 4. Flow chart showing summary of LED manufacturing process and quality screens employed.

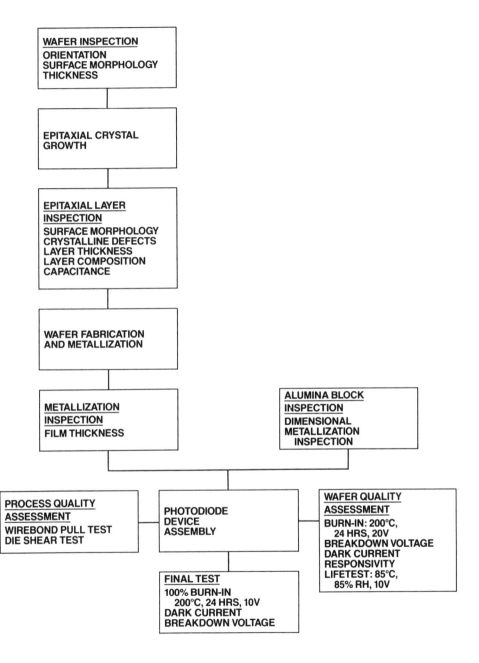

Figure 5. Flow chart showing summary of PIN manufacturing process and quality screens employed.

References

1. J. R. Jones, J. S. Kennedy, F. W. Scholl, "A Prototype CSMA/CD Local Network Using Fiber Optics", Local Area Networks '82, Sept. 1982, Los Angeles, CA.

2. D. Botez, M. Ettenberg, IEEE Trans. of Elec. Devices, 26, 1230 (1979).

3. D. Marcuse, IEEE Journ. of Quant. Elec., 13, 819 (1977).

4. D. Jenkins and A. Mabbit, IEEE Specialist Conference on Light Emitting Diodes and Photodetectors, Ottawa-Hull, September 15-16, 1982.

5. P. W. Webb, G. H. Olsen, IEEE Trans. Elect. Devices, Vol ED-30, 395 (1983).

6. R. H. Saul, F. S. Chen, P. W. Shumate Jr., AT & T Technical Journal, 64, 861 (1985).

Fiber optic edge emitting LED package for local area network applications

S. J. Anderson, P. G. Abbott and F. W. Scholl

Codenoll Technology Corporation

1086 North Broadway, Yonkers, New York 10701

ABSTRACT

A high performance edge emitting 830 nm LED has been mated with SMA and ST fiber optic receptacles using an internal lens to achieve performance comparable to a pigtailed device at substantially lower cost. The absence of a pigtail enhances reliability and simplifies handling during PC board assembly. The design goals included -7 dBm minimum launched power into a 100 micron core fiber at 150 mA drive current, compatibility with existing active device packages, electrical isolation of all pins from the package, and 0° to 50°C operation, typical of commercial Local Area Network (LAN) requirements.

These objectives are attained using a cylindrical Graded Index (GRIN) lens bonded into the package, and active alignment of the device during assembly. Where the highest power is required, as in passive star LANs, coupled powers of -3 dBm are easily achieved at 150 mA; alternatively, optical power can be traded for reduced drive current, an attractive property for active star based LANs. The devices exhibit rise times of 4 to 7 ns, with 2 ns available at reduced power output. Repeatability of coupled power for multiple connections of the same or different 100 micron core fibers is typically ±0.5 dB for SMA Type 2 connectors, and ±0.25 dB or less for SMA Type 1 and ST connectors. The devices withstand ten thermal shock cycles (-40° to +100°C) with no significant change in optical output. Accelerated life testing has demonstrated a MTTF of 2.7×10^7 hours for unpackaged devices, and preliminary results indicate comparable lifetimes for the packaged devices.

The paper will discuss the design criteria for optimum coupling in a system consisting of an edge emitting LED, GRIN lens, and connectorized optical fiber; comparison to other lens systems will be made. Data for coupled power versus drive current and fiber size (100 micron, 62.5 micron, 50 micron) will be presented. In addition, results of high pulsed current operation will be given.

1. INTRODUCTION

Active device mount fiber optic packages allow the connection of an optical fiber to an emitter or detector through a removable standard optical connector.[1] They offer a number of advantages over pigtailed devices in the design and manufacture of transmitter and receiver hybrid modules or printed circuit boards. While the maximum coupling of light from an emitter into an optical fiber is generally achieved by the direct alignment of a lensed fiber pigtail, some disadvantages are associated with the pigtail technology. Among them are the high cost for the pigtail and connector, the difficulties encountered in aligning and fixing the fiber very close to the emitting facet of the LED, the problems encountered in making a hermetic seal to the fiber, and the added effort required in handling and mounting a package with a pigtail. In an active device mount (ADM) receptacle package, a seperate lens is usually combined with the emitter in a housing with a mating optical port which allows the system fiber to be connectorized and polished in a normal manner. Some of the coupled power may be given up in exchange for the smaller size, increased ruggedness, simplicity of assembly, and lower cost which are characteristic of this design.

In the implementation discussed in this paper, a high power edge emitting LED chip, originally optimized for direct fiber coupling, has been studied under several possible active device mount configurations. The resulting package, incorporating a graded index (GRIN) lens, launches powers into a system fiber comparable to those from a direct coupled pigtail, after the loss at the pigtail-system connector is taken into account. The high coupled power achieved in these devices makes them particularly suited for implementing a passive star Ethernet local area network (LAN). They also find application in active star systems where the required optical power can be achieved at low drive currents, enhancing the design of an active hub.

2. DEVICE DESIGN

The double heterostructure edge-emitting LED chip utilized in our novel package is based on a proprietary design that has evolved to meet the following objectives, all of which are aimed at extracting and launching as much light as possible from the device into an optical fiber:

1. Light rays generated within the active layer must not be trapped there where they may be absorbed;
2. The electrical carriers injected into the active layer must be confined there to allow efficient generation of light;
3. Light rays travelling through the device must not encounter absorbing regions or interfaces on their path to the fiber core;
4. Reflections at the emitting facet must be minimized to allow the greatest amount of light to escape.

The LED device design has been adapted for both AlGaAs and InGaAs materials, utilizing relatively straightforward diffusion, photolithographic, metallization and other semiconductor processes.[2] Once cleaved and separated into discrete chips, the active surface of the LED is eutecticly attached to a BeO submount, and gold wires are bonded to effect the electrical connections to the device. The emitting facet of the assembled device is provided with an anit-reflective coating to maximize the usable output of the LED. The process has been extensively proven in use with a discrete BeO submount, used in hybrid circuit applications employing a fiber optic pigtail. Accelerated life testing data for the submounted (unpackaged) device indicates an extrapolated room temperature MTTF of over 25 million hours at 100 mA.[3]

The 830 nm AlGaAs edge-emitter is well suited for incorporation into high speed fiber optic data links, exhibiting typical transition times of 4 to 7 nanoseconds (ns). Specially doped wafers can be produced to achieve speeds of 2.5 ns or better. In addition, the device shows excellent thermal stability, with its optical output diminishing with temperature at a rate of only -0.03 dB/°C. Typical operating currents (up to 150 mA) require less than 2 volts of forward bias. The near field spot size is 35 microns by 15 microns. The far field FWHM is 45 degrees perpendicular to the junction and 120 degrees parallel to the junction. The spectral distribution is peaked at 830 nm with a FWHM of 60 nm.

To lower cost and increase handling ease and reliability, an active device package was designed and developed to meet the following goals:

1. Launch 200 microwatts of 830 nm optical power into a 100 micron core (0.29 NA) optical fiber terminated with a standard SMA connector,
2. Drive current of no more than 150 mA,
3. Operating temperature range of 0° to 60° C,
4. Package outline compatable electrically and mechanically with "standard" active device mount parts,
5. Electrical isolation of device from package exterior,
6. Adequate internal heatsinking of the LED chip to ensure that reliability is not compromised under typical operating conditions,
7. Utilize simple manufacturing processes employing as many standard or familiar steps as possible, and
8. Substantially reduce overall packaged device cost, as compared to pigtailed packages with comparable performance.

Whereas pigtailed package designs often rely on a microlens, created on the launch end of the pigtail fiber, to maximize optical coupling, such a technique cannot be used with a receptacle-type active device package, where the light is to be launched into the flat polished end of a removeable optical fiber terminated with a standard connector. Since launching the light of an edge-emitter directly into an unlensed fiber end results in very poor coupling efficiency, it was clear that a discrete lens was needed for the package design. Besides enhancing optical coupling from LED to fiber, the lens was required to be inexpensive, easy to handle and mount, offering good working distances and alignment tolerances, and able to be incorporated in a hermetic-type enclosure.

3. COMPARISON OF LENS DESIGNS

Many factors went into the selection of the lens, including overall net optical performance, mechanical alignment sensitivity, ease of handling, and cost. A number of different designs were evaluated, ranging from micro-sized sapphire spheres to larger plastic and glass lens elements.[4] Table 1 contains a comparison of several of the lens designs that were evaluated. While a very small sapphire sphere offered the highest peak coupling, a cylindrical Graded Index (GRIN) lens was selected for its good optical performance, the ease with which it can be handled and mounted in a metal package, and its relatively larger positional tolerances in alignment.

Lens Type	Lens Material	Size (mm) Dia	Length	Relative Coupling Efficiency ELED into 100μ Fiber	
Lensed F.O. Pigtail	Glass Fiber Optic	0.140	–	1.0	(0.8 with 1 dB pigtail connector loss)
Cylindrical GRIN Lens	Glass (Doped)	1.80	3.74	0.62	
Microsphere (Large)	Sapphire	1.00	–	0.50	
Microsphere (Small)	Sapphire	0.50	–	0.80	

Table 1. Comparison of size and fiber coupling efficiency for several types of lenses.

Figures 1a and 1b demonstrate this difference in mechanical sensitivity by comparing optical coupling vs. axial and radial position for the GRIN lens and for a typical sphere lens with comparable gross performance. Furthermore, it was realized that the cost of the metal receptacle body would be related to the precision required for its fabrication, and that a lens choice that permitted easier tolerances for certain key dimensions would be less expensive.

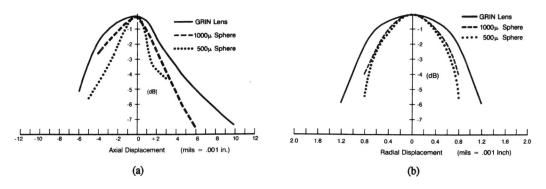

Figure 1. Comparison of lens properties: (a) axial position sensitivity of lens/device alignment; (b) radial position sensitivity of lens/device alignment

As shown in Figure 2 the GRIN lens offers far better performance than a sapphire micro-sphere when used in an off-axis or non-optimum alignment position. A modest misalignment between, for instance, the lens and fiber may be almost completely compensated for by adjusting the position of the LED, to achieve performance virtually the same as with exact optimum alignment.

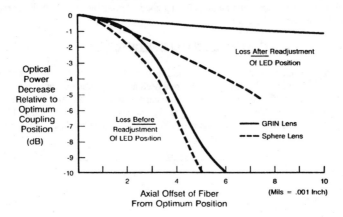

Figure 2. Active LED position adjustment to recover coupled power lost to fiber/lens misalignment.

4. PACKAGE CONFIGURATION

A Beryllium Oxide (BeO) submount was utilized as a heatsink for the LED chip, providing a flat, easy-to-metallize surface with a good thermal path for the device while also electrically isolating the device from the metal header to which the submount is soldered. Wire bonds connect the chip and submount to the leads of the header, through which the device is connected electrically to the outside world. The header assembly is tested and burned in as a subunit, prior to being incorporated in the final package.

The cylindrical GRIN lens is positioned and securely sealed in a metal receptacle package, referred to as the "body". The body locates the lens concentric to, and a specified distance from, the assumed position of the fiber core in an installed fiber optic connector. As Figures 1 and 2 indicate, the accuracy required for the concentricity and axial placement is easy to achieve with readily available and economical fabrication techniques, so the body may be a relatively low cost part.

The header, bearing the submounted LED, is positioned at the rear of the body and actively aligned with the GRIN lens, thereafter being affixed in place so as to seal the assembly, secure the placement of the LED, and obtain good thermal continuity. Provided proper materials and processes are employed, the package can readily be made hermetic; for low cost commercial applications, on the other hand, the package can easily be put together with minimum investment and in production-line quantities. Figure 3 depicts a cross section of the package configuration, along with a photograph of two versions of the package. The header and body were fabricated from metals, to meet heat sinking, sealing, mechanical stability, and other requirements. For larger production runs, further cost savings could be accrued through the use of plastic or composite materials.

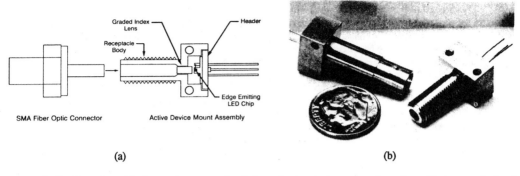

Figure 3. Configuration of Package: a) cross-sectional view of active device receptacle package; b) photograph showing ST version (left) and SMA version of active device package alongside a U.S. dime.

5. DEVICE CHARACTERIZATION

5.1. Optical/Electrical Performance.

Figure 4a shows the optical power vs. forward current profile for packaged LEDs of two power levels; the indicated power levels correspond to the optical flux coupled into a standard 100 micron core multimode fiber. Forward voltage across the device is also depicted in Figure 4a as a function of current. Coupling into smaller core fibers is enhanced by the focussing power of the lens: for 62.5 micron fiber the coupled power will typically be 75% of that for 100 micron fiber; for 50 micron core, the ratio is over 50%. Thus, with 500 microwatts (-3 dBm) into 100 micron fiber available from with better quality devices (at 150 mA), power levels of over 375 μW (-4.3 dBm) into 62.5 micron fiber and over 250 μW (-6 dBm) into 50 micron fiber are readily achievable. The linearity of the devices ensures that they will still put out relatively large levels of power even at reduced forward current (50 to 100 mA). Figure 4b shows a histogram of coupled optical power data for a typical production lot of LEDs in the active device mount package.

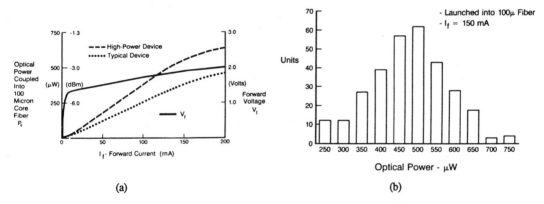

Figure 4. Electrical/Optical Performance: a) P-I and V-I characteristics for typical ELED devices; b) histogram of coupled power (100 micron fiber) for a typical assembled lot of active device mount packages.

5.2. Optical Power Reproducability.

The realized reproducability of optical power coupled into a system fiber is dependent upon the quality (length, concentricity, etc.) of the mated fiber optic connector as much as on the accuracy of the device package. Some variation in launched optical power will invariably be observed for an SMA connector as it is rotated to different positions in the device receptacle; the amount of variation can be minimized by selecting a quality connector, such as the newer ceramic types, or by using a stepless (905-style) SMA connector rather than the stepped (906-style) connector which relies on a plastic adaptor sleeve for proper mating. Figure 5 shows a histogram for optical power range obtained from rotating connectors in LED receptacles and noting the difference between the maximum and minimum powers coupled for each device. Larger core fibers offer better performance in this regard. Table 2 compares this connector rotation variance for three common fiber sizes when various types of SMA connector are used. For keyed (non-rotating) connectors, such as the ST (2.5 mm bayonet), the rotational problem is not present, and reproducability of the optical coupling is quite accurately maintained, typically within 0.5 dB, even for the smaller fiber sizes.

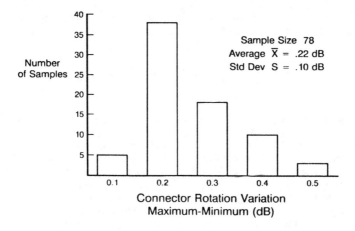

Figure 5. Coupling variation with connector rotation for 100 micron fiber and a 905-type SMA connector.

Connector Type	100 micron fiber	62½ micron fiber	50 micron fiber
SMA Type 1 (905)	± 0.3 dB	± 0.7 dB	± 1.0 dB
SMA Type 2 (906)	± 0.7 dB	± 1.5 dB	± 2.0 dB
2.5mm Bayonet (ST)	± 0.25 dB	± 0.5 dB	± 0.5 dB

Table 2. Maximum connector rotation variation for various fiber sizes and connector types.

5.3. Temperature Testing.

Assembled and tested devices have been subjected to temperature shock testing to assure the mechanical integrity of the package. Twenty-five (25) devices were each subjected to ten full cycles, each cycle consisting of one hour at -40° C followed by one hour at +100° C; data was taken for each device after each cycle (with a 30 minute stabilization period allowed before measurements were taken). None of the devices showed, at any stage in the experiment, any sign of degradation or catastrophic failure. Figure 6 shows a plot of the optical power data for the devices after each cycle, and a depiction of the temperature cycle.

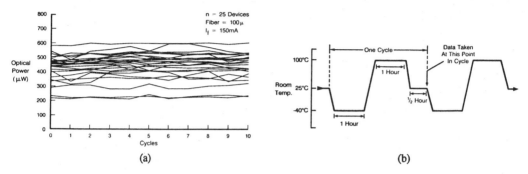

Figure 6. Temperature Shock Testing: a) 100 micron fiber coupled optical power data for 25 devices subjected to 10 temperature shock cycles; b) depiction of typical temperature shock cycle ranging from -40° C to +100° C.

5.4 Rise and Fall Times

The rise and fall times (10% to 90%) were determined for these packaged devices. A pulse generator capable of 1 ns risetime pulses was used with an impedance matching drive circuit to pulse the devices at 100 mA peak current and 50 percent duty cycle. A Si APD detector and 350 MHz oscilloscope were used in the detection circuit to measure the optical flux emitted from an optical fiber connected to the receptacle. The resulting optical waveforms are shown in Figure 7, illustrating rise and fall times of less than 5 ns. The device shown is of standard configuration with an undoped active layer. Standard devices exhibit speeds in the 4 to 7 ns range. Other devices, in which the active layer is intentionally doped to a level of approximately 10^{18} cm^{-3}, exhibit rise and fall times of less than 2.5 ns with a corresponding decrease in the optical output power.

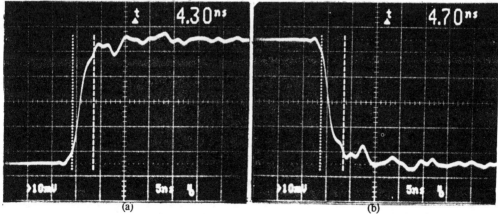

Figure 7. Rise and Fall Times of Standard 830 nm ELED in Fiberoptic Receptacle: a) rise time for typical (standard speed) device; b) fall time for the same device.

5.5. High pulsed power

In some applications it may be desireable to achieve short duration, very high power pulses from the device. In the edge emitter configuration the device is bonded to its heat sink active side down, so that the thermal impedance is very low, typically 20-50 degrees C/Watt. Therefore the pulsed and cw characteristics show very little difference for currents up to about 200 mA. The low thermal impedance allows much higher pulsed currents than could be achieved with a surface emitter. For pulse widths of 300 ns at a duty cycle of 10^{-4}, pulsed currents up to 1 Amp have been used to achieve optical power coupled into 100 micron fiber of 5 mW. The pulsed P-I characteristic of such a device is shown in Figure 8. The capability to send very high power signals may be useful in sensor applications or in networks to flag certain information, or as a means of enhancing collision detection in passive star systems.

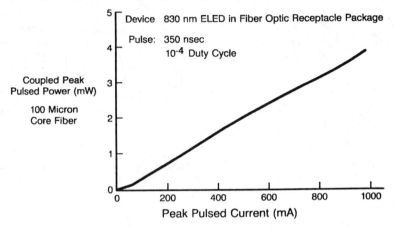

Figure 8. High Pulsed Power Output Characteristics of 830 nm ELED in Fiberoptic Receptacle Package.

5.6. Lifetesting

A program of accelerated aging was carried out on completed ADM modules.[5] Prior to lifetesting the chip headers were burned in at 100 mA and 125 degrees C for 48 hours and the assembled modules were burned in at 100 mA and 80 degrees C for 48 hours. This is the standard burn in cycle for all the production ADM receptacles. Devices from three diffferent grown wafers were tested at ambient temperatures of 60, 80, 100 and 120 degrees Celsius and 100 mA cw drive current. A total of 144 devices were characterized over a one year period. The sample was divided into groups of 16 devices for which a log-normal distribution of failure percentage versus time was established. End of life was defined to be when the optical power coupled into a 100 micron fiber reached one half of its starting value. At the time of this writing, although the longest running groups have accumulated over 8000 hours of testing, insufficient failures have occurred in any of the groups to define the mean time to failure (MTTF) from the log normal distribution. However, sufficient degradation has occurred to allow an extrapolation of the power drop versus time to the half power failure point for each group. The extrapolated MTTF for each group was plotted versus temperature to determine both the activation energy for the failure mechanism and the room temperature lifetime.

The result is shown in Figure 9, indicating a 25 degree C lifetime of three million hours. The horizontal error bars indicate the uncertainty of the extrapolation at each temperature. There are two groups represented at 80 degrees but their MTTF values are equal so that they appear as a single point. No 60 degree C data has been plotted since the degradation has been so slight after 8000 hours that an extrapolation to the half power failure time cannot be made accurately.

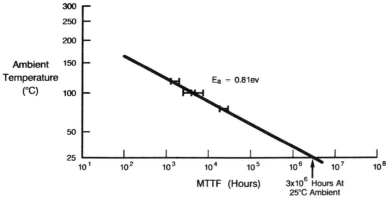

Figure 9. Accelerated Lifetest Plot (Arrhenius Plot) for 830 nm ELED in Fiberoptic Receptacle Package.

Experiments were also conducted to separate the failure rates associated with the package from those due to the LED chip itself. Extensive lifetesting on LED chips mounted on BeO blocks has been carried out over a number of years prior to the ADM package study. These tests have established a 25 degrees C MTTF of 27 million hours for the semiconductor chip. To separate package degradation effects from chip degradation effects, a measurement of total optical power emitted from the GRIN lens was made in addition to the optical power coupled into a fiber. The decrease in total power over time was found to be more gradual than that for the coupled power, indicating that the chip itself was aging at a slower rate, consistent with its previously established lifetime.

The established ADM module 25 degree C lifetime of 3 million hours is more than adequate for application to fiber optic transmitter products. The potential for even greater reliability exists in the inherent reliability of the edge emitting LED itself, allowing refinements in alignment stability to increase ADM package reliability to higher levels.

6. CONCLUSION

The design and characterization of a high performance fiber optic package for edge emitting LED's has been described. The package exhibits high coupled optical power without the disadvantages of a directly pigtailed device. The SMA and ST optical connector configurations have been described. The same package design can be extended to any commonly used type of optical fiber connector.

7. ACKNOWLEDGEMENTS.

The authors wish to thank Dr. B. Dutt and Mr. E. Salierno for their technical contributions to this project.

7. REFERENCES

1. H.M. Berg, D.L. Shealy, C.M. Mitchell, D.W. Stevenson, L.C. Lofgran, "Optical coupling in fiber optic packages with surface emitting LEDs", IEEE Trans. on Comp., Hybrids, Manuf. Tech., 6, 334-342 (1983).
2. D. Botez, M. Ettenberg, "Comparison of surface- and edge-emitting LEDs for use in fiber-optical communications", IEEE Trans., Elec. Devices, 26, 1230-1238 (1979).
3. F. Scholl, M. Coden, S. Anderson, B. Dutt, J. Racette, W.B. Hatfield, "Reliability of components for use in fiber optic LANs", in Reliability Considerations in Fiber Optic Applications, D.K. Paul, ed., Proc. of SPIE 717, 108-117, (1986).
4. K. Kawano, "Coupling characteristics of lens systems for laser diode modules using single-mode fiber", Appl. Opt.,25, (15), 2600-2605 (1986).
5. F.H. Reynolds, "Thermally accelerated aging of semiconductor components", Proc. of IEEE, 62, 212-222 (1974).

AlGaInP/GaAs red edge-emitting diodes for polymer optical fiber applications

B. V. Dutt, J. H. Racette, S. J. Anderson, and F. W. Scholl
Codenoll Technology Corporation, 1086 North Broadway, Yonkers, New York 10701

J. R. Shealy
School of Electrical Engineering, Cornell University, Ithaca, New York 14853

(Received 12 August 1988; accepted for publication 13 September 1988)

Pulsed and cw operation of AlGaInP/GaAs graded refractive index separate confinement heterostructure edge light-emitting diodes fabricated from epitaxial structures grown by organometallic vapor phase epitaxy is reported. The device consists of an active region of a single 100 Å quantum well of ternary $Ga_{0.5}In_{0.5}P$ and 1600-Å-thick lattice-matched confining layers of quaternary $(Al_xGa_{1-x})_{0.5}In_{0.5}P$. A coupled power of -10 dBm at 100 mA into a 500 μm polymer optical fiber of 0.48 NA is realized in a package consisting of a graded refractive index lens with uncoated facets from a 60×300 μm mesa-shaped stripe geometry diode. The 10–90% rise and fall times at 100 mA with a 5 mA prebias were measured to be 5 ns. Higher launched powers are expected to result from improvements in the materials growth, facet coatings, and packaging techniques.

Visible red emitters in the 640–680 nm wavelength region are of growing interest for a variety of applications such as displays, bar-code readers, compact disk players, and for transmitters in local area networks based on polymer optical fibers (POF's) with polymethyl methacrylate (PMMA) cores. These fibers have two transmission windows: one centered at 570 nm in the visible green and the second at 655 nm in the red with attenuations of 0.07–0.13 dB/m and 0.13–0.18 dB/m, respectively. Light-emitting diode (LED) transmitters are generally preferred over laser transmitters for local area networks because of their lower cost and higher reliability. Since the currently available green LED's are poor emitters, the focus is primarily on the emitters in the red region. Although AlGaAs/GaAs structures are not yet completely ruled out, the direct to indirect band-gap transition which occurs near the red wavelength of 650 nm corresponding to a composition of $Al_{0.4}Ga_{0.6}As$ may pose some limitations to realizing highly efficient LED's with high speeds. In our own experience, some commercially available surface-emitting AlGaAs/GaAs red LED's met power requirements for slower POF applications but their speed performance for 10 Mb/s Ethernet systems was limited by long 10–90% rise and fall times of 40–80 ns. Although it might be possible to improve the speed by alternate device designs incorporating restricted junction area and increased current crowding, the emphasis for fast and efficient red emitters appears to be moving towards the $(Al_xGa_{1-x})_{0.5}In_{0.5}P$ materials system grown lattice matched to GaAs substrates,[1-7] because of favorable band structure in this system.

In this letter we report the operation of an AlGaInP/GaAs edge light-emitting diode with a graded refractive index separate confinement heterostructure (GRINSCH).[8] This structure has been shown to be more efficient than the conventional double heterostructure.[8-10] The growth of the device structure by organometallic vapor phase epitaxy has been reported earlier.[7] Briefly, the epitaxial structure was grown at 76 Torr on p-type (100) oriented GaAs:Zn substrates. The structure consists of a 1.5 μm Zn-doped ($p\sim1.10^{18}$) $(Al_{0.6}Ga_{0.4})_{0.5}In_{0.5}P$ cladding layer, an undoped ($p\sim1.10^{18}$) 1600-Å-thick $(Al_xGa_{1-x})_{0.5}In_{0.5}P$ layer with a linear grading from $x=0.6$ to 0.2, a single 100-Å-thick undoped quantum well of $Ga_{0.5}In_{0.5}P$, an undoped 1600-Å-thick $(Al_xGa_{1-x})_{0.5}In_{0.5}P$ confining layer linearly graded from $x=0.2$ to 0.6, a 1.5 μm Te-doped ($n\sim7.10^{17}$) $(Al_{0.6}Ga_{0.4})_{0.5}In_{0.5}P$ cladding layer, and finally a cap layer of 2000 Å Te-doped ($n\sim5.10^{18}$) $Ga_{0.5}In_{0.5}P$. The devices were formed by Au-Ge n metallization, Ti-Pt-Au p metallization, wet etching of 60 μm mesas, scribing, and cleaving into individual chips of 300 μm cavity length. The chips were then bonded onto beryllia submounts using tin and the submount with the chip was in turn bonded onto a typical TO-can type compact disk header by soldering. Thermal compression wedge bonding completed the wire bonding to the n stripe of the mesa of the header pin electrically isolated from the base. It should be pointed out that in this scheme of packaging the junction in these experimental devices is not efficiently heat sunk. Nevertheless, these devices performed satisfactorily in both the pulsed and dc testing, although for long-term reliability better heat dissipation would be required.

The voltage-current (V-I) and the power-current (P-I) characteristics of the diodes on the header were obtained at room temperature. For power measurements, the emitter and the detector were arranged to form a 0.5 NA configuration. Figure 1 shows typical characteristics of two of the LED's. Note that the 0.5 NA powers at 100 mA were about 220 and 180 μW. The saturation behavior observed in these devices is perhaps due to the poor heat dissipation of the header-mounted chips mentioned earlier. The header with the device was then aligned with an x-y-z stage in a package consisting of a graded refractive index rod lens by using a connectorized 500 μm core POF with 0.48 NA for optimum power. Figure 2 shows this device mount package schematically.[11] The power coupled into the 500 μm 0.5 NA fiber from similar LED's so packaged versus current is shown by curves 3 and 4. At 100 mA the coupled power is ~100 μW

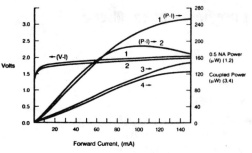

FIG. 1. Voltage-current (*V-I*) and power-current (*P-I*) characteristics of the devices numbered 1 and 2 are shown. Curves 3 and 4 show coupled power from similar devices in the active device mount package shown schematically in Fig. 2.

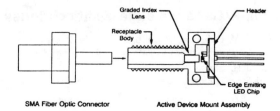

FIG. 2. Schematic of the active device mount package with the chip on the header, GRIN ROD lens, and the connectorized fiber.

(− 10 dBm) from devices with no facet coatings. Note that the degree of saturation in packaged devices is reduced from the chip on the header. We believe that this is due to the improved heat dissipation from the package. Indirect evidence for this suggestion was observed in the measurement of wavelength. The emission wavelength was found to be 665 nm with a full width at half maximum of 20 nm at both 50 and 100 mA drive currents, while a shift of ∼3–4 nm was expected based on the estimated temperature rise of the device.[12]

The 10–90% rise (t_r) and fall (t_f) times were measured with 100 mA pulses of width 100 ns with a repetition rate of 1 ms with a prebias of 5 mA. A Si avalanche photodiode served as the detector. Both t_r and t_f were found to be 5 ns. These LED's performed successfully in an Ethernet system for 10 mB/s data transmission over 80 m distances of polymer optical fibers typical of interoffice environment. These results will be reported elsewhere.[13]

In summary, pulsed and cw operation of AlGaAsP/GaAs edge light-emitting diodes, in a package with a GRIN ROD lens, with uncoated facets that couple − 10 dBm into 500 μm polymer optical fibers at 100 mA are reported. Further work is in progress to improve the performance with appropriate facet coatings and for optimizing the growth of the epitaxial structures.

The authors wish to thank D. P. Bour, M. Hatzakis, Jr., P. Lee, and P. Abbott for assistance with the device fabrication, packaging, and testing. Stimulating discussions with M. Coden and J. Helbers are gratefully acknowledged.

[1]Masao Ikeda, Yoshifumi Mori, Miromitso Sato, Kunio Kaneka, and Naozo Watanabe, Appl. Phys. Lett. **47**, 1027 (1985).
[2]Masayuki Ishikawa, Yasuo Ohba, Hideto Sugawara, Motoyuki Yamamoto, and Takatosi Nakanisi, Appl. Phys. Lett. **48**, 207 (1986).
[3]K. Kobayashi, S. Kawata, A. Gomyo, I. Hino, and T. Suzuki, Electron. Lett. **21**, 931 (1985).
[4]Masayuki Ishikawa, Yasuo Ohba, Yukio Watanabe, Hiroko Nagasaka, Hideto Sugawara, Motoyuki Yamamoto, and Gen-Ich Hatakoshi, *Extended Abstracts of the 18th International Conference on Solid State Devices and Materials, Tokyo, 1986* (Komiyama, Japan), pp. 153–156.
[5]M. Ikeda, A. Toda, K. Nakano, Y. Mori, and N. Wantanabe, Appl. Phys. Lett. **50**, 1033 (1987).
[6]Hidenao Tanaka, Yuichi Kawamura, Shunji Nojima, Koichi Wakita, and Hajime Asahi, J. Appl. Phys. **61**, 1713 (1987).
[7]D. P. Bour and J. R. Shealy, Appl. Phys. Lett. **51**, 1658 (1987).
[8]W. T. Tsang, Appl. Phys. Lett. **39**, 134 (1981).
[9]J. R. Shealy, Appl. Phys. Lett. **50**, 1634 (1987).
[10]J. Nagle, S. Hershee, M. Krakowski, T. Weil, and C. Weissbuch, Appl. Phys. Lett. **50**, 1325 (1986).
[11]S. J. Anderson, P. G. Abbott, and F. W. Scholl, presented at the SPIE conference, Boston, MA 6–10 Sept. 1988, paper No. 988-28.
[12]This estimate was arrived at by assuming that the amount of heat to be dissipated is the same and results in a similar temperature rise at various currents as in a ELED of AlGaAs of 10 μm × 125 μm size stripe, where the junction was mounted directly on the submount for more efficient heat dissipation. Since the present device is a 60 × 300 μm mesa, the current density at which it is operating is lower by an order of magnitude. It is possible that the temperature rise of the junction is also correspondingly lower and may account for the absence of a shift in emission wavelength. On the other hand, the thermal conductivity of the AlGaInP/GaInP layers is expected to be lower and the actual temperature rise may be higher. Thus, it is rather difficult to estimate the temperature of the device and careful measurements are needed for each of the packaging designs.
[13]F. W. Scholl, M. Coden, S. J. Anderson, and B. Dutt, presented at SPIE conference, Boston, MA 6–10 Sept. 1988, paper No. 991-28.

FDDI COMPONENTS FOR WORKSTATION INTERCONNECTION

S. J. Anderson, D. V. Bulusu, J. Racette, F. W. Scholl
T. Zack and P. G. Abbott

Codenoll Technology Corporation
1086 N. Broadway, Yonkers, N. Y. 10701

ABSTRACT

The purpose of this paper is to discuss cost effective components for workstation interconnection as Fiber Distributed Data Interface networks transporting data at 100 Mbps proliferate within the next few years. The 1300 nm components called for in the draft standard are relatively expensive and will probably be used in the backbone. However, 830 and 660 nm components based on AlGaAs/GaAs technology appear to be suitable and offer significant cost savings for workstation interconnections over 50 to 500 meter distances. The 830 nm interconnection still uses silica fiber, while 660 nm interconnetion could be done with cheaper polymer fiber as the medium. The discussion is centered on a comparison of the costs and performance of these alternative components with the 1300nm devices. The presently available 830 nm technology could carry the data over 500 meters, while the 660 nm technology with polymer optical fiber can be used up to 50 meters. A recently developed receiver based on a monolithic GaAs preamplifier with a sensitivity of -39 dBm average at 125 Mbps and a wide dynamic range is also described here.

1. INTRODUCTION

Within the next few years, 100 Mbps Fiber Distributed Data Interface (FDDI) networks are expected to see a rapid growth in the backbone. The distances here can go up to two kilometers. However, the workstation environment, as shown in Fig. 1, through the use of concentrators interfacing with the backbone may involve distances of the order of 50 to 100 meters. The standard is drafted around the 1300 nm components made up of quaternary InGaAsP/InP alloy system, which are relatively expensive as of this writing compared to 830 and 660 nm components. The latter devices are based on more mature AlGaAs/GaAs materials technology. It is possible that the present cost differences are due to different levels of market demands.

In the following, we compare the performance of 830 and 660 nm components with the 1300 nm devices taking into account the spectral characteristics, speed and output power of the available sources. Transmission distances for FDDI 100 Mbps data rates over glass fiber at 830 nm and over polymer fiber at 660 nm as limited by spectral width and speed of these sources are estimated. In this paper, we also describe a wideband receiver based on a monolithic GaAs preamplifier for high speed operation with a wide dynamic range of 25 dB and a high sensitivity of -39 dBm. These alternatives appear to be potential contenders for interconnections over distances of the order of a few meters to several hundred meters in the workstation environment.

2. WORKSTATION INTERCONNECTION

2.1. 830 nm Components

Optical data transmission at 830 nm is attractive because of the availability of low cost sources and detectors. As data rates increase, the distances which can be supported decrease because of pulse spreading due to modal and chromatic dispersion. However, for FDDI workstation interconnection, reasonable values of LED risetime and spectral width allow more than sufficient distance of transmission.

For accurate reception of data, the FDDI specification calls for the optical risetime exiting the fiber to be less than 5 ns. A calculation of the exit risetime can be made if the LED source risetime, spectral width, and wavelength are known along with the fiber bandwidth[1]. For a nominal wavelength of 830 nm for the LED source, the following equation relates the distance which can be achieved, maintaining a 5 ns risetime at the end of that length of fiber :

$$25 = t_{fiber\ exit}^2 = t_{LED}^2 + \underbrace{\frac{FWHM^2 \times L^2}{87.67}}_{\text{MATERIAL TERM}} + \underbrace{\frac{2.32E5 \times L^2}{f_{3\ dB\ fiber}^2}}_{\text{MODAL TERM}} \qquad (1)$$

where L is the length of the fiber in km, t_{LED} is the risetime of the LED in ns, FWHM is the full width at half maximum of the LED spectral characteristic in nm and $f_{3\ dB\ fiber}$ is the 3 dB optical bandwidth of the fiber in MHz-km. This equation results in the graphical representation shown in Fig. 2, using the bandwidth of FDDI-specified 62.5 micron fiber at 830 nm of 160 MHz-km.

Achievable LED properties, as described in the following section, allow risetimes of 3.5 ns or less and spectral widths of 60 nm or less. As seen from Fig. 2, these support a fiber length of up to 500 meters, more than sufficient for workstation interconnections.

The LED design is an edge emitting type, achieving maximum coupling of light into the fiber. This allows sufficient coupled power even with the reduced optical power characteristic of a high speed device as compared to one with a slower response. Other design criteria are shown in Fig. 3. An internally highly doped active region is incorporated in the device, which reduces the carrier lifetime and leads to the fast rise and fall time characteristics.

2.2. 660 nm Components

A potentially attractive low cost workstation interconnection appears to be the polymer optical fiber (POF) link up to 50 meters[2]. This alternative offers several benefits - the medium, the sources and detectors, the connectors, the fiber termination - are all cheaper compared to the other approaches. With the presently available polymethyl methacrylate (PMMA) fibers, the availble bandwidth of 6 MHz-km allows transmission of 100 Mbps up to 50 meters. However, with smaller numerical aperture sources such as 660 nm lasers or edge-emitting LEDs under proper launching conditions of light from the source into the fiber, which virtually obviates the modal dispersion limit to the bandwidth, this distance could be pushed much higher[3]. In addition, current research on fibers shows the potential for availability of 25 MHz-km bandwidth fibers in the near future[4], which would make it possible to realize polymer links to longer than 100 meters at FDDI data rates.

The 660 nm sources based on double heterojunction AlGaAs/GaAs alloys may pose some limitations as pointed out earlier due to band structure[5]. As discussed in reference (5), 660 nm red emitters made of AlGaInP/GaAs alloys appear to be gaining momentum because of a more favorable bandstructure in this materials system.

Notwithstanding the choice of devices based on either of these systems, the tradeoff between LED and POF properties can be calculated from equation (1), assuming the modal dispersion term dominates as is the case for step index fiber like POF. The results are shown in Fig. 4. As previously stated, 50 m systems are possible with the 6 MHz-km fiber available today and with LEDs of 3 ns risetimes.

2.3 GaAs Integrated Circuit Optical Receiver

A high sensitivity, wide dynamic range receiver has been designed around a GaAs preamplifier chip and packaged in an active device mount receptacle to achieve small size and low cost. The photodiode, which is mounted on a hybrid circuit with the GaAs preamplifier and the AGC circuitry may be be either Si or InGaAs p-i-n detector, allowing the receiver to function for all of the potential wavelengths proposed for FDDI operation. The cost savings associated with a Si detector chip useful for both 830 and 660 nm compared to a 1300 nm InGaAs detector is significant.

A graded refractive index (GRIN) lens is used to couple the greatest amount of light into a small area detector. The low capacitance of the small area detectors is important to achieve the high speed operation required of FDDI. The performance of the preamp circuit is shown in Fig. 5 and the package configuration in Fig. 6.

A combination of transmitter and receiver components described here can provide a 830 nm flux budget of at least 14 dB. Since the length of an FDDI 830 nm link is dispersion limited to 500 meters, only about

2 dB is needed to drive the cable plant (3.75 dB/km fiber loss at 830 nm). The remaining flux budget, which is greater than that available in the 1300 nm system, can be used to allow bypassing of a greater number of nodes than in 1300 nm FDDI, adding another feature to the usefulness of these alternate wavelengths. The resulting application space for 660 nm, 830 nm and 1300 nm systems is summarized in Fig. 7.

3. SUMMARY

In summary, it appears that low cost FDDI workstation interconnections can be realized with 830 and 660 nm components. The 830 nm components can run up to distances of the order of 500 meters on glass fibers, while the 660 nm components with polymer fiber can go up to 50 to 100 meters. Work is in progress to demonstrate these estimates by using modular FDDI boards.

4. ACKNOWLEDGEMENTS

We gratefully acknowledge stimulating discussions with M. Coden, E. Sztuka, R. Lefkowitz, B. Ramsey, R. Steele, M. Lynn and G. Miller.

5. REFERENCES

1. J. L. McNaughton and R. L. Ohlhaber, pp 232-237, Proceedings FOC 80, San Francisco, Sept 1980, published by Information Gatekeepers.

2. High speed polymer optical fiber networks, D. V. Bulusu, T. Zack, F. W. Scholl, M. Coden, R. Steele, G. Miller, M. Lynn, paper no. 1364-47 presented at the SPIE conference on FDDI, CAMPUS-WIDE, and METROPOLITAN AREA NETWORKS, Sept 16-21, 1990, San Jose, California, USA.

3. Unpublished work.

4. Private communication from R. Steele (Sept 1990).

5. Private communication from B. Chiron (Aug 1990).

6. B. V. Dutt, J. H. Racette, S. J. Anderson, F. W. Scholl and J. R. Shealy, Appl. Phys. Lett. pp 2091-92, 53 (21) Nov 1988.

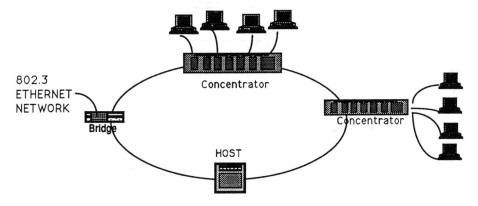

Fig. 1. Typical FDDI application illustrating the use of concentrators to attach workgroups to the ring.

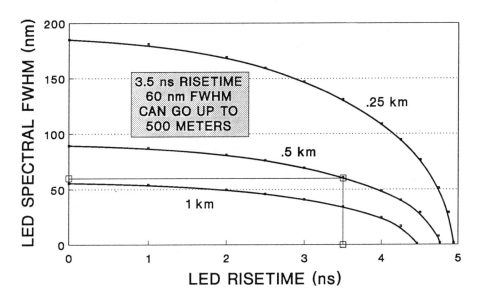

Fig. 2. FDDI at 830 nm. LED characteristics required to achieve 5 ns risetime at fiber exit.

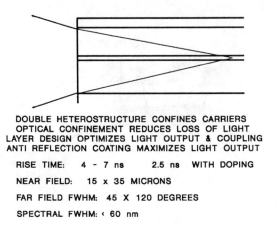

Fig. 3. Characteristics of 830 nm edge emitting LED.

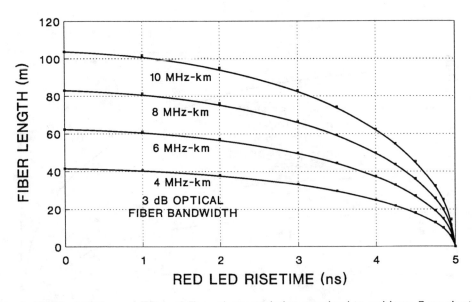

Fig. 4. FDDI at 660 nm. LED and fiber characteristics required to achieve 5 ns risetime at fiber exit.

GaAs IC PREAMP RECEIVER
HIGH SENSITIVITY - WIDE DYNAMIC RANGE

λ	BANDWIDTH	MAXIMUM DATA RATE	OPTICAL SENSITIVITY (AVG PWR)	OPTICAL DYNAMIC RANGE
1300 nm	90 MHz	135 MBPS	-42 dBm	24 db
830	90	135	-39	25
660	90	135	-38	25

Fig. 5. Characteristics of GaAs IC preamp receiver.

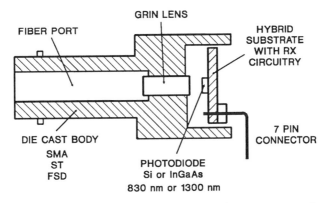

Fig. 6. Package configuration of GaAs IC preamp receiver.

λ	SOURCE STRUCTURE	SPEED ns	SPECTRAL WIDTH nm	DISTANCE meters	COST
1300	InGaAsP DH	< 3.5	~ 90	2000	HIGH
830	AlGaAs DH	< 3.0	~ 60	500	MEDIUM
660	AlGaInP GRINSCH	< 5.0	~ 20	50	LOW

Fig. 7. Summary of component properties for FDDI.

Making Fiber Optic Terminations — Step-by-Step Part One

By John Cocomello and Richard Focht, Codenoll Technology Corp., Yonkers, N.Y.

This is the first of a two-part article on fiber optic terminations. Part Two, which will be featured in the March issue of Connection Technology, will focus on a compatibility study that was recently performed to measure the mating and insertion loss of the 906 ceramic optical connector as compared to the standard metal 906 connector.

Fiber-optic cable has become cost competitive with coaxial cable. Better manufacturing processes, higher volume and less costly materials, such as combinations of stainless and ceramic rather than all stainless construction in connectors, has reduced the costs of a fiber network.

The fiber optic connector, one of the most important parts of a fiber-optic installation, influences the price of the cable plant and how effectively it functions.

The most commonly used connector is the screw-on SMA, even though it has several disadvantages. For instance, it can unscrew as a result of vibration. In addition, the actual connectorization is a time-consuming laborious affair that does not lend itself to automation, although several recent innovations, the ceramic-steel "906" type connector and automated multiple connector polishing equipment, have breathed new life into the SMA-type connector.

Also widely used is the ST, a BNC-type positive-locking connector, which is unaffected by environmental factors and rotational variations. The ST's ease of connectorization and installation, in addition to low cost, are rapidly making it the standard connector of choice.

The following is a description of the connectorization process for the SMA style stainless steel optical connectors.

The connector is assembled in the following sequence:

Step 1: Cable Preparation. The outer jacket is stripped with jacket strippers, removing 1-5/8" in short 1/2" sections. This will expose the "Kelvar" strands and the buffered fiber (see figure 1). The Kelvar fibers then are trimmed to 3/8" long and the buffered fiber is stripped using "No-Nik" strippers. Small sections are removed until 5/8" of buffered fiber remains.

Step 2: Epoxy Preparation. The two-part epoxy is mixed until a uniform color is obtained (see figure 2).

Step 3: Attaching the Connector. The heat-shrink tubing is slid over the cable jacket (see figure 3) and the eyelet is carefully slid over the jacket, the flared end toward the end with the fiber. The fiber then is cleaned using a soft cloth or tissue soaked with isopropyl alcohol. If the cable jacket diameter is less than 2.7 mm (0.106"), the spacer tubing can be slipped over it for added support.

The connector body is filled half full with epoxy (see figure 4). Two or three drops of epoxy are applied to the outside of the knurled support sleeve. (A thin coat is advised.) The primary ferrule is slowly inserted into the connector body. (Note: Some connectors do not use a primary ferrule.) Slow, careful insertion ensures that air is displaced by the epoxy.

The fiber is inserted, with a rotating motion, as far as possible into the con-

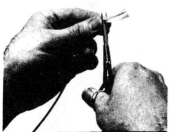
Figure 1, cable preparation.

Figure 2, epoxy preparation.

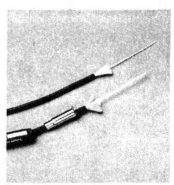
Figure 3, applying heat-shrink tubing.

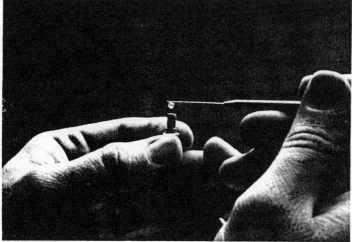

Figure 4, filling the connector with epoxy.

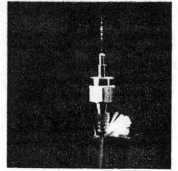

Fiber 5, the fibers must spread over the outside of the knurled support sleeve.

nector body. The Kelvar fibers must spread over the outside of the knurled support sleeve. The optical fiber will pass through and extend past the connector body (see figure 5).

A hemispherical bead of epoxy is applied to the front tip of the connector and optical fiber and the eyelet is brought up and over the Kelvar fibers and bottomed onto the body shoulder behind the "C"-ring of the connector body. The eyelet is crimped to the connector body and the heat-shrink tube is slid up over the eyelet as far as possible.

Heat is applied from the heating gun by moving it slowly from the top of the connector to the tip and then back. This process is repeated until the epoxy turns dark brown (see figure 6).

After the epoxy is cured, the heat is directed to the heat shrink tubing. With a cleaving tool, the fiber is lightly scribed and the fiber is pulled straight away from the connector (see figure 7).

Step 4: Polishing. Polishing is performed by hand in three stages, using 30 μm lapping papers in the first stage, 3.0 μm lapping film in the second and 0.3 μm lapping film for the third stage.

A polishing fixture (see figure 8) maintained at 0.3860 to 0.3863 is used. A clean, flat surface is necessary for polishing (Plexiglass works well) and the polishing surface must be flooded with water (figure 9).

Polishing is performed in an elongated "figure 8" (see figure 10). During the first stage, 3 μm lapping paper is used initially to lightly polish and remove the excess glass. Polishing is continued until all the epoxy is removed from the front of the connector. During stage 2, 3 μm lapping film is used making approximately 35 figure 8s. In stage 3, 0.3 μm lapping film is used to make approximately 45 figure 8s.

Step 5: Inspection. Under magnification, the metal portion of the connector face is checked for scratches. Scratches indicate the need for further polishing. Small chips are acceptable on the outer edge, but not in the center. Large chips either require further polishing or re-termination (see figure 9).

SMA 905 and 906 style connectors come in two versions, all stainless steel and metal with a ceramic insert.

The all-stainless steel connector is manufactured in a way that the metal barrel is overlength and then is polished to size along with the optical fiber. It has a short land area that guides the fiber due to the difficulties associated with drilling a 125 micron dia. hole in stainless steel. It requires

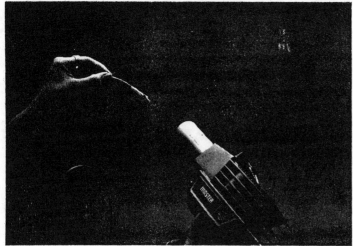

Figure 6, heat-curing the epoxy.

Figure 7, the fiber is lightly scribed.

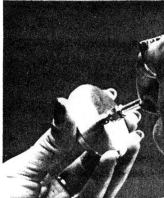

Figure 8, polishing.

Figure 10, polishing is performed with an elongated "figure 8."

constant monitoring during polishing to insure intolerance length. If the faces of two (2) connectors abut one another during mating, one or both connectors generally are destroyed. This is because

Figure 9, flooding the polishing surface with water.

as it is polished, steel, which is softer than glass, polishes off first, allowing the glass to protrude. It also has a large rotational variation in coupling due to the short land area and inherent problems with drilling.

Consequently, if two (2) connectors are mated face to face, the off center placement, as well as the downward angle, can cause a greatly varying optical coupling as the connector is rotated.

The ceramic ferrule connector comes with the overall length to finished tolerance. This has several benefits. During polishing, only the glass is removed to the surface of the ceramic ferrule. The ceramic being harder will not polish so that monitoring and measuring is not required to maintain the preset length.

The ceramic SMA is a PC-type, physical contact connector. In addition to mating the stainless steel type, this connector can be mated to the point that it actually comes in contact.

The rotational variations are improved over the steel type because of the much longer land area that guides the fiber and because of better concentricity between the ID and OD of the ferrule (see figure 2).

The ST type of connector is similar to the ceramic 906 in that it employs a ceramic ferrule. It differs in that it is engaged by a BNC type of captivator and is spring loaded to always mate in a physical contact mode. Another difference is that it is keyed to always couple in the same position, which virtually eliminates rotational variation.

Making Fiber Optic Terminations — A Compatibility Study And Evaluation, Part Two

By John Cocomello and Richard Focht, Codenoll Technology Corp., Yonkers, N.Y.

This is Part Two of a two-part article on fiber optic terminations. Part One, featured in the February issue of Connection Technology, *provided a step-by-step description of a termination procedure.*

A test was performed to measure the mating and insertion loss of the 906 ceramic optical connector as compared to the standard metal 906 connector. The test also measured insertion losses for the mating of the ceramic connector to the standard 906 connector for the purpose of studying compatibility. The types of mating combinations were drill thru to drill thru, ceramic to ceramic, ceramic to drill thru, drill thru to ceramic, four rod through to drill thru and four rod to ceramic. The attachment structure (adapter) used was of the conventional type. All of the conventional connectors were terminated using the method established by Codenoll for production of terminated optical cable.*

Test Procedure

The procedure for measuring the insertion loss for the connector under test utilized a stable optical power source to which the master connector was terminated. For accuracy, a half-sleeve was used when measuring the reference power out of the master connector. Once the reference power was established, the master connector was mated to the test connector with a standard bulk head adapter and an alignment sleeve. The opposite end of the test fiber was also terminated and fitted with a half sleeve for accuracy. This end of the fiber was inserted into the optical power meter. The difference between reference power and mated power was the insertion loss in dB.

Five mate-unmate cycles were made for each test connector resulting in 30 measurements of test power for each mating combination. The mean (average) power was calculated for each test connector and the deviation between mate-unmate cycles was derived from the measurements.

Drill Thru to Drill Thru

The typical insertion loss for the SMA 906 drill thru connector has been 1.0 dB and the manufacturer guarantees 1.5 dB. Test results for the mating of two drill thru connectors enforced the previous typical connector loss for this style connector of 1.0 dB with a standard deviation of ±0.12 dB. Currently manufacturing passes any termination that is below 1.5 dB of loss. Depending on the polishing process, the loss for this connector varies anywhere from 0.5 to 1.5 dB. This connector is assembled with epoxy protruding out of the end of the connector along with the fiber. During polishing, the connector is polished down along with the glass and the epoxy. The operator must periodically stop and measure the length of the connector with gages, which must be calibrated each day. This connectorization process is tedious and subjects individual connector lots to vary greatly in insertion loss from day to day. Polishing of this connector includes wet polishing and three types of polishing for the grinding of the metal body of this connector.

Ceramic to Ceramic

The typical insertion loss for the SMA rapid termination (rt) ceramic connector, when it is coupled through a conventional adapter, is specified by the manufacturer to be 0.73 dB with a standard deviation of ±0.08 dB. This connector does not require the body to

---

*The ceramic connectors were terminated using the Amphenol termination procedure 949-1002 for SMART connectors.

be polished since it arrives at optimal length. This relieves manufacturing from using wet sanding and periodic gaging of connector length during polishing. The connector is assembled with epoxy. (The difference from the standard SMA connector is that no epoxy must protrude out of the end of the connector.) The polishing of the cleaved fiber is accomplished by the use of a single type of polishing paper and dry sanding, which eliminates the operator's having to stand over a sink. Since only the fiber is polished and the connector is at optimal length, the insertion loss does not vary greatly from lot to lot. Also, the precision of this ceramic allows a deviation loss of only ± 0.08 dB as compared to 0.12 for the standard 906 connector. Needless to say, this connector is preferred by manufacturing and quality assurance departments.

Compatibility — Ceramic to Drill Thru Mating

Experimental test results for the mating of a drill thru connector to the new ceramic Type 906 were conducted by using the ceramic 906 as the master connector and by changing the drill thru connectors each time. The measurements revealed that the insertion loss varied between 1.1 to 0.88 dB with the mean valve being 0.97 dB loss. The standard deviation was measured to be ±0.16 dB.

The insertion loss of ceramic to drill thru is comparatively better than the insertion loss for drill thru to drill thru — 0.97 dB versus 1.04 dB. The rotational sensitivity for ceramic to drill thru is close to the same as for drill thru to drill thru, ±0.16 dB versus ± 0.12 dB.

Considering the accuracy of the ceramic connector with rotational sensitivity as was measured in the ceramic to ceramic test, it leads me to believe that the drill thru introduces the high rotational sensitivity numbers.

Compatibility — Four Rod to Ceramic

The four rod 906 was dropped from production because of the high cost of the connector. The termination process and polishing is identical to the standard drill thru connector. On occasion, manufacturing does terminate cables with the four-rod type connector because of its accuracy.

Experimental test results for the mating of a four rod SMA connector to the new ceramic type 906 were conducted by using the four rod 906 as the master connector and by changing the ceramic connectors each time. The measurements showed this combination to have a high coupling efficiency with the mean insert loss to be 0.39 dB with a standard deviation to be ±0.16 dB.

The reason for the high coupling efficiency was later discovered to be caused by the under-polished four rod connector, which shows the vulnerability of metal connectors to the accuracy of the polishing methods and the length gaging. Another important factor is the measured deviation of ±0.16 dB. This seems to be typical when using a metal connector as can be seen for ceramic to drill thru also with a deviation of ± 0.16 dB; ceramic to ceramic being ±0.08 dB.

Compatibility — Four Rod to Drill Thru

In the test to measure the mating of a four rod connector to a drill thru, the master connector was the four rod 906 and the drill thru 906 was changed each time. The measurements showed this combination to have a mean insertion loss of 1.03 dB with a standard deviation of ±0.3 dB.

The four rod connector used here was the same four rod found to be oversized in the previous test. The difference in insertion loss between the four rod to drill thru versus four rod to ceramic is 1.03 dB versus 0.3 dB.

The rotational sensitivity of the four rod to drill thru is the worst case of all mating combinations. This is most likely due to the inaccuracy of the drill thru and the variations of the lengths.

Study Conclusions

The standard 906 drill thru connector has a mean insertion loss of 1.04 dB. The connector length has a direct effect on this loss. Maintaining the optimal length during polishing requires gaging of its length throughout the polishing process. This has made the metal 906 connector vulnerable to under-polishing and over-polishing. The inaccuracy of the drill thru also contributes to a high rotational sensitivity in which the operator must tweek the connector for maximum performance. In addition, the polishing of metal connectors requires the use of water because of the grinding of the metal ferrule, and three types of lapping paper.

The ceramic 906 connector has a mean insertion loss of 0.73 dB. It is supplied with an extremely precise "critical length," which eliminates most of the rough grinding and polishing. With this connector, only the glass is removed during polishing; no ceramic ferrule material is removed. The accuracy of the ceramic connector allows it freedom from rotational sensitivity, giving it excellent repeatability.

The ceramic connector clearly is an acceptable replacement for the drill thru 906. This connector will increase the yield of termination and throughput. In addition, the ceramic connector offers better coupling and repeatability. Also, this precise critical length will lessen the load on quality assurance of measuring each connector on incoming inspection, eliminating the inaccuracies of gaging and effects of the metal 906 connector.

Automatic Polishing Machine

An automatic polishing machine** was recently evaluated by means of a trial production run. The total time spent on the machine was 12 hours. This amounts to a throughput of 25 connectors per hour or a time of 2.5 minutes to polish one SMA metal connector. These numbers would have increased if a second polishing bushing were used in the operation. It requires 10 minutes to unload and load 12 connectors onto the polishing bushing. This amounts to 20 minutes lost on an hour of polishing. Certainly, another 12 connectors an hour could be polished by loading a second bushing during each polishing cycle.

The total polishing potential with a second fixture is 36 connectors an hour with a weekly total of 1,260 connectors based on a seven-hour day. Manual polishing stands at about six to eight minutes per connector with a weekly total of 350 connectors maximum.

In comparison, it requires 3.6 man hours of manual work to match one man hour of an operator using the polishing machine. It is important to note that the metal SMA is the worst case in reference to polishing times. Ceramic connector times would be dramatically faster.

Loss Measurements

All the connectors polished with the machine during the trial production run were measured for insertion loss using the identical method for manually polished connectors. Every one of the 296 connector insertion losses was below 1 dB, with 2% falling on the 1 dB loss.

After the polishing cycle is complete, each connector must be gaged to ensure that it is within the specified length tolerance. Most of the connectors required a manual touch-up as they were slightly under-polished. This was done intentionally in order to reduce the amount of failures that could result from over-polishing. The amount of touch-up could be reduced with more careful control over the automatic polishing process, although inspection is always necessary. The amount of time needed to touch-up the overlength connectors averaged at one minute.

The cost of polishing 12 connectors with the machine is 75 cents. Three different pads are required at 25 cents each. These pads are then discarded. The cost difference in polishing pads between machine and manual polish is negligible. This cost may be slightly less for the machine pads.

References

1. G. J. Sellers, M. Margolin, J.B. Salzberg, R. Essert and I. Grois, *Multi-Mode SMA Connector Performance*, Allied Amphenol Products, 1986.

2. *SMA rt Termination Procedure 949-1002*. Allied Amphenol Products, 1986.

SAE Technical Paper Series

890202

A Fiber Optic Connection System Designed for Automotive Applications

Dominic A. Messuri, Gregory D. Miller and Robert E. Steele

Packard Electric Div.
General Motors Corp.

International Congress
and Exposition
Detroit, Michigan
February 27–March 3, 1989

"Reprinted with permission © 1989 Society of Automotive Engineers, Inc."

The appearance of the code at the bottom of the first page of this paper indicates SAE's consent that copies of the paper may be made for personal or internal use, or for the personal or internal use of specific clients. This consent is given on the condition, however, that the copier pay the stated per article copy fee through the Copyright Clearance Center, Inc., Operations Center, P.O. Box 765, Schenectady, N.Y. 12301, for copying beyond that permitted by Sections 107 or 108 of the U.S. Copyright Law. This consent does not extend to other kinds of copying such as copying for general distribution, for advertising or promotional purposes, for creating new collective works, or for resale.

Papers published prior to 1978 may also be copied at a per paper fee of $2.50 under the above stated conditions.

SAE routinely stocks printed papers for a period of three years following date of publication. Direct your orders to SAE Order Department.

To obtain quantity reprint rates, permission to reprint a technical paper or permission to use copyrighted SAE publications in other works, contact the SAE Publications Division.

All SAE papers are abstracted and indexed in the SAE Global Mobility Database

No part of this publication may be reproduced in any form, in an electronic retrieval system or otherwise, without the prior written permission of the publisher.

ISSN 0148 - 7191

Copyright © 1989 Society of Automotive Engineers, Inc.

Positions and opinions advanced in this paper are those of the author(s) and not necessarily those of SAE. The author is solely responsible for the content of the paper. A process is available by which discussions will be printed with the paper if it is published in SAE Transactions. For permission to publish this paper in full or in part, contact the SAE Publications Division.

Persons wishing to submit papers to be considered for presentation or publication through SAE should send the manuscript or a 300 word abstract of a proposed manuscript to: Secretary, Engineering Activity Board, SAE.

Printed in U.S.A.

A Fiber Optic Connection System Designed for Automotive Applications

Dominic A. Messuri, Gregory D. Miller and Robert E. Steele

Packard Electric Div.
General Motors Corp.

ABSTRACT

Functions controlled by electronics in automobiles are projected to continue to increase throughout the 90s'. As the sophistication of the functions performed increases, data communications will play an increasing role. Accordingly, electromagnetic compatibility will become increasingly difficult to attain both functionally and economically with all-conductor-based data transmission. The need for alternatives such as fiber optics becomes apparent. This paper presents a fiber optic connection system designed specifically to meet automotive requirements.

AS AUTOMOTIVE ELECTRONIC CONTENT and sophistication continue to increase, the need for electromagnetic compatibility becomes pronounced. In addition, the costs of routing high data rate signals around the vehicle are prompting designers to look for alternatives to conductor-based media. Fiber optic technology offers a potential solution to these problems. The major factors holding back the application of fiber optics in the past have been lack of functional requirement and cost.

To date, the majority of resources in the fiber optics industry have been focused on the issues surrounding telecommunications. To successfully apply fiber optics to the needs of the telecommunications industry, ultra-low loss connectors and cable have been developed. High power, narrow line-width lasers and extremely sensitive wide-bandwidth receivers have also been needed. The increase in functionality obtained by far offset the cost associated with the system components. As fiber is moving into metropolitan and local area networks, networking components are being implemented. Currently, connection systems for industrial and office automation are being developed and marketed. These components are compatible with the manual, on-site assembly and associated systems that divorce the power distribution network from the signal distribution network. Automotive applications, on the other hand, require compatibility with high volume, off-site assembly of systems integrated into the total vehicle power and signal distribution system which is then installed on an assembly line.

In this paper application requirements will be defined, potential solutions will be reviewed and a connection system designed to meet these requirements will be presented.

AUTOMOTIVE CONNECTION SYSTEM REQUIREMENTS

DIMENSIONAL CONSTRAINTS - The connection system must meet the dimensional constraints of its application. The application of fiber optics must not result in an increase in the dimensions of the connector.

CIRCUIT BOARD DESIGN AND ASSEMBLY - Automotive fiber optic connection systems should be compatible with circuit board design and assembly requirements. Header lead spacings, lead lengths, and mounting techniques must all be compatible with high-

0148-7191/89/0227-0202$02.50
Copyright 1989 Society of Automotive Engineers, Inc.

density circuit board design standards while providing the level of mechanical integrity required in the automotive application. Materials and packaging must be compatible with circuit board processing and handling techniques.

HIGH VOLUME ASSEMBLY COMPATIBLE - Product and process development for automotive fiber optic connection systems must address the need for high volume production.

HARNESS ASSEMBLY - Fiber optic terminations and connectors must be designed for high volume, low cost and excellent quality. Terminations must use low cost components which can be automatically assembled. Connector assembly operations should not be significantly different from the electrical connection assembly, allowing both hand and automated assembly.

VEHICLE ASSEMBLY - The fiber optic system must be designed to facilitate installation on an assembly line. Connection systems must be designed for single operation installation. Screw-on connections that can be partially mated are not acceptable. It is undesirable to complicate the assembly process with an increase in the number of connections to be mated, or to have restrictions on harness routing.

COMPATIBILITY WITH AUTOMOTIVE ENVIRONMENT - The fiber optic connection system must meet the rigors of the normal operating conditions encountered: thermal, chemical, and physical.

SERVICEABILITY - The components and system must last the life of the vehicle and fit into the user and service environments. Components and systems that are beyond the capability of the service environment to repair are unacceptable.

DESIGN CONSIDERATIONS

FIBER SELECTION - Generally speaking, there is a relationship between connector cost, fiber cost and fiber diameter (see Figure 1). As fiber diameter increases, the cost of the connector components decreases. The cost of the connection components is driven largely by the tolerances required to achieve low optical losses. As fiber diameter increases, these tolerances relax. On the other hand, the cost of optical fiber increases roughly as a square law function of the diameter of the fiber due to the additional material. In addition, as fiber diameter increases, the fiber becomes less flexible and vehicle assembly costs can be driven upward. Fiber diameter then is seen as a cost driver in connector design.

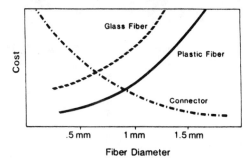

Figure 1. Costs as a Function of Fiber Diameter

By adding these costs (see Figure 2), an optimal fiber diameter can be determined. Allowing for pricing differences between manufacturers, a diameter of approximately seven hundred fifty microns results in a minimum datalink cost when using plastic fiber. Approximately five hundred microns results in a minimum datalink cost when using glass fiber.

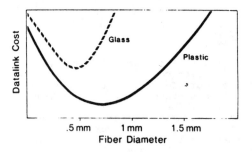

Figure 2. Datalink Cost Optimization

Note that glass datalinks will usually cost more than plastic datalinks in automotive applications. The proposed design utilizes one thousand micron fibers which are currently a standard product, but it is compatible with smaller fiber diameters.

ELECTRO-OPTICAL DEVICE PACKAGING - Electro-optical device packaging can impact device cost, connector tolerances, and connector configuration. There are two basic types of packages: header and lead frame.

Typical header and lead frame optical device packages are shown in Figure 3. The header types tend to be much more expensive than the lead frame type. Also, the lead frame parts can be made with integral fiber lead-in. This reduces the tolerance requirement on the connector, since fiber alignment is dependent only on the device package, rather than both the package and the connector. For these reasons, the lead frame package was chosen.

Figure 3. Header and Lead Frame Optical Device Packages

LED DEVICE SELECTION - There are two potential devices to select from: 660nm (red) and 820nm (infrared). The 660nm part typically has less power launched, but the attenuation in the plastic fiber is significantly less at 660nm than at 820nm. Fiber attenuation at 660nm is ~0.2 dB/m, while at 820nm it is ~2 dB/m. This factor can impact the system optical power margin. Figure 4 illustrates receiver optical power vs. link length for the two types of LEDs. For short links the infrared part will result in more power at the receiver. Typically, infrared parts are specified to operate at higher data rates. Since there may be applications for both 660nm and 820nm LEDs and since this does not impact the connector design, the type of LED is determined by the application.

OPTICAL RECEIVER SELECTION - Optical receivers are available as discrete silicon detectors with external amplification or with the detector and amplification integrated into a single package. Since the current out of the detector is very small, the integrated device has the advantage of being significantly less susceptible to electromagnetic interference due to the reduced lead length between the detector and the amplifier. For this reason, the receiver with the integral amplification was chosen. The integrated device also simplifies the PC board.

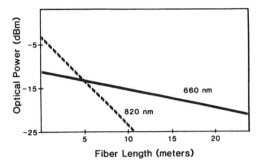

Figure 4. Optical Power versus Link Length as a Function of Wavelength

PACKAGING THE ELECTRO-OPTICAL DEVICES - There are three possible options for packaging the electro-optical devices:
(1) in the harness connector, mating electrically;
(2) on the PC board (pigtailed), mating fiber to fiber;
(3) in the header connector, mating fiber to device.

Harness Connector Packaging - Packaging the devices in the harness connector would require the creation of a second protected environment, outside the controller (Figure 5). The mated connection would be electrical rather than optical. The electrical lead length would be longer than the other options which would result in increased electromagnetic radiation from the LED because of the relatively high current. In addition, the harness connector is subjected to much more physical abuse during shipping and handling than the controller module.

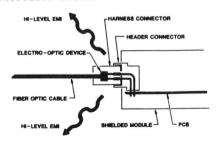

Figure 5. Harness Connector-Mounted Devices

On the PC board (Pigtailed) - Mounting the devices to the printed circuit with a fiber pigtail to connect to the module connector is an improvement over the harness connector mounting due to improved electromagnetic compatibility and there is no requirement for a second protected environment (Figure 6). There are two main disadvantages. First, there is an added optical interface involved. The module connection would be a fiber-to-fiber connection but the device requires a pigtail connection which increases the device cost. Second, the fiber pigtail would require special attention to minimize the opportunities for damage.

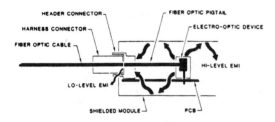

Figure 6. Fiber Pigtailed, Board-Mounted Devices

Header Connector Packaging - The third option, mounting the electro-optical devices in the module header connector, offers all the advantages of the pigtailed version without the disadvantages (Figure 7). Device lead length is minimized, reducing electromagnetic interference. The devices are not pigtailed, so the device cost is reduced; and since the devices are in the header connector, occupied PC board space is reduced.

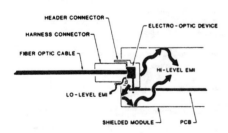

Figure 7. Header Connector-Mounted Devices

FIBER ALIGNMENT - Fiber alignment will have a significant effect on the amount of coupled optical power. To minimize the amount of tolerance stack-up, the fiber alignment should be dependent on as few components as possible. To optimize this, the previously described device package with integral lead-in was chosen, and an independent spring-loaded ferrule was selected. This minimizes the tolerance requirements on the harness connector and header connector.

FIBER TERMINATION - Optical fibers are terminated in two ways, crimping or epoxying. Crimping has the advantage that it is similar to current processes and is simple and quick, which leads to easy application of high volume assembly techniques. Epoxying tends to be time consuming since the epoxy must set up before the fiber end can be finished, it is potentially messy, and it requires another control in the process to monitor epoxy condition. Crimping also is compatible with prefinished fiber ends, which allows another processing option.

HARNESS BUILD - Since the fiber optics will be a part of the power and signal distribution system, assembly into the total package must be facilitated. This can be accomplished in two ways: first, by terminating and connectorizing the fiber as a subassembly or second, by designing the connector to accept the fiber after it is terminated in a fashion similar to todays wiring. Figure 8 illustrates how the preterminated fiber is assembled into the harness connector.

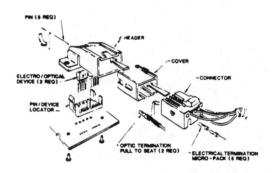

Figure 8. Components of Electro-Optic Connection System

VEHICLE BUILD - During the vehicle assembly process, when the vehicle systems are being installed, each additional operation adds time/cost to the assembly process. If the fiber optic connections were discrete connections, separate from the electrical connections, this would add a significant number of operations to the process (one or two per module). For this reason, the optical connections are integrated with the electrical connections in a single connector. An additional benefit of this construction is that the wires provide the strain relief for the fibers to keep them from being pulled away from the electro-optical devices by normal routing. Figure 9 shows a completely assembled combined electrical/optical connection system in the standard snap lock configuration.

Figure 9. Assembled Electro-Optic Connection System

ENVIRONMENTAL COMPATIBILITY - All automotive connection systems must be capable of withstanding the extreme environmental conditions encountered in the vehicle, and must perform satisfactorily in the intended application.

In order to evaluate the performance of the Electro-Optic Connection System, samples of the prototype assembly were subjected to laboratory environmental testing. Parameters were monitored during the testing to determine any variation in the performance of the connection system resulting from the environmental conditions. The test plan for this evaluation is shown in Figure 10. Per the test plan, entire fiber optic data links were subjected to the following environmental conditions:

(1) High Temperature Exposure: +85°C, 1000 hours;
(2) Thermal Cycling: -40°C to +85°C, 125 cycles;
(3) Dust Exposure: coarse grade dust, 5 hours;
(4) Sequential Environmental Cycle: 15 cycles
    16 hours, 98% relative humidity, +38°C,
    2 hours, -40°C,
    2 hours, +85°C,
    4 hours, +25°C;
(5) Mechanical Shock: 50 G peak, 5 shocks, 3 axes;
(6) Random Vibration: +25°C, -40°C, +85°C, 3 axes.

Evaluation of test results indicates that this Electro-Optic Connection System can perform satisfactorily in the automotive environment with little variation in parameters.

APPLICATIONS

AUTOMOTIVE - Fiber optics can provide increased performance, higher reliability, and potential cost savings, relative to coax and twisted-shielded cable, in a number of automotive applications.

Point-to-point fiber optic data links provide significant advantages for high speed data transmission such as video and audio data. Transmitting these high data rate signals via fiber optics reduces electromagnetic radiated emission concerns associated with conductor-based data transmission systems. Similarly, data signals with fast transition times can be safely transmitted optically to control the radiated emissions caused by the high frequency harmonics associated with the transitions.

Fiber optics can be very beneficial as a means of protecting the integrity of a system from electromagnetic susceptibility. For example, anti-lock brake systems, suspension control systems, and other critical automotive systems require careful design considerations to protect from electromagnetic interference. The use of fiber optics in these systems could directly address some design considerations.

With the ever increasing number of computer controlled systems in automotive vehicles, a high-speed data transmission network will become a necessity. Conductor based network systems can require costly termination techniques in order to maintain shield integrity and constant line impedance throughout the network structure. Fiber optics provides a high speed data transmission medium which has potential as a network structure in automotive applications primarily because of the ability to maintain electromagnetic compatibility. Automotive fiber optic networks will become viable with the development of low-cost optical networking components.

NON-AUTOMOTIVE - Fiber optics is finding increased use in areas such as factory automation and intra-office communications. In the medical field, fiber optics is being used in various sensor applications. Audio equipment suppliers are incorporating fiber optic links for transmission of digital audio signals among the various components in the audio system.

The environmental considerations in these non-automotive applications are much less severe than the automotive environment. The Electro-Optic Connection System can meet the performance requirements of many of these applications and offers a low-cost, high-reliability connection system. Field terminations can be made easily and the connection system is robust, by design, to withstand abusive handling. Use of this connection system in non-automotive applications provides a means of increasing the volume usage, which in turn results in reduced costs and spurs increased development activity of products which can capitalize on the combined applications of automotive and non-automotive markets.

CONCLUSIONS

Fiber optics will find applications in the automotive market. The benefits will be increased vehicle functionality without the electromagnetic compatibility difficulties usually encountered. A key factor in the introduction and proliferation of fiber optics is the development of a cost effective connection system that addresses both the functional and non-functional needs of the automotive industry. The connection system presented was designed to address the manufacturing, assembly, and functional issues associated with automotive applications.

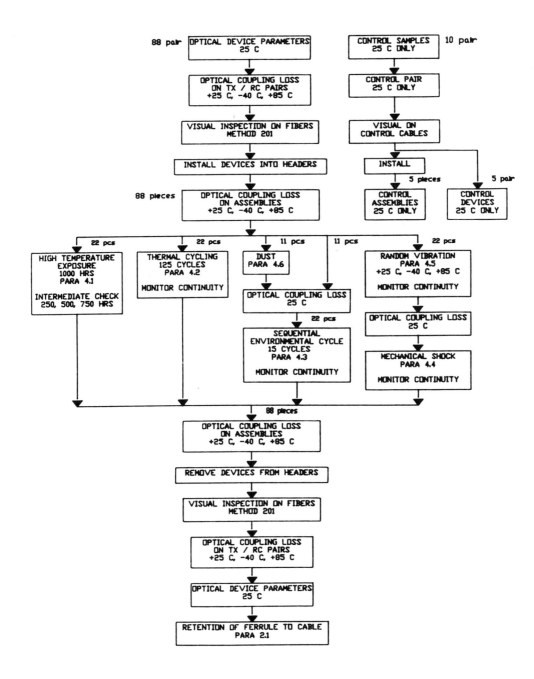

Figure 10. Electro-Optic Connection System Test Plan

Fiber Optic Data Links

Designing a Fiber Optic Communications System, p. 171
E. M. Raasch, W. B. Hatfield

III

OPTICAL NETWORKS

Designing a fiber-optic communications system

Compared to copper cable, glass fiber offers impressive benefits, including higher speed over greater distances, better noise immunity and lower costs. Here's how to choose a fiber-optic data link for your applications.

By Ernest M. Raasch and W. Bryan Hatfield

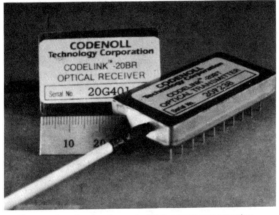

In fiber optics, tiny equipment yields enormous capacity.

Today there is a new emphasis on fiber optics in communications systems. It's a quiet revolution, probably best stated by a longtime telecommunications manager of a Fortune 100 oil company who said, "What's going on? Suddenly, all these computer people are talking about fiber."

The fiber-optic bandwagon is picking up speed. And, as with bandwagons everywhere, people who don't want to miss the parade are jumping aboard.

Almost daily, there are announcements of more and more companies basing their data communications systems on glass fiber, rather than copper wire. The list is impressive. It ranges from telephone companies like Southwestern Bell to manufacturers like Polaroid, from brokers like Merrill Lynch to insurance companies like Aetna, from universities like Stanford to government agencies like the Bureau of the Census. Optical fiber is no longer the exclusive province of long-haul telephone systems. Many people feel that fiber optics is a must in today's and tomorrow's communications systems for voice, data and video.

A decade before fiber became commercially available, mainframe computer users knew that their computers' transmission speeds were being constricted whenever data had to be sent over common carrier lines. In the 1960s, for example, when computer manufacturers were selling "channel couplers" that typically operated at speeds of 1.5M bps between mainframes, common carrier speeds through private branch exchanges (PBXs) were typically less than 64k bps. Furthermore, the quality and access times of common carrier lines, including "conditioned" lines, left much to be desired.

Why fiber optics?

To achieve greater data rates and higher transmission quality, MIS executives began to move from traditional carriers to microwave and satellite links. But these two higher-tech solutions had their own limitations and costs. Eventually common carriers of all types—from twisted pair to coaxial cable to microwave to satellite—learned of the benefits offered by fiber-optic cable:

Higher bandwidth/low attenuation. The higher bandwidth of fiber offers uniformly low attenuation from the lowest to the highest data rates, even above 200M bps. While the bandwidth of wire is drastically reduced and the attenuation drastically increases over longer distances (as the square of the distance), fiber allows any standard interface to operate at its maximum bandwidth at any distance. Fiber's large bandwidth is typically 10 to 100 times that of coaxial cable ordinarily used in local-area networks.

Radio frequency interference and electromagnetic interference immunity. Fiber-optic links are virtually immune to noise in any electrical environment, with typical bit error rates of less than 10^{-9}. Fiber links are free from ground loops, cross talk and lightning strikes. Fiber is preferred for secure communications systems.

Lower cost. Fiber costs less per meter than shielded data and coaxial cables and is less expensive to install because of its light weight and small size. Fiber also saves upgrading and office-moving costs. For example, a fiber cable originally installed to serve a 327X computer will also support token-ring, Manufacturing Automation Protocol, PC Network and Ethernet networks as well as RS-232, RS-422, V.35, MIL-188C, T-carriers, transistor-transistor logic and emitter-coupled logic or any communications standard, either serial or parallel, synchronous or asynchronous.

Greater distances and more users. Fiber optics offers data rates of over 200M bps at distances up to 60 kilometers without repeaters. In contrast, with low data rate RS-232 signals, copper cable is limited to just 50 to 200 feet. A 327X fiber cable network can support up to several thousand users with multiplexing.

Given these benefits, you can see that a data communications system based upon fiber optics has some inherent advantages over other technologies—at the time of installation and as the system grows in years to come.

The components

Most discussions of fiber optics tend to focus on the glass fiber itself. Equally important, however, are the transducing devices that convert electrical signals to optical signals and back again. There are two wavelengths of commercial importance in fiber-optic communications, 830 nanometers and 1300 nanometers. These wavelengths relate primarily to the availability of optical emitters and optical detectors. The light loss and dispersion in a fiber decreases at the longer wavelength; however, transmitter power generally increases at the shorter wavelength. Choosing which wavelength to use is a matter of optimizing all aspects of system design, including cost.

A second factor in system design is the nature of light propagation in the fiber. Depending on the transmitter/fiber combination, the light propagates either in a single mode or in many modes (multimode). Although the light loss is lower in the former case, the technology is more complex and expensive. Single-mode fiber-optic cores measure about 8 microns in diameter. There are a variety of "standard" multimode fiber core diameters. The most important are 50, 62.5, 85 and 100 microns.

Building a fiber-optic system

A fiber-optic communications system can be designed and built step-by-step through three levels: 1) point-to-point links; 2) local-area networks; and 3) wide-area networks. System design might at first seem arcane, but with some background knowledge and aids such as the accompanying tables, the job turns out to be quite manageable. For now, we'll discuss only point-to-point links using multimode fiber. Networks and single-mode fiber will be discussed in a later article.

A point-to-point link interconnects one computer or peripheral to another. Such a link consists of a multimode fiber-optic cable together with a matched pair of optical transmitters and optical receivers.

Optical transmitters convert electrical energy from the computer or terminal circuits into light pulses that are coupled into and transmitted

> "*The fiber-optic bandwagon is picking up speed and people who don't want to miss the parade are jumping aboard.*"

TABLE 1
Optical Receiver Characteristics

		PINFET Receiver	AC Coupled LED/PIN Links			DC Coupled LED/PIN Links				DC Coupled Laser/PIN
Parameter	Units	Codeamp-50A	Codelink-3	Codelink-20A	Codelink-100B	Codelink-1	Codelink-10	Codelink-20B	Codelink-50	Codelink-100
Data Rate (maximum)	Mbps	80	3.2	20	125	1	10	20	50	150
(minimum)	Mbps	1.0	0/0.2	0/1.5	20	0	0	0	0	0
Codedata Manchester Coding	—	—	—	Internal Standard	Internal Standard	—	—	—	—	—
Clock Rate (maximum)	MHz	40	1.6	10	60	0.5	5	10	25	60
(minimum)	MHz	0.5	0.2	1.6	10	0.01	0.01	0.01	0.01	10
Receiver—Detector Type	—	PIN	PIN	PIN	PIN	PIN	PIN	PIN	PIN	PIN
—Sensitivity	dBm	−40	−38	−30	−27	−33	−30	−30	−20	−12
—Dynamic Range	dB	34	27	27	>23	23	18	15	15	12
Optical Flux Budget	dB	(Subtract receiver sensitivity from transmitter output power in Table 2)								12

Detailed specifications such as these simplify choosing the correct mix of products for any given application. The text gives an example of how to use this table.

TABLE 2
Optical Transmitter Characteristics

Wavelength (nanometers)	Core (microns)	Numerical Aperture	Optical Power* Minimum	Optical Power* Typical
830	50	0.20	125/ – 9.0 100/ – 10.0 80/ – 11.0	140/ – 8.5 110/ – 9.5 90/ – 10.5
830	85	0.26	290/ – 5.4 230/ – 6.4 180/ – 7.4	335/ – 4.7 270/ – 5.7 210/ – 6.8
830	100	0.29	445/ – 3.5 355/ – 4.5 280/ – 5.5	500/ – 3.0 400/ – 4.0 315/ – 5.0
1300	50	0.20	50/ – 13.0 30/ – 15.0	60/ – 12.2 40/ – 14.0
1300	62.5	0.29	60/ – 12.2 35/ – 14.3	65/ – 11.7 50/ – 13.2
1300	100	0.29	75/ – 11.2 45/ – 13.5	90/ – 10.5 60/ – 12.2

*microwatts/decibels below 1 milliwatt

Many factors affect a system's optical power. Here, you can see how wavelength, fiber core size and aperture interact to produce the range of powers shown.

through the optical fiber. The light sources commonly are either light-emitting diodes (LEDs) or injection laser diodes (ILDs). LEDs come in two types, edge and surface emitting, with the former generally considered to be more reliable and powerful. Laser diodes are used for longer distance links. ILDs are more expensive than LEDs and their service life is generally less by a factor of 10. Edge-emitting LEDs are estimated to be capable of lasting over 1000 years. Both LEDs and ILDs are available in 830-nanometer and 1300-nanometer wavelengths. LEDs can deliver coupled power (into a fiber) of up to 1000 microwatts, while ILDs typically supply power levels of 1000 to 3000 microwatts.

Optical receivers detect and amplify the light pulses coming from the source and convert them back into electrical energy, which is fed into the data terminal equipment. Because the receiver must have high sensitivity and low noise, a choice of receivers based on two types of detectors exists: the PIN detector or the APD detector.

PIN receivers contain a photodiode with unity gain (a positive-intrinsic-negative photodiode with low noise bipolar or field effect transistor amplifiers) coupled with a high-impedance front-end amplifier. This device offers low-voltage operation, high reliability and low sensitivity to temperature variations.

The APD receiver (avalanche photodiode) produces a gain of 100 or more, but also produces noise that may affect the sensitivity of the receiver. Typically, APD receivers are extremely sensitive, requiring as few as 200 photons to be detected at the receiver per bit transmitted at data rates of 200M to 400M bps.

Obviously, the choice of receivers and transmitters can significantly impact the cost and operation of the data communications link. To choose the correct optical data link terminal equipment for a particular application, you must know the minimum and maximum data rates as well as the data format. Some fiber-optic receivers accept optical inputs having arbitrary duty cycles. But many receivers, particularly those of the most sensitive design, will function properly only if their input has a 50% duty cycle averaged over a fairly short interval. Devices accepting arbitrary inputs are said to be dc coupled while those that require a 50% duty cycle are ac coupled.

Manchester encoding technique

One way to take advantage of the generally greater sensitivity of an ac coupled receiver is to encode the incoming electrical signal at the transmitter in a way that will achieve a 50% duty cycle optical signal. The most common means of doing this is the so-called Manchester encoding technique. Of course, the encoded optical signal must be decoded at the receiver before the signal is passed on to the terminal equipment. If desired, the data can be encoded directly with a synchronous clock signal, allowing both the data and the synchronous clock to be transmitted over a single optical fiber. Some fiber-optic transmitters and receivers

> *"There's more to fiber optics than glass fiber: Devices which convert electrical signals to optical signals are just as important."*

offer this encoding function as a built-in feature.

In a fiber-optic point-to-point link, the optical power delivered to the photodetector must equal or exceed the sensitivity limit of the receiver. The system gain or "flux budget" available to the system designer can be found either by consulting data tables supplied by equipment manufacturers or by:

Available System Gain = $10\log(P/S)$ (dB)

where P is the power output (in microwatts) of the transmitter and S is the sensitivity in microwatts of the receiver. (Sensitivity is generally that incident optical power level at which the receiver exhibits an error rate of less than 10^{-9}.) The available system gain defined in this manner is expressed in decibels. The total optical losses in the link, made up of fiber attenuation, connector and splice losses, together with any desired system margin, cannot exceed the available system gain.

On the other hand, all fiber-optic receivers have a limiting dynamic range. That is, if the incident power exceeds a certain level, the error rate will exceed the specified limit. (Information on dynamic range can also be found in manufacturers' literature.) This then implies that a point-to-point link must have both minimum and maximum losses given by:

Available Gain – Dynamic Range
< Loss < Available Gain

A number of aids, such as the accompanying tables, simplify system design and make it relatively easy to juggle the factors involved.

An example

A hypothetical case study, using IEEE standard 802.3, will help illustrate some of the points already discussed. Suppose you need to connect two pieces of terminal equipment located in separate buildings 6 kilometers apart; the fiber will enter each building through a wall-mounted fiber-optic connector. The transmission rate will be 10M bps, and you want a 6-decibel operating margin. You also need to transmit a synchronous 10-MHz clock signal. And to minimize cost, you want to use LED transmitters (preferably 830 nanometers), and 50-micron multimode fiber. The first step in building this system is to figure the signal losses: At 830 nanometers, a typical 50-micron multimode fiber loss might be 3 decibels/kilometer, so the loss over 6 kilometers would be 18 decibels.

The fiber link as described would contain three connectors. The definition of receiver sensitivity usually includes an allowance for connector loss between the system fiber and the receiver. In optical transmission, connector loss might be 1.5 decibels. For the link in question, then, the total connector losses for three connectors would be 4.5 decibels.

To calculate the total system loss, you'd add the fiber attenuation plus the optical connector losses plus the desired margin. In this case, 18 + 4.5 + 6 = 28.5 decibels.

> "*System design might at first seem arcane, but it turns out to be quite manageable.*"

Now all that's left is to select the products that meet the requirements. Table 1, Optical Receiver Characteristics, shows that the data rate requirements could be met by the Codenoll Technology Codelink-20A LED/PIN transmitter/receiver, for example. These devices provide synchronous transmission with integral Manchester encoding/decoding. The table also gives the rated sensitivity of the CL-20A as −30 dBm (decibels below one milliwatt).

Table 2, Optical Transmitter Characteristics, shows that the highest 830-nanometer LED power that can be coupled into a 50-micron fiber with a 0.20 numerical aperture is −9 dBm. Thus, the available system gain in this case would be −9 − (−30) = 21 decibels. This, however, is well below the required system gain of 28.5 decibels. This combination of fiber, transmitter and receiver would not meet system design.

What's the next step? Evaluate the possibility of using a 1300-nanometer LED/PIN device. Although this class of transmitter/receiver is somewhat more expensive than the 830-nanometer devices, the attenuation in a 50-micron optical fiber can be considerably less: 1 decibel/kilometer rather than 3 decibels/kilometer; the system loss in this case would then be reduced to 16.5 decibels. Because the 1300-nanometer LED technology is not as advanced as the 830-nanometer technology, a Codelink-20A operating at 1300 nanometers has a maximum power output of approximately −13 dBm. Receiver sensitivity is unchanged. Therefore, the available gain would in this case be −13 − (−30) = 17 decibels, which does exceed the required 16.5 decibels. Thus, the requirements of this hypothesized link would be satisfied by a 1300-nanometer transmitter/receiver pair such as Codenoll Technology's Codelink-20A.

For any point-to-point fiber-optic link, the same relatively simple procedure for calculation of the flux budget is used to choose the proper fiber-optic link for transmitter and receiver. In upcoming articles, we'll examine other issues of fiber-optic system design and implementation, including optical local-area networks, Ethernet-compatibility, and wide-area networks utilizing multiplexers. **TPT**

Ernest M. Raasch is executive vice president of operations at Codenoll and also a director and member of the Executive Committee. Raasch was previously president and CEO of Gestetner Corp—USA.

W. Bryan Hatfield is vice president of quality engineering at Codenoll Technology Corp, Yonkers, NY. Prior to joining Codenoll, Dr Hatfield held executive positions in research and development with firms including A-M Printer Systems, Singer Corp, and Bell Laboratories.

III

III
176

Fiber Optic Local Area Networks

Implementation of a Fiber Optic Ethernet Local Area Network, p. 179
M. H. Coden, F. W. Scholl

Reliability of Fiber Optic LANs, p. 185
M. H. Coden, F. W. Scholl, W. B. Hatfield

Passive Optical Star Systems for Fiber Optic Local Area Networks, p. 193
F. W. Scholl, M. H. Coden

Development of Fiber Optics for Passenger Car Applications, p. 205
R. E. Steele, H. J. Schmitt

Fiber Optic Ethernet in the Factory Automation Environment, p. 215
W. B. Hatfield, F. W. Scholl, M. H. Coden, R. W. Kilgore, J. H. Helbers

Fiber Optic LANs for the Manufacturing Environment, p. 221
W. B. Hatfield, M. H. Coden, F. W. Scholl

Investment Bank Trades up to Fiber Optic Ethernet, p. 227
S. J. Anderson, F. W. Scholl

High-Rise Optical Communications Networks, p. 237
D. E. Stein

A Few Fiber Optics Applications, p. 243
D. W. Maley

Passive Star Based Optical Network for Automotive Applications, p. 245
L. K. DiLiello, G. D. Miller, R. E. Steele

Introduction to CSMA/CD Network Design, p. 259
W. B. Hatfield

A Migration Strategy to FDDI for the Fiber Optic LAN Cable Plant, p. 279
W. B. Hatfield, M. H. Coden

IV

178

Implementation of a fiber optic ethernet local area network

Michael H. Coden and Frederick W. Scholl

Codenoll Technology Corporation
1086 North Broadway, Yonkers, New York 10701

Abstract

A fiber optic transceiver and network compatible with all Ethernet local area network hardware and software are presented. The description, theory of operation, and feasibility analysis of a 1024 node system are given. Several installations of fiber optic Ethernet systems are described.

Introduction

By replacing the coaxial cable of the Xerox, Digital, Intel, Hewlett-Packard, IEEE-802 adopted Ethernet local area network, with fiber optic cable, a network compatible with all Ethernet hardware and software has been realized. The fiber optic network covers greater areas, at lower cost with greater data security, safety, reliability of operation and greater effective throughput.

Theory of operation

Ethernet is a broadcast type network. Each node listens for a clear network (Carrier Sense) then transmits or broadcasts (Multiple Access) a packet. If two nodes transmit at one time, they must stop and retransmit after a delay (Collision Detection). This contention algorithm gives Ethernet its generic name: Carrier Sense Multiple Access with Collision Detect (CSMA/CD). The CSMA/CD broadcast type local area network is ideally suited to a star topology (Figure 1) which minimizes the distances, transit times, and hence probability of a collision between nodes. Figure 1 also shows that under the Ethernet specification of a fixed round trip propagation delay of less than 45 µs the star topology allows much greater areas to be covered at lower cost.

The fiber optic Ethernet is facilitated by a transmissive or reflective optical star coupler. This allows a passive star topology which is immune to the reflection and termination problems found in coaxial systems. Optical benefits of the star coupler approach include:
1) a minimum number of connectors between all nodes, usually 0, 2 or 4 depending on the use of splices (Figure 2);
2) an increase in optical attenuation of only 3.0 to 4.5 dB each time the number of nodes is doubled (Table 1).

Fiber optic transceiver

The Codenet-2020 (Figure 3) fiber optic Ethernet transceiver consists of: transmitter, receiver, collision detection, power regulation and transceiver cable driver elements (Figure 4). The transmitter, receiver, collision detection and optional Manchester encoder/decoder are contained in two PC card mountable hybrid integrated circuit modules occupying 1.17 cubic inches. The modules have TTL input/output signals and operate from +5 VDC with average supply current less than 390 mA from $0^{\circ}C$ to $70^{\circ}C$. They may be integrated directly into the Ethernet node equipment or combined with +12 VDC power regulation and twisted pair cable drivers to simulate identically the function of a coaxial transceiver as in the Codenet-2020 transceiver boxes (Figure 3).

Transceiver LED optical power coupled into a 100 micron core 0.29 NA fiber is typically greater than 350 uW (-4.6 dBm). Values as high as 470 uW (-3.3 dBm) can be obtained. Transceiver PIN detector sensitivity is typically 1.5 uW (-28.2 dBm) for bit error rates of 10^{-9}.

System configuration

Referring to Figure 2 it may be noted that system flux budget must satisfy the following inequality:

$$2C_T + 2C_S + 2 \cdot a_f \cdot l_{max} + E_S + 10 \, LogN + M \leq 24.0 \, dB \qquad (1)$$

where:

C_T = connector or splice losses at transceiver in dB.
C_s = connector or splice losses at star in dB.
a_f = attenuation of fiber in dB/km.
l_{max} = length of longest cable from star to any node in km.
E_s = excess loss of star in dB (See Table 1).
N = number of nodes in star.
M = system flux margin.

The maximum network allowed by the Ethernet specification has 1024 nodes with no more than 2500 meters between any two nodes. The maximum Codenet optical Ethernet system might consist of 32 cluster stars each with 32 nodes as shown in Figure 5. For equal node-to-star spacing we have l_{max} = 417 meters, so that 6 x 417 = 2500 meters. Using a selected 100 um core fiber, $a_f \leq 4$ dB/km, and with $C_T \leq 1.0$ dB and $C_s \leq 0.1$ dB we have a worst case flux budget of 0.2 + 2 + 3.3 + 18.5 = 24.0 so that a 32 x 32 = 1024 node system is feasible using Codenet-2020 Transceivers. Other combinations of nodes per star may be obtained by using splices or different connectors, and varying the length and attenuation of the fiber. The area covered in the system in Fig. 5 is 5 km^2, much greater than is possible with coaxial cable Ethernet.

Reliability and throughput

If any cable in a fiber optic star configured Ethernet network is damaged or severed, only the node on that cable is affected. If any part of the coaxial cable system is damaged, or severed, the resulting reflections will usually cause continuous data collisions rendering the entire network useless until the damaged cable is located and repaired. Furthermore, from Figure 1 we see that a typical Codenet fiber optic Ethernet network has many fewer active parts to fail than the coaxial system.

In the fiber system distances and transit times between nodes are shorter so there is less probability of a collision. Furthermore, the unique Codelogic collision detection circuitry developed for the Codenet-2020 transceiver detects a collision in less than 75 ns, 7 times faster than the 500 ns that may be used in coaxial systems.

Codenet implementations

At the time of this writing Codenet fiber optic Ethernet local area networks have been implemented at several cites with several different types of Ethernet hardware including: Ungermann-Bass Network Interface Units, ARPANET, INTEL Multibus and others. Networks have been configured with both fiber and coax, using standard Ethernet repeaters as well as fiber alone. In all cases, no modification to the Ethernet hardware or software was required and operation was immediate.

No. of Nodes	E_s(dB)	10LogN(dB)	Max Loss(dB)
4	2.0	6	8.0
8	2.0	9	11.0
16	3.5	12	15.5
32	3.5	15.0	18.5
64	5.0	18.0	23

Table 1. Codestar Losses

Excess loss E_s includes loss as well as output power non-uniformity. Numerical values are worst case results for presently available Codestar.

References

1. The Ethernet Local Area Network Specification, Digital, Xerox, Intel, Version 1.0, September 30, 1980.
2. J. R. Jones, J. S. Kennedy, F. W. Scholl, "A Prototype CSMA/CD Local Network Using Fiber Optics," Local Area Networks '82, September, 1982, Los Angeles.
3. M. H. Coden, F. W. Scholl, "High Flux Budget Budget System for High Speed Computer Interfaces and Local Area Networks," International Conference on Optical Communications, September, 1982, Raleigh, NC.
4. M. H. Coden, "Fiber Optics and Local Area Networks," Local Area Networks '82, September 1982, Los Angeles.
5. H. O. Erwin, M. H. Coden, F. W. Scholl, "Communications, Tracking, and Docking on the Space Station," IEEE National Telesystems Conference, November, 1982, Galveston, Texas.

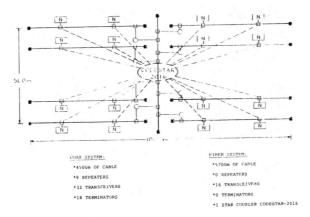

Figure 1. Comparison of CSMA/CD Ethernet fiber optic star vs. coax

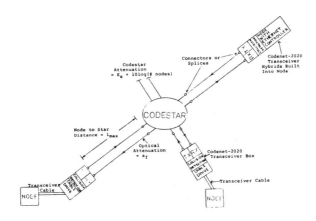

Figure 2. Codenet configuration

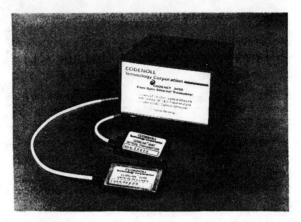

Figure 3. Codenet-2020 fiber optic Ethernet transceiver

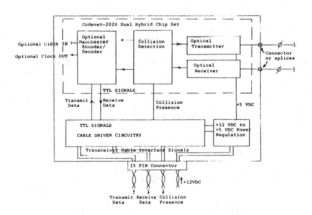

Figure 4. Block diagram of Codenet-2020 transceiver

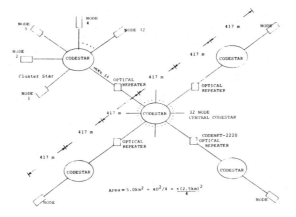

Figure 5. 1024 node system covering 5 square km

IV

184

ABSTRACT

"Reliability of Fiber Optic LANs"
Michael Coden, Frederick Scholl, W. Bryan Hatfield
Codenoll Technology Corporation
1086 North Broadway
Yonkers, New York 10701
(914) 965-6300

Fiber optic Local Area Network Systems are being used to interconnect increasing numbers of nodes. These nodes may include office computer peripherals and terminals, PBX switches, process control equipment and sensors, automated machine tools and robots, and military telemetry and communications equipment. The extensive shared base of capital resources in each system requires that the fiber optic LAN meet stringent reliability and maintainability requirements. These requirements are met by proper system design and by suitable manufacturing and quality procedures at all levels of a vertically integrated manufacturing operation. We will describe the reliability and maintainability of Codenoll's passive star based systems. These include LAN systems compatible with Ethernet (IEEE 802.3) and MAP (IEEE 802.4), and software compatible with IBM Token Ring (IEEE 802.5). No single point of failure exists in this system architecture.

At Codenoll, a MIL-I-45208A quality system is used for both military and commercial fiber optic manufacturing. Aspects of this system and other manufacturing quality controls will described.

RELIABILITY OF FIBER OPTIC LANS

INTRODUCTION

A local area network or LAN is a method by which distributed computing systems can be interconnected locally, typically within a building or a small set of buildings, in order to achieve a number of benefits such as resource sharing and common access to an expanded data base. Of particular importance are the standards-based LANS now in the process of definition by committees of the IEEE. These include the Carrier Sense Multiple Access with Collision Detection, the so-called Ethernet (IEEE 802.3); the Token Passing Bus, or MAP standard (IEEE 802.4); and the Token Ring Access Method (IEEE 802.5). LAN environments include factory as well as office and laboratory installations.

Reliability refers to the ability of a device or system to perform to established specifications over an extended period of time. In the case of Local Area Networks the concept of reliability should be considered in a broader context than simply the MTBF of the various system components. Experience has shown that the perceived reliability of a LAN system will depend upon both the topology and media used to implement the LAN as well as the inherent component reliability.

LAN TOPOLOGY

A number of topologies have been developed for Local Area Networks. These include the ring or loop, bus, active star and passive star configurations illustrated in Figure 1. Another way of looking at these topologies is that they can be grouped into two basic types of technologies: broadcast and non-broadcast. In a non-broadcast network data must pass through "uninterested" equipment (equipment for which it is not intended) before reaching the appropriate destination equipment. For example, in a ring or loop data is passed sequentially from node to node. In a conventional (active) electronic star or broadband network, the data is passed from the source node to an intelligent central controller which then sends it to the destination node. Both of these configurations have the potential for a single point of failure, a node in the ring, the ring medium, or the active star, which can disable the entire network. Redundancy or by-pass switches can alleviate but never entirely eliminate this potential for catastrophic failure.

In broadcast systems, the source node transmits or broadcasts its data on the network and it is received by all the nodes on the network. Each node examines the data that has been broadcast, accepting and using the data only if the node recognizes its own destination address. The functionally equivalent bus and passive star configurations are examples of broadcast systems. A particular attraction of the passive star LAN implementation is that failure of either a node or a segment of the LAN medium will have but a limited effect on overall network operation. As a passive device the star itself should have a very high reliability. As noted previously, this characteristic is not shared by any other of the basic LAN topologies. The passive star LAN topology displays a number of additional attractive features vis-a-vis the other LAN topologies (Figure 2). The star topology tends to minimize cabling costs, it is the most easily maintained configuration and it offers maximum flexibility, an important consideration in office and factory installations where relocation rates for work stations of up to 30 percent per year are not unusual. Support for these assertions is provided by long-standing building wiring practice in the telephone industry as well as the Token Ring Network cabling scheme proposed by IBM (Figure 3). The star topology network is also easily expanded to very large sizes, as shown in Figure 4, through the use of a tree structure and repeaters. All of these factors may be considered reliability related under a somewhat dynamic re-definition of system reliability that includes equipment relocation and system expansion as environmental parameters in addition to the more usual temperature, humidity, vibration, etc.

A final attraction of the star LAN topology is its relative ease of maintainence. Through a sequence of logical trouble shooting procedures, most network failures can be quickly isolated to a particular branch of the tree and then to a particular star. The offending branch or star can them be disconnected from the remainder of the network which can then continue to function while maintenance is performed on the isolated offending element.

LAN MEDIUM

Until recently all local area networks had been built around coaxial cable or twisted wire technology. The most used and supported network in today's market is Ethernet. Conforming to IEEE standard 802.3, Ethernet is a broadcast type network in which each node listens for a clear network (Carrier Sense) then transmits or broadcasts (Multiple Access) a packet of data. If two nodes transmit at one time, they must stop and re-transmit after a delay (Collision Detection). This contention algorithm gives Ethernet its generic name, Carrier Sense Multiple Access with Collision Detection (CSMA/CD). Ethernet systems have been installed in business offices, research laboratories and in banking and financial institutions. Its applications in industry include plant process computer control, robotic control and data acquisition. Although Ethernet has numerous current uses, there have been problems with its implementation using coaxial cable. Operation on electrical cable is impaired by inductive interference such as that seen in such electrically noisy environments as power plants, substations, machine works, robotic controls, data systems in factories and even fluorescent light fixtures. Improper termination or "kinking" of the coaxial cable during system installation or relocation can cause reflections which inhibit network operation. Fiber optics as a transmission medium is immune to these problems. In large systems, ground loops can be a serious problem in electrical conductors. Use of fiber optics not only avoids ground loops entirely but allows much larger systems to be implemented. Additionally, fiber optics provides a very secure, virtually untappable medium of communication, a very desirable feature in many applications. Since almost all the factors listed above, and summarized in Figure 5, relate to the integrity or, reliability, of the transmitted data, it would seem that a fiber optic LAN implementation would serve to optimize LAN reliability.

Optical fiber has a very high bandwidth. Therefore, once it is installed and operational, it need never be replaced if the network is upgraded to a higher performance standard. Such is not the case with copper based networks. As illustrated in Figure 6, whenever a network is modified, new cabling must be installed thereby re-introducing the start-up problems attendant with every new system installation.

FIBER OPTIC LAN

As it happens the passive, fiber optic star coupler is a readily fabricated device. A number of bare fiber sections are brought into intimate contact by twisting. This region is then melted by heating and carefully pulled apart creating a uniform biconical taper. Optical power incident in a fiber on one side of the taper is divided approximately equally among all the fibers on the other side of the taper. Cross talk between fibers on the same side of the taper is virtually non-existent.

Since early 1983, Codenoll Technology has marketed a fiber optic Ethernet system compatible with IEEE 802.3[1]. This system, which includes fiber optic Ethernet transceivers and repeaters, is based upon the Codestar series of fiber optic star couplers. To date thousands of Codenet fiber optic Ethernet systems have been installed in office, factory, campus and military environments around the world. User experience in this wide variety of environments serves to validate the reliability-through-design concepts outlined earlier in this paper. Codestar couplers are manufactured with N x N input/output ports where N=4, 8, 16, 32, 64. A three tiered, tree structure network as shown in Figure 4 could then contain over 250 thousand nodes if implemented with 64 port star couplers. In coaxial Ethernets, reflections on the line limit the number of node connections to one every five meters.

In addition to the potential for extremely high node densities, fiber optic Ethernets also offer large area coverage. Timing considerations in the IEEE 802.3 standard limit the maximum distance between any two nodes to 2.5km. Therefore, a coaxial system could cover only about 0.25 square kilometers. Limitation on the maximum length of coaxial Ethernet cable, necessary to assure signal waveform integrity, would require the use of four repeaters in this case (A repeater is an electronic "blackbox" used to interconnect Ethernet segments). A Codenet fiber optic Ethernet system with a maximum distance of 2.5km between radial nodes could cover an area of 5 square kilometers without a single repeater and still maintain all standard Ethernet timing considerations.

In 1986 the Codenet series of standards based fiber optic LANS was expanded to include passive star based, tree structure systems which implement IEEE 802.4 (Manufacturing Automation Protocol or MAP) and are software compatible with IEEE 802.5 (token ring).

MANUFACTURING SCREENS

The point has been made that LAN reliability is enhanced through proper choice of a network topology (passive star) and transmission medium (fiber optics). Equally important are the process control and quality screens employed during the manufacture of the LAN components.

Codenoll Technology is a vertically integrated company with design and manufacturing capability at all levels of fiber optic technology. This capability begins (see Figure 7) with high radiance, edge emitting LEDs which are designed to provide the high fiber coupled powers required to implement large LANs. Incorporation of LEDs and PIN photodiodes into high performance fiber optic transmitters and receivers is the next step on the integration ladder which is completed with incorporation of the link (transmitter and receiver) elements into active LAN system components (transceivers, repeaters, etc.). Paralleling the active component manufacturing ladder is the production of passive components, that is, connectorization of fiber optic jumpers and cables as well as passive fiber optic star coupler assemblies. Control over virtually the entire LAN system component manufacturing process allows reliability to be achieved through implementation of quality control procedures at all levels of product manufacture.

Although the primary application of LAN products is commercial, Codenoll has established an overall Quality Assurance System which conforms to the requirements of the Military Specification MIL-I-45208A. Under this system, a number of critical functions including vendor qualification and inspection of all incoming materials, production drawing control, documentation of manufacturing processes, traceability of production lots, and calibration of test equipment are controlled.

The quality screens incorporated into the LED manufacturing process were described in detail in a companion paper presented at this conference.[2] Briefly, before release to final LED device production, each wafer is subjected to an extensive array of tests. These tests include measurement of spectral characteristics; radiation patterns; such electrical and mechanical properties as breakover voltage, resistance and thermal impedance; optical output power and fiber coupling; and, finally, temperature accelerated life testing. Failure to meet established criteria result in rejection of the wafer. Before being passed on for the next level of production, each LED is burned in, under power, for 48 hours at $135^{\circ}C$ following which its electrical properties are again verified and the device is sorted according to fiber coupled optical power.

Codenoll fiber optic transmitters and receivers are assembled using thick film hybrid microelectronics. The devices are packaged in metallic, printed circuit board mountable, dual-in-line packages. Optimum optical performance is achieved through a proprietary technique in which a fiber optic "pigtail" is aligned directly to either the LED, in the case of a transmitter, or photodiode. Completed links are burned in for 48 hours at $70^{\circ}C$ under power after which they are tested to specified performance standards.

In the case of transmitters, measurement of the optical power both before and after burn-in assures device reliability. The long term reliability of the transmitter and receiver assembly processes is monitored by subjecting devices, on a sample basis, to temperature cycling and extended, high temperature life testing.

LAN system components, such as transceivers, which incorporate Codenoll fiber optic transmitters and receivers are tested both statically on the bench and dynamically in an actual LAN test bed. Again, extended testing of sample units under both ambient and high temperature conditions provides a monitor of LAN component level reliability.

As a passive device, the reliability of the Codestar fiber optic star coupler is determined solely by its mechanical reliability as there are no potential sources for electronic failure. Designed primarily for commercial applications, the Codestar package has been qualified to withstand mechanical vibration far in excess of that normally encountered by commercial products. Prior to final testing, each unit is temperature cycled between $-50^{\circ}C$ and $+65^{\circ}C$ several times in order to identify potentially weak units.

SUMMARY

In this paper we have evaluated the reliability of Local Area Networks from two perspectives. First, we have developed the thesis that the inherent system reliability can be greatly enhanced through proper choice of network topolgy and method of implementation. The fiber optic passive star network was shown to be superior in that it offers immunity from electromagnetic interference and ground loops while offering low cost both upon installation and through subsequent system upgrades, ease of maintainence and troubleshooting and greatest flexibility as regards system reconfiguration. The second point made was that during the manufacturing process production controls and screening processes insure that the quality and reliability of Codenoll Codenet and Codestar fiber optic LAN products are commensurate with the high conceptual reliability of Codenoll fiber optic LANS.

REFERENCE

1. M. H. Coden and F. W. Scholl, SPIE 1983 Technical Symposium East, Arlington, Virginia, April 4-8, 1983.

2. S. J. Anderson, M. H. Coden, B. Dutt, W. B. Hatfield, J. Racette, F. W. Scholl, SPIE Fiber/LASE '86 Conference on Fiber Optics, Optoelectronics and Laser Applications in Science and Engineering, Cambridge, Massachusetts, September 14-26, 1986.

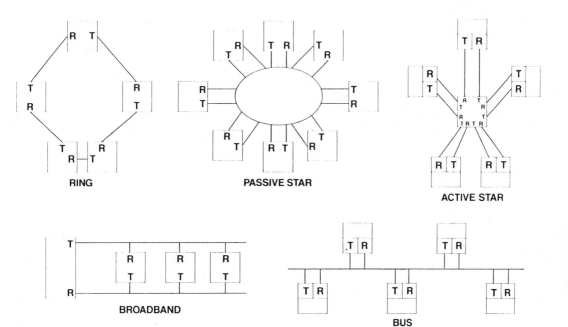

Figure 1—LAN Topologies

Figure 2—Features of the Passive Star LAN Topology

- No single point of failure
- Failures easily diagnosed and isolated while major portion of LAN remains in operation
- Minimum cabling cost
- Highly flexible and expandable for system reconfiguration and growth

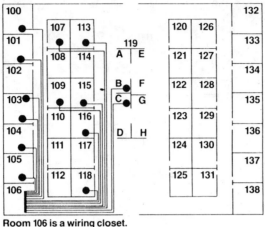

Room 106 is a wiring closet.
● = Ring Member (Telephone)

Figure 3—Floor Plan of a Typical Telephone or Token Ring Cable Installation

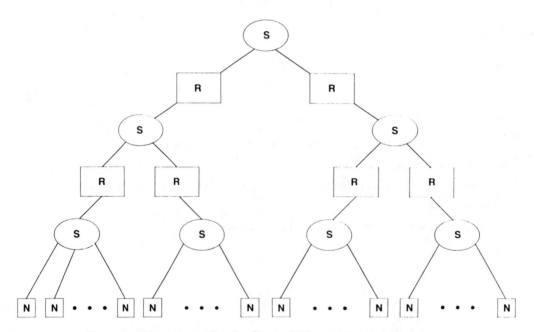

Figure 4—Expansion of Passive Star LAN Through the Tree Structure

Figure 5—Advantages of Fiber Optics as a LAN Medium

- Immune to inductive interference
- Immune to electrical noise
- Immune to ground loops
- No electrical fire hazard
- Secure communication
- Compatible with all networks and data rates
- Lightweight

NETWORK	DATA RATE	CABLING
ASCII	19.2 Kbps	Twisted Pair
IBM 5251	1.0 Mbps	Twin-Ax
PC NETWORK	2.0 Mbps	75 ohm Coax
IBM 3278/9	2.35 Mbps	92 ohm Coax
TOKEN RING	4.0 Mbps	Dual Twisted Pair Shielded Data Cable
IEEE 802.4 MAP (G. M.)	5/10 Mbps	75 ohm Coax or Fiber Optic
IEEE 802.3 ETHERNET	10 Mbps	50 ohm Coax
ADVANCED TOKEN RING	16 Mbps	Fiber Optic
IEEE – 802.6	50 Mbps	Fiber Optic
ANSI – FDDI	100 Mbps	Fiber Optic

Figure 6—LAN Cable Specifications

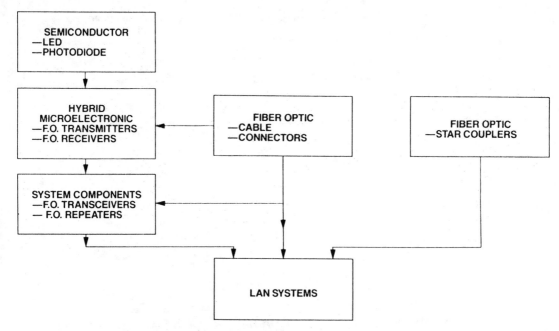

Figure 7—Codenoll Technology Integrated LAN Manufacturing

Passive Optical Star Systems for Fiber Optic Local Area Networks

FREDERICK W. SCHOLL AND MICHAEL H. CODEN

*Abstract—*The first commercial passive star-based fiber optic Ethernet local area network was described in 1982. Since then several improvements on this system have been made and passive star implementation of three other previously described networks have been developed. The thesis of this paper is that passive optical star systems represent an attractive and versatile architecture for implementation of many fiber optic LAN's. The passive star toplogy, we believe, best achieves the system goals of versatility, reliability, noise immunity, and data security. These networks are built around transmissive optical power splitters that are inherently reliable and contain no electronic components. Each node on the network is served by a duplex fiber cable and an appropriate optical transceiver. Several important technologies have been developed to support the above systems capabilities. The passive optical star topology clearly mandates high optical gain for the transceivers.

In addition, cost drivers in the market place have led to introduction of integrated circuits, surface assembly technology, high performance LED packages, and improved star coupler manufacturing processes. A variety of system installations in office, government, and factory settings have been completed using the above passive star systems. The largest of these is an Ethernet at Southwestern Bell's 44 story headquarters in St. Louis, MI. This all-optical system has 540 nodes, over 4000 computer terminals, and 92 mi of fiber.

I. Introduction

THE first commercial passive star-based fiber optic Ethernet local area network was described in 1982 [1]. Since 1982 several improvements have been made and passive star implementation of three other previously described networks have been developed: ARCNET, MAP, and PC Net. The thesis of this paper is that passive optical star systems represent an attractive and versatile architecture for implementation of many fiber optic LAN's.

The passive star topology, we believe, best achieves the system goals of versatility, reliability, noise immunity, and data security. These networks are built around transmissive optical power splitters that are inherently reliable and contain no electronic components.

Each node on the network is served by a duplex fiber cable and an appropriate optical transceiver. Noise immunity and data security are ensured by appropriate shielding and packaging of the transceivers. Single fiber breaks result in loss of only the associated terminal, and not in more extensive network failures. Typically, up to 32 nodes can be attached to a single star; larger numbers of nodes can be accommodated using repeaters. Radial length of fiber from star to transceiver can be up to 1 km; again, larger distances can be covered using repeaters. These network sizes are suitable for work station clusters, intrabuilding systems, and campus installations.

Fiber optic networks compatible to several network standards have been built using appropriately designed optical transceivers. The optical transceiver implements the physical layer (layer 1) of the ISO network model. All of the transceivers share a common high output power LED transmitter and a high sensitivity PIN diode burst mode receiver. The fiber optic systems implemented to date are compatible to IEEE 802.3 (Ethernet), MAP and IEEE 802.4, PC Net, and ARCNET.[1]

Our 802.3 compatible fiber optic system was described first in 1983. Since that time over 20 000 optical nodes have been installed. These transceivers are AUI (attachment unit interface) compatible to 802.3; a proposal describing a full fiber optic system is under discussion within IEEE 802 standards groups.

A passive star system developed by NTT is similarly under discussion in the Japanese LAN Standardization Committee. Ethernet transceivers operate at 10 MHz baseband, provide optical transmit, receive, and collision detection.

PC Net was originally a CSMA/CD broad-band system operating at 2 MHz. Our fiber optic version has been developed using passive star technology; the bandwidth available allowed a data rate increase to 10 MHz. MAP, 802.4, and ARCNET all refer to token bus networks, operating at 10 MHz and 2.5 MHz, respectively. We have developed electrically compatible baseband fiber optic systems again using passive star topology.

Several important technologies have been developed to support the above systems capabilities. The passive optical star topology clearly mandates high optical gain for the transceivers. In addition, cost drivers in the market place have led to introduction of surface assembly technology, high performance LED packages, and improved star coupler manufacturing processes. Edge emitting 830 nm LED's with minimum launched power of 0.5 mW in 100 μ-graded index fiber were developed and have been manufactured for the past five years. Reliability measure-

---

Manuscript received November 12, 1987; revised February 18, 1988.
The authors are with Codenoll Technology Corporation, Yonkers, NY 10701.
IEEE Log Number 8821341.

[1]ARCNET is a registered trademark of the Datapoint Corporation.

ments indicate MTTF of 2.0×10^7 h at 25°C. High performance burst mode receivers (packet operation) now have -37 dBm peak sensitivity, 22 dB dynamic range, and operate with 10 MHz Manchester coded data.

A variety of system installations in office, government, and factory settings have been completed using the above passive star systems. The largest of these is an Ethernet at Southwestern Bell's 44 story headquarters in St. Louis, MI. This all optical system has 540 nodes, over 4000 computer terminals, and 92 mi of fiber.

II. Passive Star Topology

Passive optical star-based fiber optic systems have been under development for over 15 years. Early work was on passive star LAN's for avionics systems was carried out by Biard [2] while Rawson and Metcalf [3] later sought to apply passive star technology to the emerging Ethernet LAN standard. More recent work on passive optical star networks [1], [4]-[8] has continued to emphasize application to standards-based LAN's as these emerge as the systems of choice. At present, two separate proposals for passive star systems standards are being reviewed, respectively, by IEEE 802 Committee and by the Japan LAN Standardization Committee.

The physical topology of the passive optical star system is shown in Fig. 1. This network can operate as a logical broadcast bus as in Ethernet CSMA/CD applications or as a token bus in 802.4 or ARCNET applications. The hardware components consist of an optical transceiver at each node, a passive optical star power splitter at the network hub, and sections of duplex fiber optic cable interconnecting each transceiver to the hub. The passive star contains no electronics and serves to optically divide the power from each transceiver to the others in the network.

The size of this network segment is limited by flux budget, dynamic range, and timing considerations for the network protocol being implemented. Flux budget and dynamic range are also affected by the number of nodes attached to a particular segment. Two simple equations describe the optical flux budget and dynamic range. First, define the following parameters:

TX = Peak transmitter power launched into system fiber, dBm.
RX = Peak receiver sensitivity, dBm.
L = Radius of fiber segment, $L_{min} \leq L \leq L_{max}$, km.
α = Optical fiber loss, dB/km.
M = System optical power margin, dB.
LS = Star coupler loss, including bulkhead optical connectors at star, dB.
FB = Flux budget, or allowable loss between transmitter and receiver, dB.
DR = Receiver optical dynamic range, dB.

Using these parameters the optical flux budget is expressed as follows:

$$FB = TX - RX = LS + 2L_{max}\alpha + M. \quad (1)$$

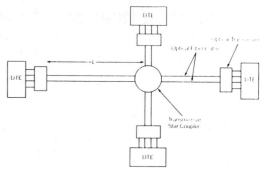

Fig. 1. General layout of a passive star fiber optic network. Individual nodes are optically interconnected by duplex fiber optic cable and a transmissive star coupler. The optical transceivers provide transmit and receive function and other information particular to each network. The case of 802.3 networks is illustrated where the four lines to the DTE include transmit, receive, collision presence, and electrical power.

The receiver dynamic range is described as follows:

$$DR \leq FB - LS - 2\alpha L_{min}. \quad (2)$$

Given a particular receiver, transmitter, and star coupler, these equations can be used to determine L_{min}. For maximum installation flexibility, we want $L_{min} = 0$.

These equations can be applied to a 10 MHz Ethernet CSMA/CD system as an example. The following parameters are assumed:

TX = -7 dBm peak power (100 μ core fiber).
RX = -37 dBm peak power.
α = 5 dB/km.
M = 3 dB.
DR = 22 dB.

Therefore, the relation between star coupler loss and network radius is

$$30 = LS + 2 \times 5 \times L_{max} + 3$$
$$27 = LS + 10 L_{max}.$$

Using the equation for receiver dynamic range

$$FB - DR \leq LS + 2\alpha L_{min}$$
$$8 \leq LS + 2\alpha L_{min}$$
$$8 \leq LS + 10 L_{min}.$$

In Fig. 2, these relations are shown graphically using the above parameters and others of interest. For a typical case using a 16-port biconic-fused star coupler, a network radius of 1 km can be achieved. This is equivalent to 2 km distance between optical transceiver or attached DTE's. It is significantly greater than the 500 m maximum segment length that can be attained with conventional coaxial Ethernet. Using biconic-fused couplers the receiver dynamic range is sufficient to permit operation to $L = 0$ for all but the four-port star coupler. An alternate family of star couplers is based on mixing fiber technology (described in Section IV). As shown in Fig. 2, these

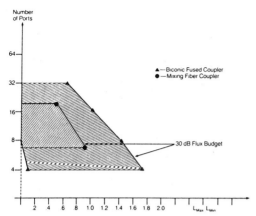

Fig. 2. Operating range for single segment passive star networks. Parameters used: 30 dB flux budget, 5 dB/km fiber, 3 dB margin, and 22 dB dynamic range. Operating range is within the shaded areas, limited on the right by flux budget and on the left by dynamic range.

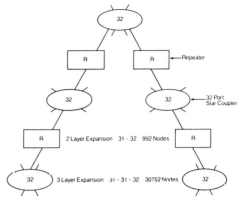

Fig. 3. Three layer expansion of network size beyond the 32 ports of a single segment.

couplers operate with seven and 19 ports and permit operation to $L = 0$.

Passive star fiber optic systems fill a well-defined application niche in an overall LAN marketplace that is served by coaxial and other wire systems, as well as other fiber optic LAN systems. Some useful points of comparison can be made by limiting consideration to 10 MHz CSMA/CD networks. In comparison to coax 802.3 system, the above discussion shows that larger segments can be built using fiber. Up to 100 nodes can be accommodated per segment using 802.3 coax technology, while a single passive star will presently support up to 32 ports using available LED and PIN receiver technology. A recent survey, however, indicated that 44 percent of all LAN's contain 20 nodes or less [9]; a single passive star segment will adequately support most work group clusters. A second key consideration involves network reliability.

From Fig. 1, it is clear that any cable break or damage will affect only the DTE locally connected to the damaged cable. Connection of additional terminals does not involve intrusion or disruption of the existing network. These features have proved to be very important for networks installed in factory and financial institutions where network availability is critical. Two other fiber optic CSMA/CD networks have been developed and commercialized in addition to the passive star. They are described in [11] and [12]; the topologies adapted are, respectively, an active star and a ring.

Practical networks may involve more than the 32 terminals supported by a single passive star hub; these may be distributed as well over a campus-like installation of many separate buildings.

For example, the backoff algorithm of 802.3 permits up to 1024 nodes per network, while 802.4 token bus has a virtually unlimited number of nodes (limited only by 48 bit address field). These extended networks are created by joining together passive star segments using repeaters. Repeaters also allow interconnection of coaxial systems and fiber optic star segments. Each repeater will regenerate and retime the signals on an input port and transfer these to the output port. An 802.3 repeater is a one-way valve that transmits in one direction at a time unless a collision (simultaneous transmission by two or more nodes) occurs on the network. In this case, the repeater will notify both sides of the network by transmitting a jam signal. A three layer hierarchical topology using passive star hubs is shown in Fig. 3.

Two layers will accommodate 992 nodes while 3 layers permits up to 30 752 (Ethernet limit is 1024, however). The geographical size of these CSMA/CD networks is limited by the roundtrip propagation delays including cables and optical transceivers. The passive star introduces no propagation delay. Reference [9, Appendix A] describes an 802.3 compatible network of 4.5 km distance from DTE to DTE.

Fig. 4 illustrates the four types of optical repeaters that have been developed to interconnect passive star systems into hybrid coax–fiber networks. Further discussion of repeater function will be given in Section III to follow.

III. Fiber Optic Passive Star Systems

Our emphasis to date has been to develop fiber optic implementations of previously defined network standards. Passive star systems developed to date include the first Ethernet [1], MAP, and PC Net fiber optic implementations as well as a passive star ARCNET compatible system. This approach makes fiber optic LAN's available to users now; other fiber optic networks such as FDDI [13] will likely play an increasingly important role in the future. To date the most widely implemented fiber optic LAN is Ethernet compatible; our discussion will emphasize these systems. Table I summarizes some basic properties of the four network protocols.

IEEE 802.3 compatible Ethernet hardware is shown in Fig. 5. This hardware includes a stand alone optical trans-

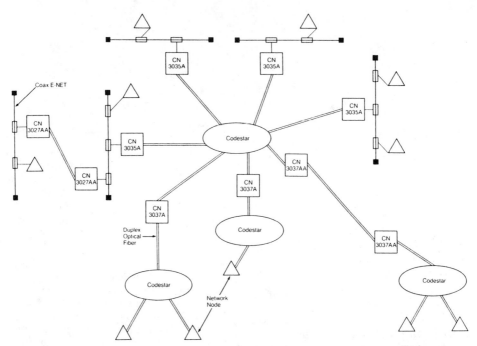

Fig. 4. Optical hybrid network. Repeaters are used to extend the geographical extent of the network as well as to interconnect coax and fiber media.

TABLE I
SUMMARY INFORMATION ON NETWORK SYSTEMS AVAILABLE IN PASSIVE STAR IMPLEMENTATION

Network	Access Protocol	Line Rate	Maximum No. of Stations	Maximum Network Size	Optical Line Coding
802.3 (Ethernet)	CSMA CD	10 MHz	1024	4.5 km	Manchester
802.4	Token Bus	10 MHz			Manchester
ARCNET	Token Bus	2.5 MHz	255	6.4 km	Miller
PC Net	CSMA CD	10 MHz	1000	4.8 km	Manchester

ceiver, a board level optical transceiver that interfaces to a common personal computer bus, a passive optical star power splitter, and network repeaters, both electrical to optical and optical to optical. The stand alone optical transceiver implements the physical layer of 802.3 including transmit, receive, and collision detection. The board level optical transceiver implements in addition, the MAC sublayer within the data link layer (layer 2). A gate array implements the interface to the PC bus. System design is facilitated by making use of the optical design parameters listed in Table II.

An important feature of all CSMA/CD networks is collision detection. A collision occurs if two or more nodes attempt to transmit simultaneously. The goal of transceiver design is 100 percent detection of collisions. In practice, a small collision detect inefficiency will not affect network performance. For example, a bit error rate of 1×10^{-9} will cause a certain percentage of frame check errors and missed packets. This percentage will range from 1.2×10^{-5} for the maximum frame length of 12 144 bits to 0.5×10^{-6} for minimum frame length of 512 bits. An undetected collision will also cause a certain number of lost packets depending on the product of the probability of missed collision and the probability of a collision. Since a heavily loaded network may experience large numbers of collisions, a collision detect inefficiency of $< 1 \times 10^{-5}$ would seem to be a suitable practical limit. Several techniques have been proposed for collision detection in passive star systems; see [6], [9], [14]-[19].

We have successfully implemented the pulse width error detection method described in [17]. In this technique, a timing circuit is used to detect the presence of pulses greater than a preset threshold t_{CD}. If only one station is attempting to transmit, then the 10 MHz Manchester code prescribed by 802.3 allows only 50 or 100 ns pulses on the optical fiber. The presence of pulses greater than $t_{CD} > 100$ ns indicates that two or more stations are attempting to transmit. A collision between two packets is shown in Fig. 6. This signal is obtained by connecting an optical waveform analyzer to an unused port on the star coupler and observing the resulting electrical signal on a LeCroy 9400 digital oscilloscope. The 5 MHz signal in the expanded scale trace shows the preamble of the first packet.

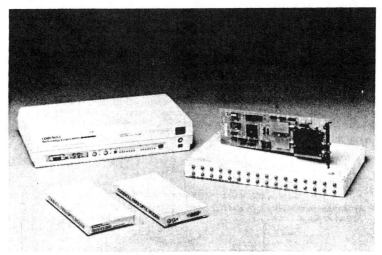

Fig. 5. Typical 802.3 hardware consisting of (clockwise, from foreground): stand-alone optical transceiver, coax-to-optical repeater, transceiver card, and 16-port passive star coupler.

TABLE II
OPTICAL DESIGN PARAMETERS FOR PASSIVE STAR NETWORKS

Transmitter power	7 dBm (peak) (100 micron core, 0.29 na)	
Operating wavelength	830 nm	
Receiver Sensitivity	37 dB (peak)	
Receiver Dynamic Range	22 dB	
Typical fiber attenuation	5 dB·km	
Star coupler losses (maximum)	4 port	10 db
	8 port	13 dB
	16 port	17 dB
	32 port	21 dB
	7 port	18 dB
	19 port	22 dB

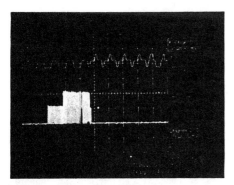

Fig. 6. Optical collision between two packets. Upper trace shows collision event to expanded waveform at 200 ns per division; lower trace shows entire packet at 5 μs per division, with collision occurring 5 μs from start of first packet. While preamble of one packet has only 100 ns between transitions, two colliding packets generate wave with 200 ns between transitions.

The collision with the second packet produces a composite waveform with pulses about 200 ns in width. This collision results from optical pulses with nearly equal amplitude. For large differential optical signal levels the strong signal begins to dominate the weaker, and collisions may not be detected successfully.

These phenomena have been studied in the experimental setup shown in Fig. 7. This three-node test bed for CSMA/CD networks was developed to measure collision detection efficiencies under a variety of real network traffic and topology configurations. Three programmable traffic generators are connected through optical transceivers to a passive star hub. The multiport transceiver units and internetwork bridge units are used for experimental convenience. A logic circuit was developed to monitor, for each optical transceiver, the total packets transmitted, the total sensed collisions (as indicated by collision detect output from the transceiver), the actual collisions occurring (simultaneous transmission by the transceiver in question and one or more of the other transceivers), and false collisions (the transceiver in question signals that a collision has occurred when in fact it is the only unit transmitting). Fig. 8 shows collision detect efficiency as a function of optical power difference. The loopback power was adjusted to be equal at the start; then transceiver $T1$'s power was increased in 1 dB increments. The differential power measured at a particular receiver would be higher by the variations in the star coupler (typically 1–2 dB).

We have conceived and analyzed an improved collision detect method [20] that predicts collision detection inefficiency of less than 1.9×10^{-6} over 6 dB optical differential signal level. The two principal features of this technique are transmission of large amplitude optical pulses and random encoding of 8 of these pulses within the first

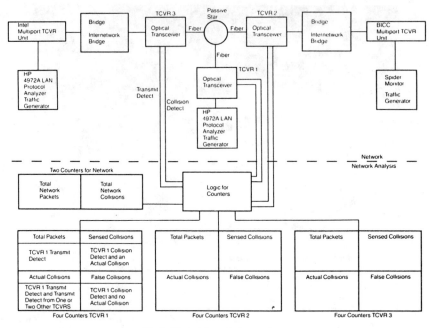

Fig. 7. Testbed for optical networks.

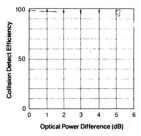

Fig. 8. Collision detection efficiency measured on network testbed. Fiber length was 500 m to star for each transceiver. Transceivers set to generate no false collisions. 64 byte packets were used with 18 percent total network traffic.

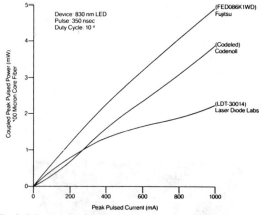

Fig. 9. Pulsed power output of three commercially available LED's. "Codeled" device is edge emitting, the others are surface emitters.

16 bits of the preamble. The $4x$ amplitude large optical pulses are detectable in the presence of a second optical data stream of 6 dB ($4x$) larger amplitude. Since these pulses are 25 ns in width, they are distinguishable from the normal 50 ns and 100 ns intervals. A collision is counted if any transceiver, upon receiving a packet, counts more than eight 25 ns pulses. The probability of a missed collision can be calculated to be $< 1.9 \times 10^{-6}$. The large amplitude optical pulses are generated by edge emitting and surface emitting LED's; Fig. 9 shows pulsed power output characteristics of several commercially available LED's.

An alternate collision detection technique has been reported using coding rule violations detected at the output of a partial response circuit [6], [15], [29]. In [29], a measured collision detection inefficiency of 10^{-7} with optical power difference of 8 dB is reported. Fig. 10 shows experimental data [29] for collision detection time. Future work in this field will involve continued testing of the above described wide dynamic range collision detection techniques, also needed is an international standard using one of these methods.

Optical repeaters were mentioned in Section II and are illustrated in Fig. 4. The repeaters designated 3035A and 3037A are used for locally interconnecting two optical LAN's or an optical and a coaxial LAN. For both re-

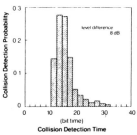

Fig. 10. Collision detection time reported for a CRV technique. Total collision detection inefficiency was reported to be 1×10^{-7}.

TABLE III
SUMMARY OF FIBER OPTIC SPECIFICATIONS IN 802.4, DRAFT H

Topologies	Passive Star Active Star Active Bus Active Headend	
Signalling	5, 10, 20 Mbs Differential Manchester Coded Packet Transmission	
Optical Media	Networks may be implemented in any multimode fiber provided all parameter specifications are met. A standard test fiber (62.5/125, na .275) is defined for measurement of fiber optic parameters	
Optical Connector	As specified by ANSI X3T9.5-84-88	
Emitter Characteristics	Center Wavelength FWHM Launch Power Rise/Fall Time	800-910 nm 60 nm 9 ± 2 dBm defined by wave shape template in draft standard (Fig. 16.3)
Receiver Characteristic	Moderate Sensitivity 11 dBm to −31 dBm High Sensitivity 21 dBm to −41 dBm	
Optical Power Budget	Moderate Sensitivity 20 dB Receiver High Sensitivity 30 dB Receiver	

peaters, the optical transceiver is packaged internal to the repeater housing. For remote interconnection of coaxial or optical LAN's, the 3027AA or 3037AA repeaters are used. The point-to-point fiber link between the repeater units has additional collision detection via detection of simultaneous transmission and reception on the fiber cables. A recent draft standard describes the fiber optic signaling on remote repeater links [21]. This will ensure interoperability of vendor equipment.

A second CSMA/CD network is IBM PC Net [22]. As originally developed this was a broad-band coaxial cable system operating at 2 Mbits. Our optical transceiver replaces the RF modem with an optical transceiver. The board level transceiver is compatible to the IBM PC bus interface and measures $13\frac{1}{8}'' \times 3\frac{7}{8}''$.

This transceiver contains not only the physical layer interface (layer 1) but also four higher level interface layers: data link (layer 2), network (layer 3), transport (layer 4), and session (layer 5). Layers 3-5 are implemented in firmware executed by an on board 80188 microprocessor.

Board level optical modems essentially compatible to the IEEE draft standard 802.4 have been developed. IEEE 802.4 is a token bus system that originally was based on coaxial cable; the new revision [23] contains two chapters describing fiber optic systems options and fiber optic media. The new revision allows active and passive systems as shown in Table III.

Present designs allow a 20 dB flux budget with a 62.5 μ core fiber. Electrical signals supplied by the optical modem to the MAC layer hardware on a separate card include; receive data, receive clock, and three lines of receive data type information. The MAC layer provides the optical modem with: transmit data, transmit clock, and three lines of transmit data type information. Optical modem cards are compatible to VME bus and multibus backplanes.

A passive star fiber optic ARCNET transceiver card[2] was recently described [24]. The baseband coaxial cable system was developed in 1977 and has more recently been described in detail [25]. This system uses a bipolar line code transmitted over RG 62A/U cable. An active hub approach is used. Our ARCNET transceiver card measures $4.2'' \times 13.3''$ and is compatible to the IBM PC bus. A number of points of comparison exist between the fiber system and the coax system. First, the maximum hub-to-node distance for the coax system is 610 m limited by cable losses. The optical fiber system can support a node to star distance of 1 km limited by roundtrip propagation time of the optical signal. For 1 km this is $2 \times 1 \text{ km}/2 \times 10^8 \text{ m/s} = 10 \text{ } \mu\text{s}$. The optical receiver output is blanked for 10 μs following end-of-transmission so that a particular station does not see its own signal returning. This blanking interval is compatible to the minimum 12 μs gap between ARCNET packets.

The coax system uses a bipolar line code with average voltage of zero. For the fiber system, we have adopted a Miller code with constant average duty cycle of 50 percent. The maximum network size of 6.4 km (with repeaters) is the same for fiber and coax system and is limited by internode timing considerations of the ARCNET protocol.

IV. TECHNOLOGIES FOR PASSIVE STAR LAN SYSTEMS

The passive star LAN systems' needs have driven optoelectronic and optical technologies in three main areas; LED transmitter output power, burst mode receiver sensitivity and dynamic range, and optical star coupler excess loss and loss uniformity.

We have developed high brightness 830 nm edge emitting LED's with 0.5 mW output power in a pigtail configuration and 0.2 mW output power in an active device mount package. Sensitivity for burst mode 10 MHz Manchester operation has been improved to −37 dBm peak optical power. Low cost seven and 19 port star couplers have been fabricated using large core mixing fibers and etched cladding design to reduce insertion loss. These technologies will be described in what follows.

---

[2]Joint development of Codenoll Technology Corporation and Standard Microsystems Corporation.

TABLE IV
BURST MODE RECEIVER SPECIFICATIONS

Peak Optical Power Sensitivity	37 dBm
Dynamic Range	22 dB
Power Supplies	+12 (20 MA) −12 (Photodiode bias) 5.2 (70 mA)
Design	Si FET with high Z transimpedance
Output	Analog with limiting

TABLE V
SPECIFICATIONS FOR 19 PORT MIXING FIBER STAR COUPLER

Minimum loss	17 dB
Maximum loss	22 dB
Input output fiber	100 micron core, 0.29 na
Mixing fiber	600 micron core, 22 na, step index
Outside Diameter of etched fibers	120 micron

High power edge emitting LED's were designed, (US patent applied for), to couple maximum power into 100 μ core fiber. This is the most commonly used fiber size for passive star systems. Fig. 11 shows a BeO package configuration suitable for direct fiber pigtailing. Launched powers are routinely 0.500 mW measured through a bulkhead connector in 100 μ, 0.29 nA fiber.

More recently an active device mount (ADM) package has been developed for these LED's. An approach utilizing active alignment and GRIN lens has been chosen. Figs. 12 and 13 show the package design and the actual hardware. Coupled powers of 0.200 mW into 100 μ fiber and 0.130 mW in 62.5 μ fiber are obtained at 100 MA current.

Burst mode receivers are essential parts of all passive star packet communications systems. The specifications for our most recent receiver design are shown in Table IV. This receiver is currently implemented in the low cost surface mount technology as shown in Fig. 14. Board size is $1.2'' \times 3.15''$; the PIN detector is mounted in an ADM package.

Development of passive star couplers has been motivated by needs to minimize insertion loss, minimize port-to-port variations in loss while maintaining low manufacturing costs. The biconic fused coupler [26] meets only the first two goals well. We have chosen to pursue the development of the mixing rod approach following the earlier developments of others. To date, seven port and 19 port stars have been fabricated with losses for the seven port star of 16.5 dB ± 1.5 dB and for the 19 port star, 19.5 ± 2.5 dB. These losses include bulkhead connector loss at the input and output of star. The seven port is not optimized for lowest loss. The 19 port star is optimized by chemically etching the fiber cladding to reduce packing fraction loss. Specifications for the 19 port star are shown in Table V.

Further work is continuing to reduce the loss and improve port-to-port uniformity. The advantage of the mixing fiber approach is that only two straightforward and easily automated manufacturing technologies are needed: connector polishing and etching of silica.

A third approach to multimode coupler technology has recently been described [27]. This technology uses a monolithic glass substrate with multimode waveguides fabricated by ion exchange. For the 1×8 coupler reported, excess loss ≤ 2.5 dB, port-to-port uniformity = ±0.75 dB.

V. INSTALLATIONS OF PASSIVE STAR LAN's

Installations of passive star Ethernet systems have been carried out since 1983—this network system currently dominates over all other types of fiber optic LAN's.

These systems are installed to provide assured network availability, to provide network security, to provide a communication backbone suitable for migration to higher data rates, and to provide a communication medium free from EMI problems. Three installations will illustrate the application of passive star fiber optic LAN's in office and factory settings.

The largest passive star Ethernet LAN is located at the 44 story Southwestern Bell facility in St. Louis, MI [28]. This system contains 540 nodes, 92 miles of fiber, and supports over 4000 computer terminals and workstations.

The architecture uses a three-layer star hierarchy with three stars on the 22nd floor fiber distribution center, eight port stars serving groups of five floors (two spare ports), and each floor served by two 16 port stars. The fiber optic system was installed for the reasons mentioned above and, in addition, because it made the best use of the limited under floor duct space.

Two other installations illustrate the use of passive fiber systems in factory LAN's The layout of the first is shown in Fig. 15. This grinding compounds manufacturer had initially attempted to network equipment in six buildings by threading an Ethernet coaxial cable through them. The exposed portions of this network were very susceptible to electrical storms. In several instances, equipment damage resulted from lightening strikes. The fiber optic implementation shown in Fig. 11 solved this problem.

Coaxial Ethernet segments were retained in three of the buildings connected to a central fiber optic star coupler through repeaters. Individual computers in the remaining buildings connect directly to the star coupler.

A second factory LAN application is illustrated by Fig. 16. This system was installed in an airplane manufacturing plant where noise problems had been encountered with a coaxial cable LAN. This, together with the unique requirement that some computers needed to physically move over distances beyond that afforded by coaxial cable, led to implementation of the hybrid coax/fiber network depicted.

Here a centralized VAX computer is connected to a number of Codenet coax-to-fiber repeaters through a

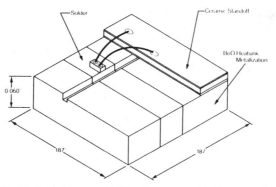

Fig. 11. Physical structure of LED packaged on heat sink. Step in ceramic heat sink is provided for fiber alignment.

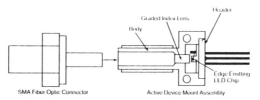

Fig. 12. Active device mount (ADM) design. This package makes use of compact disk laser header and graded index (GRIN) index lens.

Fig. 13. Active device mount located on optical transceiver card.

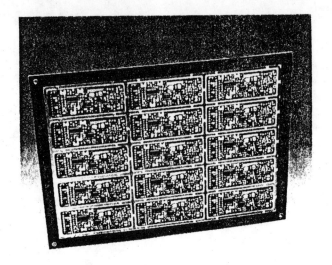

Fig. 14. Surface mount PC board technology provides low capacitance carrier for *Si* FET burst mode receiver.

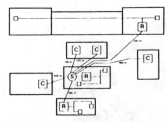

Fig. 15. Fiber optic network implementation for campus-like factory application.

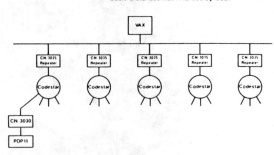

Fig. 16. Fiber optic network implementation for airframe manufacturing application.

coaxial cable LAN segment. Beyond that point, the network is fiber optic consisting of several passive star segments to which are attached the peripheral computers.

VI. Conclusion

Passive star-based LAN systems have been successfully installed since 1983. They provide an attractive alternative to wire systems for both CSMA/CD and token bus LAN's. In the future, we can expect to see increasing use of integrated circuits and ultimately OEIC's to reduce the manufacturing costs of the optical transceivers. We expect to see fiber optic transceivers integrated directly into work stations much as RS 232 modems are a standard accessory at present.

Introduction in the future of international standards covering passive star systems will make these systems more accessible to users and installers of LAN's. Application driving these installations will be: backbone

LAN's, individual work station connections where duct space or EMI is an issue, and factory LAN's of all types.

ACKNOWLEDGMENT

Major contributions to the research and development reported herein were made by S. Anderson, W. Hatfield, B. Dutt, J. Helbers, and H. Jayawickrama.

REFERENCES

[1] J. R. Jones, J. S. Kennedy, and F. W. Scholl, "A prototype CSMA/CD local network using fiber optics," presented at Local Area Networks '82, Los Angeles, CA, Sept. 1982.
[2] J. R. Biard, "Optoelectronic aspects of avionic systems," AFAL-TR-73-164, June 1973.
[3] E. G. Rawson and R. M. Metcalf, "Fibernet: Multimode optical fiber for local computer networks," *IEEE Trans. Commun.*, vol. COM-26, pp. 983-990, July, 1978.
[4] S. Moustakas and H. H. Witte, "Passive fiber optic star bus for local area networks: System design and performance," *Siemans Forsch.*, vol. 12, pp. 182-187, 1983.
[5] E. E. Bergmann, "Fiber optics as a physical layer for MAP," Presented at IEEE 802.4H Stand. Meet., Irvine, CA, Feb. 17, 1986.
[6] K. Ozawa, E. Doi, and N. Yokota, "CSMA/CD fiber optic LAN," Presented at IEEE 802.8B Stand. Meet., Vancouver, B.C., July 1987.
[7] W. B. Hatfield, F. W. Scholl, M. H. Coden, R. W. Kilgore, and J. H. Helbers, "Fiber optic Ethernet in the factory automation environment," presented at SPIE O-E/Fiber '87, San Diego, CA, Aug. 1987.
[8] C. H. DeGennaro, "1553 compatible multiple speed fiber optic data bus," Presented at SAE-9C Committee, San Diego, CA, Apr. 13, 1987.
[9] M. H. Coden, W. B. Hatfield, and R. W. Kilgore, "Proposal for passive fiber optic star configured LAN standard for IEEE 802.3 CSMA/CD," Dec. 23, 1986.
[10] *Office Products Analyst*, vol. 11, no. 4, New York, Apr. 1987.
[11] "Joint proposal for active star configured fiber optic LAN standard," Presented to IEEE 802.8A Committee, July 3, 1987.
[12] "Fiber optic CSMA/CD ring network architecture," Presented to IEEE 802.8A Committee, Dec. 22, 1986.
[13] J. Hutchinson and D. Knudson, "Developing standards for a fiber optic LAN-FDDI," *SPIE*, vol. 715, pp. 72-79, 1986.
[14] F. W. Scholl, "Method and apparatus for detecting the collision of data packets," U.S. Pat. 4 560 984, Dec. 24, 1985.
[15] K. Oguchi and Y. Hakamada, "New collision detection techniques and its performance," *Electron. Lett.*, vol. 20, pp. 1062-1063, Dec. 1984.
[16] N. Tokura, Y. Hakamada, and K. Oguchi, "Burst signal receiving apparatus," U.S. Pat. 4 462 582, Dec. 31, 1985.
[17] F. W. Scholl and M. H. Coden, "Data receiver," U.S. Pat 4 561 091, Dec. 24, 1985.
[18] J. W. Reedy and J. R. Jones, "Methods of collision detection in fiber optic CSMA/CD Networks," Advances in Local Area Networks, pp. 264-276, 1987.
[19] J. Helbers, "General requirements for collision detection using large pulse technique," Codenoll Technol. Corp., unpublished memorandum, Mar. 1986.
[20] F. W. Scholl, "Passive star Ethernet systems," Presented at 802.8A Committee, Chicago, IL, June 2, 1987.
[21] "Medium attachment unit and baseband medium specification for a vendor independent fiber optic inter repeater link," IEEE 802.3, draft H., Aug. 1987.
[22] T. Sammons, S. C. T. Schnetlagy, and J. Head, "An inside look at IBM's LAN," *PC Mag.*, vol. 136, Feb. 5, 1985.
[23] "Token passing bus access method and physical layer specification," IEEE 802.4. draft H., Aug. 1987.
[24] G. Karlin and C. Tucker, "ARCNET on fiber: A viable factory automation LAN," presented at SPIE O-E/Fibers '87, San Diego, CA, Aug. 16, 1987.
[25] A. J. Malinger, "The ARCNET local area network design and implementation of a local area network," presented at FOC/LAN 83, Atlantic City, NJ, Oct. 10, 1983.
[26] E. G. Rawson and A. B. Nafarrate, "Star couplers using fused biconically tapered multimode fibers," *Electron. Lett.*, vol. 14, p. 274, 1978.
[27] E. Paillard, "Recent developments in integrated on moldable glass," presented at Fiber Opt. '87 Conf., London, England, Apr. 29, 1987.
[28] D. E. Stein, "High-rise optical communications networks," presented at FOCLAN '85, San Francisco, CA, Sept. 18, 1985.
[29] N. Yokota, "Proposal for IEEE 802.3 AUI Compatible CSMA/CD fiber optic passive star local area network," Presented to IEEE FOSTAR group, Irvine, CA, Jan. 14, 1988.

Frederick W. Scholl received the Ph.D. degree in 1974 from Cornell University, Ithica, NY, where he developed and characterized new ternary nonlinear optical materials.

From 1974 to 1977, with Rockwell International Science Center, he developed new GaAsSb avalanche photodiodes and AlGaAs photodiodes for study of ps phenomenon. From 1977 to 1979 at Optical Information System Division of Exxon Enterprises, he was responsible for semiconductor laser research and production. Following a year at Columbia University, in 1980, he joined Mr. Coden in founding Codenoll. In addition to his responsibilities as Senior Vice President and Director of Codenoll, he has been Adjunct Associate Professor at Polytechnic Institute of New York, Brooklyn. He also has 18 publications and eight patents.

Michael H. Coden received the Bachelor's degree in electrical engineering from the Massachusetts Institute of Technology, Cambridge, the Master's degree in business administration from Columbia University, New York, NY, and the Master's degree in mathematics from the Courant Institute of Mathematical Sciences at New York University, NY.

He has taught Computer Science at Columbia University Graduate School of Business (1974-1975) and Semiconductor Physics at Fairleigh-Dickinson University, Department of Electrical Engineering (1979).

He has been Chairman of the Board, President and Chief Executive Officer of Codenoll since its incorporation in 1980. From 1975 to 1979, he had been employed by Exxon Enterprises, Inc. From 1977 to 1979, he was the Manager of Exxon Enterprises Optical Information Systems, a manufacturer of semiconductor laser and related products, and from 1975 to 1977, he was Program Manager with responsibility for evaluating venture capital investments and providing management assistance. Prior to 1975, he held management positions at Digital Equipment Corporation and Hewlett-Packard Company, where he managed the design and marketing of computer hardware and software and was an Officer of Maher Terminals, Inc., where he designed and implemented on-line computer control of all operational and financial functions of the corporation.

Mr. Coden has received an Award of Merit from the American Chemical Society and The Laser Institute of America. He has presented more than 30 papers (20 of which have been published) in computer science, local area networks, fiber optics, electro-optical technology, and high technology management. He has obtained three patents. He is a member of Beta Gamma Sigma Business Honorary Society and professional societies. He is also listed in American Men and Women of Science, Who's Who in Business and Finance, International Business Mens Who's Who, and Men of Achievement.

IV

PROCEEDINGS
Of SPIE-The International Society for Optical Engineering

Volume 840

Fiber Optic Systems for Mobile Platforms

Norris Lewis, Emery L. Moore
Chairs/Editors

Sponsored by
SPIE—The International Society for Optical Engineering

Cooperating Organizations
Applied Optics Laboratory/New Mexico State University
Center for Applied Optics/University of Alabama in Huntsville
Center for Applied Optics Studies/Rose-Hulman Institute of Technology
Center for Electro-Optics/University of Dayton
Center for Optical Data Processing/Carnegie Mellon University
Georgia Institute of Technology
Institute of Optics/University of Rochester
Optical Sciences Center/University of Arizona

20-21 August 1987
San Diego, California

FIBER OPTIC SYSTEMS FOR MOBILE PLATFORMS

Volume 840

Session 1

Automotive Fiber Optic Applications

Chair
Marek Wlodarczyk
General Motors Research Laboratories

DEVELOPMENT OF FIBER OPTICS FOR PASSENGER CAR APPLICATIONS

R. E. Steele

Packard Electric Division, General Motors Corporation
P.O. Box 431, Warren, Ohio 44486

H. J. Schmitt

Kabelwerke Reinshagen GmbH
Wuppertal, Germany

ABSTRACT

The benefits of fiber optics for telecommunications and Local Area Networks (LANs) are well documented. The benefits to passenger car applications are not as clearly defined. This paper examines the differences between Telecommunications, LAN, and automotive point to point and network applications. Current production automotive applications of optics and fiber optics, automotive data communications trends, and both functional and non-functional requirements and constraints will be described.

1. INTRODUCTION

On an electronically well equipped vehicle with a data communication network, the copper-based electrical system consists of power, signal, and data circuits. The largest portion, 78%, is for power and ground distribution for lamps, motors, and solenoids. Low current sensor and switch status signals account for 20% of the total circuits. The remaining 2% is for data distribution.

As the vehicle electronics content continues to increase through the addition of new electronic features and the replacement of mechanical controls with electronic control, the use of serial data links and networks will proliferate. The data rates which started out low, will increase as the applications change from diagnostics and data sharing to control applications. As the system requirements increase, the copper-based approaches will become much more difficult to fabricate. As more uniform transmission structures are required, fiber optics will become a viable alternative. On board fiber optic network architecture will be determined by not only functional requirements such as fault tolerance, power moding, and latency but also other considerations such as total cost including piece cost, vehicle buildability, and serviceability. Other considerations such as elasticity -- how easily the network can be expanded and contracted -- must also be addressed.

The objective of this paper is to evaluate potential network architectures using both functional and non-functional automotive considerations. This paper will review the current automotive applications of fiber optics, look at automotive data communications trends, compare telecommunications, LAN, and automotive applications, examine the advantages of fiber optics with the automotive application in mind, define vehicle considerations, and evaluate and rank various network architectures based on the defined vehicle considerations.

2. CURRENT AUTOMOTIVE APPLICATIONS OF FIBER OPTICS

Fiber optics has been in use in automobiles since the mid 1960's. The primary uses have been illumination and sensors.[1] The illumination applications provide light to remote areas of the vehicle where it would be difficult to locate and service a standard light bulb. In this way, a single incandescent light bulb can service many such areas. Fiber ribbons also provide a means to back light switch legends without requiring the space for the typical incandescent lamp or light bar.

Production fiber applications in optical sensors include lamp monitoring and windshield washer fluid level. When used for lamp monitoring, one end of the fiber is positioned to receive light from the lamp to be monitored. The other end is mounted inside or outside the vehicle where the driver can see it. This provides a visual indication when the lamp is on. Another optical sensor is a windshield washer fluid level monitor. The fiber transmitted a green color when the fluid level was acceptable and a red signal when the fluid was low. This was accomplished with a simple float and an incandescent lamp.

Data communications via fiber optics has had very limited applications. The Toyota Century, introduced in the 1982 model year, uses fiber optics to communicate information between the control modules located in the doors and a central processing unit. The data is transmitted at 2000 bits per second.[2] Another production application is the Mitsubishi Debonair which uses fiber optics to transmit switch information from a rear seat remote radio control unit to the radio chassis. Both applications are on high option content low volume vehicles. Proliferation of these types of fiber optics applications has been at a stand-still due to cost. Low cost wired options operating at the same data rates with signal waveshaping are quite effective.

While many have looked at the possibility of fiber optics to replace much of the copper-based power distribution system using optical multiplex,[3,4,5] the main area where fiber optics offers some potential benefit is in the area of on board data communications. Since the late 1970's, the number of microprocessor based systems in automobiles has increased significantly. There are production vehicles with 15 separate microprocessor based systems. This number is expected to continue to increase as functions such as Anti-lock brakes, Electric Power Steering, Electronic Transmission Control are added.

3. AUTOMOTIVE DATA COMMUNICATIONS TRENDS

Automotive applications of serial data communications started to appear in the early 1980's. Point to point, function specific applications were introduced followed by networks connecting the various vehicle computer systems. As can be seen from Figure 1 , data rates started out low but have grown rapidly. Projecting this growth out to the 1990's, shows the potential need for data rates in the Mega bit/sec. region. This growth will not be linear, but will occur in steps dictated by system functional requirements and cost. As data rates grow, the transmission media complexity will increase. Today, most applications are implemented with single randomly laid wire. Fiber optics or any other uniform transmission media cannot compete with a single wire on the basis of cost alone. This situation will change as data rates increase to the point where single randomly laid wire does not function acceptably. At higher data rates, twisted pair or twisted shielded pair will be required depending on data rate and vehicle EMC requirements. As more stringent vehicle requirements evolve, fiber optics will become a viable alternative.[6]

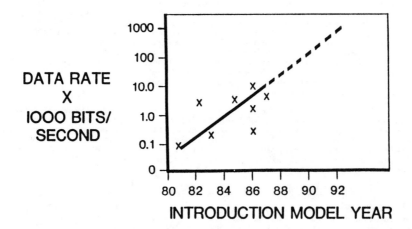

Figure 1   Data Rate Trend

4. TELECOMMUNICATIONS VS LAN VS AUTOMOTIVE

Telecommunications, LAN, and automotive applications have very distinct differences. The main reasons fiber optics is used in telecommunications applications are the long distances and high data rates that can be achieved. Fiber optic links minimize the requirements for repeaters. Repeaters are a high cost portion of the total system. They represent not only an initial cost but also an ongoing repair cost, especially when located at the bottom of the ocean. Telecommunication applications are fiber intensive. The use of connectors is minimized in order to maximize transmission distance.

LANs are significantly shorter in length than telecommunications links, but are still quite long compared to an automobile. A LAN can cover several kilometers. Connectors are used for the final distribution points.

Automotive applications can be characterized as connector intensive. A vehicle designed with a 12 node fiber optic network will have from 30 to 60 meters of fiber with the typical fiber length of 1.5 meters.

Another unique characteristic of automotive applications lies in the build techniques. Telecomm and LAN start on-site with a reel of cable and connector components. The fibers are routed and the connector components are assembled into a system on-site. The few terminations that do exist are done manually one at a time. Automotive power and signal distribution systems are assembled off-site and brought to the vehicle assembly line as complete units for assembly into the vehicle. This off-line assembly allows for the vehicle specific system to be prebuilt and tested prior to installation into the vehicle. Since all the terminations are completed at one location, an efficient automated termination process can be developed to keep cost low.

5. ADVANTAGES OF FIBER OPTICS

The typical advantages of fiber optics are weight, size, electromagnetic compatibility (EMC), dielectric nature, and bandwidth. When applied to the automobile these advantages must be reassessed using the automotive application as the criterion.

5.1 Weight
A fiber replacement for a cable containing 3200 twisted pair is a clear winner, but a 0.35 mm^2 (22 Ga.) single copper conductor is not heavier than an automotive grade glass or plastic fiber.

5.2 Size
Individual fibers are quite small, but by the time a fiber is strengthened to the point it can survive the total automotive environment, it will be bigger than the data wires in the vehicle today. This condition will change as the data rate increases and the need for a twisted, twisted shielded, or coaxial transmission media becomes necessary.

5.3 EMC
Due to the many electronic systems including a sensitive radio receiver being packaged in the close confines of the automobile, EMC is a very significant factor. With fiber optics, the fiber can be routed anywhere without causing a problem in another system or having another system interfere with its operation.

5.4 Dielectric
One of the common faults in the automobile is a short to ground. Since a fiber cannot be electrically shorted, it is possible for the majority of the bus to continue to function. This will depend heavily on the network architecture.

5.5 Bandwidth
Data communications in the automobile are still very slow compared to telecommunications and LANs. Automotive communications data rates today are less than 10K bits/sec. Projections of past growth would put this at 1M bit/sec. in the not too distant future. Copper-based systems are still more than adequate.

The typical advantages of fiber optics do not all apply to the automotive application. If the capability of the fiber is taken as the requirement, the fiber implementation has many advantages over the copper implementation. In this case, the automotive application requirements are the goal and many of the advantages of fiber optics do not apply or are not very significant. As data rate requirements increase or EMC becomes an issue, the advantages of fiber optics will increase in significance.

6. VEHICLE CONSIDERATIONS

There are a number of possible architectures that could be implemented in an automotive network. For the purpose of this study, the single and double ring, the active and passive star, and the linear tapped bus will be used. In order to evaluate their effectiveness, a typical vehicle will be developed to compare the capability of each architecture to address automotive issues. The issues to be compared are: cost (which will be derived from number of terminated fiber ends, fiber length, receiver transmitter pairs and any other items the network will require), complexity, power moding impact, fault tolerance, expandability, serviceability, and latency.

6.1 Cost

Total system cost (not just initial piece cost) will determine which network architecture will prevail. Initial cost will be the combination of the fiber ends, receiver transmitter pairs, and fiber length. Total cost will include assembly, warranty, and service costs. A relative comparison will be made to determine system advantage or disadvantage. (The best system will have no extra components, one receiver transmitter pair per node, and one fiber per node.)

Cost factors that can be readily quantified are fiber ends, receiver transmitter pairs, fiber length, and any other significant system components.

6.1.1 Fiber ends - Due to the large number of fiber ends to finish in an automotive network, the number of fiber ends will be proportional to a significant part of the system cost and be inversely proportional to reliability (the more mated ends, the more opportunities for a fault to occur.

6.1.2 Receiver transmitter pairs - Active devices are costly components. Minimizing receiver transmitter pairs will minimize cost and also result in improved system reliability.

6.1.3 Fiber length - Although the amount of fiber per vehicle will be small, it still represents a cost factor that is easily quantified.

For the purposes of this evaluation, a single vehicle implementation will be used for the comparison. The vehicle has 12 nodes which are distributed around the vehicle as they might be based on modular build trends. To be consistent with modular build, connections are required between the main body and the engine, instrument panel, and rear compartment. The distances involved are approximate but are consistent across implementations.

Applying each of the 5 network architectures to the vehicle implementation, shown in Figure 2 a, b, and c, results in the fiber ends, fiber length in meters, and receiver transmitter pair count shown in Table 1. Also listed under misc. are other significant items that need to be considered. A relative cost scale (last column on the right) is generated using a scale of -10 for the highest cost to +10 for the lowest cost.

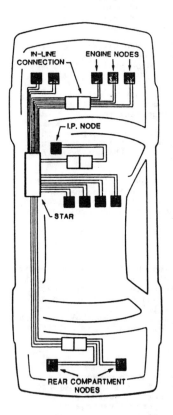
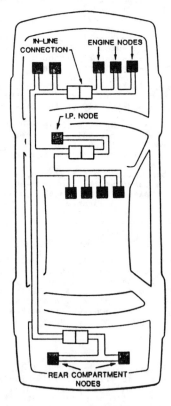
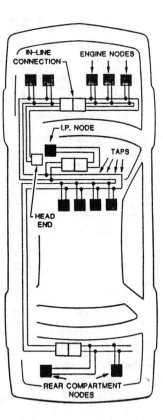

a) Star - Active or Passive   b) Ring   c) Tapped Bus

Figure 4 Vehicle Network Architecture Alternatives

7. Architecture Evaluation

In order to compare the five architectures based on the vehicle considerations a scale from -10 for worst to +10 for the ideal will be applied to the five network architectures. Assuming that the considerations have equal weighting, the network architectures can be relatively compared. Table 2 summarizes the results of the comparisons.

7.1 Active Star

The active star based system, shown in Figure 2a, consists of the 12 nodes to be networked and a central active node which has a receiver transmitter pair for each node in the network and logic to allow retransmission of received data to all nodes.

COST - Additional active node with associated electrical content and requires two receiver transmitter pair per node. (-10)

COMPLEXITY - The added node and it's associated electrical complexity with two fibers per node connecting to it. (-10)

POWER MODING - Star must be powered up at all times. (-5)

FAULT TOLERANCE - The central node offers many internal and external fault opportunities that could cause the network to be inoperative. (-5)

EXPANDABILITY - Limited by the original hardware design and may require many design levels. (-10)

SERVICEABILITY - Some fault identification is possible. Fault will be isolated to a specific spoke on the star or the central node. (0)

LATENCY - Central node retransmission will increase latency. (-5)

7.2 Transmissive Star

In the transmissive star based system, shown in Figure 2a, each node has a transmit fiber and a receive fiber that connects to the passive star input and output ports respectively. The light transmitted by any node enters the star and is divided equally among the output ports of the star. The star output ports are connected to each node's receiver.

COST - Has minimum number of receiver transmitter, but added passive star will add cost. (0)

COMPLEXITY - Requires 2 fibers per node to route to the star, but no electrical connections. (0)

POWER MODING - Nodes can be powered in any order or combination. (+10)

FAULT TOLERANCE - Fault will be limited to only one function. Star could be a single point fault, but could be hardened. (+5)

EXPANDABILITY - Limited by star port count and optical flux budget. (-5)

SERVICEABILITY - Some fault identification is possible. Fault will be isolated to a specific spoke on the star. (0)

LATENCY - All nodes will see the data when it is transmitted. (+10)

7.3 Single Ring

In the single ring, shown in Figure 2b, the data is passed around the ring from node to node until it completes a circuit. Data is received and retransmitted by every node.

COST - The fewest fiber ends and requires no special hardware. (+10)

COMPLEXITY - Complexity is minimized since there are no added nodes and the complexity is evenly distributed. (+10)

POWER MODING - All modules must be powered up for operation. (-10)

FAULT TOLERANCE - A fault anywhere in the system will cause the network to be inoperative. (-10)

The ring is the least expensive because it has the minimum number of fiber ends, no excess fiber, the minimum number of receiver transmitter pairs and no misc. components. The most expensive is the active star which has a large number of receiver transmitter pairs and an additional complex central node. The double ring is given a -5 due to the number of fiber ends and the large number of receiver transmitter pairs required. The tapped bus is rated -10 due to the number of taps and the added, although simple, head end node. The transmissive star is rated 0 due to the number of fiber ends and the added transmissive star element.

	FIBER ENDS	FIBER METERS	OPTICAL PAIRS	MISC.	REL. COST
ACTIVE STAR	72	60	24	+1 NODE	-10
TRANS. STAR	72	60	12	+1 STAR	0
SINGLE RING	36	28	12		+10
DOUBLE RING	72	56	24		-5
TAPPED BUS	68	42	13	+22 TAPS +1 NODE	-10

Table 1. SUMMARY OF COST FACTORS AND RELATIVE COMPARISON

6.2 Physical Complexity

Physical complexity is an issue that will impact the vehicle assembly plant operation. A physically complex system will be difficult to assemble reliably. This could add to assembly costs, dealer warranty cost, or owner maintenance cost. (The best system will not have concentrations of complexity or have extra fibers, nodes, or receiver transmitter pairs.)

6.3 Power Moding

The electronic systems on the vehicle are not all powered up at all times. Some systems may be purposely powered down during a portion of the time. Some systems will be maintained on for a time after the vehicle is turned off. Times such as ignition off, crank, or accessory operation require only a portion of the vehicle system to be operational to conserve battery charge. The network should allow for this reduced level of network operation. (The best system would allow network operation with any combination of nodes powered.)

6.4 Fault Tolerance

Continued network operation under a fault condition is desirable. The tolerance of a system to faults could impact the type of functions that could be included on the network. The more tolerant the network is to faults, the more critical functions that can be implemented on it. (The best system will continue to function with any single point fault.)

6.5 Expandability

Expandability for vehicle variability, future vehicle improvements, and dealer installed options must be allowed for in the initial design. Easy network expansion and contraction is a desirable feature. (The best system will be expandable without cost to the lesser system.)

6.6 Serviceability

Second only to initial functionality, serviceability is a very important issue. The vehicle must be serviceable within the existing service network. This will include local service stations. (The best system will be self-diagnosing.)

6.7 Latency

The time between data updates will impact the type of function performed on the network. Control applications will require rapid data exchange updates. (The best system will have minimum transmission latency.)

EXPANDABILITY - Easily expanded by adding nodes into the ring. (+10)

SERVICEABILITY - Faults are not easily isolated. (-5)

LATENCY - Each module must retransmit the data, adding to the latency between the first and last node. (-10)

7.4 Double Ring

In the double ring, which is the same as the single ring in Figure 2b except with redundant fibers, data is passed around in two directions on two sets of fibers. If a fault occurs, the node preceding the fault, sends the data around the ring in the other direction on the second loop, completing the communication. In this manner, faulty nodes or fiber interconnects can be bypassed.

COST - Redundant receiver transmitter pairs and fiber end count. (-5)

COMPLEXITY - Complexity is evenly distributed, but, each node requires four optical connections and two receiver transmitter pairs. (-10)

POWER MODING - Only one module could be powered down and the rest of the network function. For multiple modules to be powered down, they would have to be positioned in line on the ring which would impact the physical complexity. (-10)

FAULT TOLERANCE - very tolerant of single faults (nodes or fiber interconnect). The remainder of the network continues to function. (+10)

EXPANDABILITY - Easily expanded by inserting a node into the ring. (+10)

SERVICEABILITY - Network is capable of determining the location of a fault. (+10)

LATENCY - Each module must retransmit the data, adding to the latency between the first and last node. (-10)

7.5 Linear Tapped Bus

The linear tapped bus can be implemented in many forms depending on the capabilities of the tap. If the tap is assumed to be unidirectional and symmetrical the most probable configuration would consist of a transmit fiber, a receive fiber and an active head end as was shown in Figure 2c. Each node has a transmit tap on the transmit fiber and a receive tap on the receive fiber.

COST - Large number of taps and head end node with no offsetting fiber length or fiber end reduction will cause cost to be high. (-10)

COMPLEXITY - Taps and head end node will impact complexity. (-5)

POWER MODING - Head end node must be powered up continuously. (-5)

FAULT TOLERANCE - A fault will always cut off some portion of the network. Certain faults such as the head end node or portions of the fiber distribution close to the head end will disable the network. (-5)

EXPANDABILITY - Expandability will be limited by the tap characteristics and the original optical flux budget. (+5)

SERVICEABILITY - A number of conditions can cause the network to be inoperative making diagnosis difficult.(-5)

LATENCY - Head end retransmission will add some latency. (-5)

	ACTIVE STAR	TRANS. STAR	SINGLE RING	DBL. RING	LIN. TAP
COST	-10	+0	+10	-5	-10
COMPLEXITY	-10	0	+10	-10	-5
POWER MODING	-5	+10	-10	-10	-5
FAULT TOLERANCE	-5	+5	-10	+10	-5
EXPANDABILITY	-10	-5	+10	+10	+5
SERVICEABILITY	0	0	-5	+10	-5
LATENCY	-5	+10	-10	-10	-5
TOTAL SCORE	-45	+20	-5	-5	-30

Table 2. SUMMARY OF ARCHITECTURE AND VEHICLE CONSIDERATIONS

8. CONCLUSIONS

Current automotive applications of fiber optics take advantage of the unique properties of optical fibers to address specific functional challenges in the areas of illumination and optical sensors.

Though current automotive data communications needs are adequately met with standard single wire implementations, trends indicate that the single wire will not meet the future vehicle requirements. Comparison of telecommunications, LAN, and automotive applications indicate significantly different characteristics which will require different implementation considerations. In automotive, the connection becomes the primary factor as opposed to the fiber.

Based on the assumptions made, the factors considered, and the vehicle implementation, the following conclusions about network architecture can be made:

1. The single ring is inexpensive, but not very fault tolerant
2. The double ring is costly and complex.
3. Both ring configurations have functional deficiencies in latency and power moding.
4. The active star is very costly, complex and susceptible to single point faults.
5. The linear tapped bus is costly with no offsetting benefits.
6. The transmissive star would appear to be the best option especially if the cost of the star with some expansion ports is reasonable.

As the automotive electronics system evolves, data communications will play an increasingly significant role. It is the opinion of the authors that the unique characteristics of fiber optics will lead to significant automotive data communications applications in the mid 1990's.

9. REFERENCES

1. A.L. Harmer, Fibre Optics in Automobiles, pp. 174-185 SPIE Vol. 468 Fibre Optics '84. (1984)
2. Y. Matsuzakiand K. Baba, Development of Multiplex Wiring System with Optical Data Link for Automobiles, SAE Technical Paper Series 840492. (1984)
3. K. Sekiguchi, Fiber Optic Application in Automobile, pp. 27/6 1-9, Electro/82 (1982)
4. N. Yumoto, H. Ikeda, T. Sugimoto, K. Hayashi, F. Sakamoto, Optical Data Link for Multiplex Wiring, SAE Technical Paper Series 830320. (1983)
5. Y. Himono, M. Miyahara, K. Nishimura, Y. Enomoto, J. Nakajima, S. Inao, Optical Link for Automotive Multiplex Wiring, SAE Technical Paper Series 840494. (1984)
6. R.E. Steele, H.J. Schmitt, Electromagnetic Compatibility Considerations in Optic Signal Transmission Systems, International Journal of Vehicle Design Vol. 6, No. 6, Nov. 1985 pp. 737-747. (1985)

Fiber Optic Ethernet in the Factory Automation Environment

W.B. Hatfield, F.W. Scholl, M.H. Coden, R.W. Kilgore, J.H. Helbers

Codenoll Technology Corporation
1086 North Broadway, Yonkers, New York 10701

ABSTRACT

The requirements for a Local Area Network (LAN) suitable for use in factory environments are discussed. In many cases these requirements are met by fiber optic Ethernet (IEEE 802.3 compatible) LANs. Implementation of a passive fiber optic star coupler based Ethernet LAN is presented in terms of hardware and system design rules. Application of the passive star coupler topology LAN to a number of factory applications is described.

1. INTRODUCTION

Factories have certain unique characteristics which must be considered when selecting a data communications network. Chief among these are their large physical extent, harsh physical environment and the presence of different types of industrial systems, many having real time requirements. The Manufacturing Automation Protocol (MAP) standard was developed specifically to meet these requirements. MAP incorporates a token passing protocol to satisfy the perceived need for a deterministic system to control factory equipment and a broadband transmission medium for improved noise immunity and greater transmission distances.

For various reasons, MAP has not yet evolved into a stable standard. This has caused many potential users of factory Local Area Networks (LANs) to consider alternative networking standards. A frequently considered alternative because of its relative maturity as a standard and its large installed base is the so-called Ethernet or IEEE 802.3 standard.

The major theoretical argument against the use of Ethernet in the factory environment is that rather than a deterministic token passing protocol, Ethernet utilizes the Carrier Sense Multiple Access with Collision Detect (CSMA/CD) access method.

Under this non-deterministic media access method, any station on a common bus can transmit whenever it detects a quiet period on the medium. If two stations attempt to transmit simultaneously (collision) both must stop, back off and re-transmit at a later time.

Although Ethernet allows more rapid access to a network than a token passing protocol under most circumstances, it was developed primarily for the office environment where strict determinism is not a necessity.

The applicability of Ethernet to a factory environment will depend upon the specific equipment involved; however, it should be noted that the explosive growth of distributed processing has tended to eliminate the need for determinism in a factory LAN.

The need for longer distances and greater noise immunity than is inherent in coaxial cable, baseband Ethernet implementations is addressed by IEEE 802.3 compatible LANs now available which utilize optical fiber as the transmission medium.

2. A FIBER OPTIC IEEE 802.3 COMPATIBLE LAN

Fiber Optics brings a number of advantages to LAN design. It provides greater distance capability than baseband coax, and lower cost and less maintenance than broadband coax. Optical fibers are immune to electromagnetic noise, lightning and ground potential differences. In many instances, fiber optics offers greater flexibility in system installation and re-configuration. Because of its extremely high bandwidth, a single fiber installation can accommodate the anticipated increase in LAN data rates well into the future. The history of electrical LANs, on the other hand, has been that each new LAN standard requires a new cable installation.

One implementation of an IEEE 802.3 (Ethernet) compatible fiber optic LAN is based upon the passive fiber optic star topology. The fiber optic star coupler (Figure 1) is a totally passive device having some number of INPUT ports and an equal number of OUTPUT ports. The basic principle of the star coupler is that an optical signal coupled into any of the INPUT ports is divided approximately equally among all the output ports. The optical star coupler is, therefore, functionally equivalent to the tapped coax cable used in the electronic Ethernet. In its Codestar series of optical star couplers, Codenoll Technology offers couplers manufactured by the fused biconical process having 4, 8, 16 and 32 ports (see Figure 2). Also available are 7 and 19 port couplers manufactured by an alternative process which are of somewhat higher loss and significantly lower cost. The loss associated with transmission from any INPUT port to any OUTPUT port is an important parameter in passive star topology network design. Loss parameters for the Codestar Couplers are summarized in Figure 1.

A fiber optic transceiver is required to interface between any equipment designed to IEEE 802.3 and the passive star network.

The functions of the transceiver are to:

- Transduce transmit data into optical signals and launch them onto the fiber optic transmission media.
- Detect optical receive data signals and convert them to electrical signals.
- Detect the simultaneous presence of more than one transmission (Collision Detect) and notify the associated station controller.

Codenoll Technology manufactures two types of fiber optic transceivers:

Codenet 3030A/3030B (Figure 3) — a standalone unit which interfaces via standard transceiver cable to any 802.3 port.

Codenet 3051/3051A (Figure 4) — an integrated unit which mounts in an expansion slot of any IBM PC, XT, AT or compatible computer.

Specifications for Codenet Transceivers are given in Figure 5.

A fiber optic LAN is configured by connecting each transceiver to a pair of INPUT/OUTPUT ports on the star coupler using multimode duplex optical fiber as indicated in Figure 6. Star couplers are available in all the popular mulimode fiber sizes. The only variation will be in the amount of optical power which can be coupled from the LED source into the multimode system fiber. Standard power options for various fiber diameters are shown in Figure 7.

The passive star based fiber optic LAN has a number of unique characteristics. Since the star coupler is a totally passive optical device, it exhibits none of the failure modes common to electronic equipment. In this sense, there is no single point of failure in the passive star LAN. The passive star can be placed in any convenient location, it requires no power or controlled environment, it is not subject to EMI nor does it emit any detectable radiation. A break in any cable will affect only the connected node, thereby simplifying troubleshooting. Nodes can also be added to or removed from a network without affecting overall network operation.

3. PASSIVE STAR LAN DESIGN

Like all fiber optic systems, passive star based LANs must satisfy two design criteria in order to assure reliable optical performance.

3.1. Sensitivity Limit

The optical losses (attenuation) due to the passive star coupler, optical fiber attenuation, any connectors or splices, together with a designated operating margin should not cause the signal launched into a fiber (at the transmitter) to be attenuated below the level at which it can be detected by a receiver (this is known as the sensitivity level). Expressed analytically, and noting that each transceiver must detect its own "reflected" signal (for collision detection):

$$\text{Available System Gain} = 10 \log \frac{\text{Transmitter Output}}{\text{Receiver Sensitivity}} \geq \text{Coupler Loss} + \text{Connector Loss} + 2\beta L_{max} + \text{Operating Margin}$$

where β = Fiber attenuation (dB/Km) and

L_{max} = the longest coupler-to-node length.

3.2 Dynamic Range Limit

Optical receivers have an upper limit to their detection range, above which they saturate. The optical circuit must contain sufficient loss to avoid this effect.

For the fiber optic receivers used in the Codenet-3030B and 3051A, this operating (or dynamic) range is approximately 21 dB. Therefore;

$$\text{Available System Gain} - 21 \text{ dB} \leq \text{Coupler Loss} + \text{Connector Loss} + 2\beta L_{min}$$

Where L_{min} = the shortest coupler-to-node fiber length.

Two examples of passive star LAN designs are shown in Figures 8 and 9. Figure 8 shows the calculation for a maximum extent segment. This is achieved with a minimum loss (4 port) coupler and a maximum gain (100/140 micron fiber) transceiver.

Assuming a 3 dB operating margin, the maximum coupler-to-node distance is found to be 1.7 Km giving a total segment coverage of 9.1 Km². If, however, there are nodes situated less than 200M from the coupler, some attenuation must be added to the line to avoid exceeding the dynamic range limit of the optical receivers.

The second design example (Figure 9) assumes a larger number of network nodes. With a 32 port coupler and 62.5/125 fiber, the maximum coupler-to-node distance, again allowing 3 dB margin, is 375M for a total segment coverage of 0.44 Km^2. In this instance, the coupler loss is sufficient to assure adherence to the dynamic range limit for all coupler-to-node distances.

4. NETWORK EXPANSION

A fiber optic star coupler and the network nodes connected to it through fiber optic transceivers are referred to as a LAN segment. Large fiber optic networks are created by interconnecting two or more LAN segments through a fiber optic Ethernet repeater. A fiber optic repeater (see Figure 10) contains two integral Codenet fiber optic transceivers and joins two LAN segments (Star Couplers) by attaching through one port on each coupler. The repeater retimes and amplifies all signals it receives from one LAN segment and passes (repeats) the signal to the other LAN segment. Codenet fiber optic repeaters have the same optical specifications as Codenet transceivers. Hybrid repeaters having one port for connection to a fiber optic segment and the second port for connection to a coaxial segment are also available. The IEEE 802.3 standard allows up to four repeaters in any signal path. Consequently, very large Ethernet networks can be created using the three level tree structure indicated in Figure 11.

5. PASSIVE FIBER OPTIC STAR COUPLER FACTORY LAN APPLICATIONS

In this section we describe several installations in which factory LAN applications which could not be adequately addressed by copper based technology have been satisfied using the passive fiber optic star coupler LAN technology discussed in the preceding sections.

5.1 Steel Mill

Here it was desired to connect two microprocessors controlling the production of steel plate with a VAX computer. The distance from computer room to mill pulpit was too great, 900 meters, for a single coaxial cable LAN segment and the hostile electrical environment made reliable operation of any electrical conductor based network doubtful. The solution implemented was to interconnect the two controllers and the VAX through a four part Codestar coupler as indicated in Figure 12.

5.2 Airframe Manufacturer

As is so often the case in factory environments, this user had encountered noise problems with a coaxial cable LAN. This, together with the unique requirement that some computers needed to physically move over distances beyond that afforded by coaxial cable, led to implementation of the hybrid (coax/fiber) network depicted in Figure 13. Here a centralized VAX computer is connected to a number of Codenet coax-to-fiber repeaters through a coaxial cable LAN segment. Beyond that point, the network is fiber optic consisting of several Codestar segments to which are attached the peripheral computers.

5.3 Power Generating Station

Here the problem was to interconnect computers in three buildings separated by as much as 1900 meters in the extremely electronically hostile environment of a power generating station. The passive fiber optic star coupler solution shown in Figure 14 provides this user with reliable error free operation.

5.4 Grinding Compounds Manufacturer

This user had initially attempted to network equipment in six building by threading an Ethernet coaxial cable through all six buildings as indicated in Figure 15. The exposed portions of this network were very susceptible to electrical storms. In several instances equipment damage resulted from lightning strikes. Additionally, overall network operation was slowed by the excessive length of the Ethernet Cable. The fiber optic implementation that solved both of these problems is shown in Figure 16. Coaxial Ethernet segments were retained in three of the buildings connected to a central fiber optic star coupler through repeaters. Individual computers in the remaining buildings connect directly to the star coupler.

6. SUMMARY

Although the Ethernet or IEEE 802.3 network standard was not developed specifically for factory use, a fiber optic implementation compatible with this standard does offer features which make it an attractive solution for many factory applications. Hardware now exists which allows a user to implement a standards based network in demanding applications requiring the unique distance, noise immunity and fault tolerant properties inherent in fiber optic based LANs.

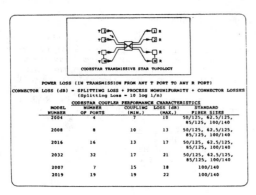

Figure 1. Characteristics of the Passive Fiber Optic Star Coupler

Figure 4. Codenet 3051 Ethernet Board for IBM PC or Compatible Computers

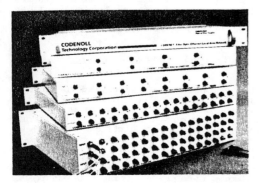

Figure 2. 4, 8, 16 and 32 Port Codestar Passive Fiber Optic Star Couplers

Specifications

Light Source:	Codeled high radiance edge-emitting Aluminum Gallium Arsenide LED.
Light Detectors:	Silicon PIN-type Photodiode.
Optical Wavelength:	830nM ± 20nM
Optical Sensitivity:	0.2μW (−37dBm)
Optical Saturation:	30μW (−15dBm)
Fiber Optic Cable:	Codenet duplex CFOC-XXX-2—or equivalent
Optical Connector:	C-10 Series SMA-type standard (other connector types available on special order)
Electrical Interface:	IEEE 802.3 standard transceiver cable—CN-8010—or equivalent
Electrical Connector:	15-pin, male subminiature D-connector
Indicator:	Power on
Storage Temperature:	−55° C to +80° C (−67° F to 170° F)
Operating Temperature:	0° C to +55° C (32° F to 131° F)
Humidity:	5% to 90% (non-condensing)
Radiation Suppression:	Part 15, Subpart J for computing devices and VDE (Class B)
Dimensions:	9.875 in. L × 4.55 in. W × 1.00 in. H (25.08 cm. L × 11.56 cm. W × 2.54 cm. H)

Figure 5. Codenet Fiber Optic Ethernet Transceiver Specifications

Figure 3. Codenet 3030A Fiber Optic Ethernet Transceiver

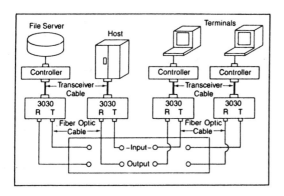

Figure 6. Illustration of a Passive Fiber Optic Star Coupler Ethernet LAN Segment

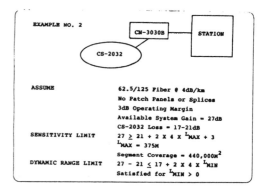

Figure 9. Design of a 32 Node Passive Star Coupler LAN Segment using 62.5/125 Optical Fiber

AVAILABLE OPTICAL POWER OPTIONS		
OPTION NUMBER	FIBER SIZE	OPTICAL POWER BUDGET
013	100/140	30
012	100/140	26
033	62.5/125	27
032	62.5/125	23
003	50/125	25
002	50/125	21

Figure 7. Codenet Fiber Optic Ethernet Transceiver Optical Power Options

Figure 10. Codenet 3037A Fiber Optic Ethernet Repeater

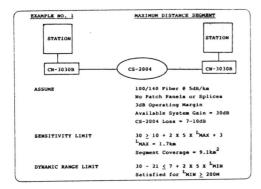

Figure 8. Design of Maximum Length Passive Star Coupler LAN Segment

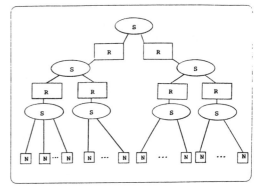

Figure 11. Fiber Optic Network Expansion in Three Layer Tree Structure

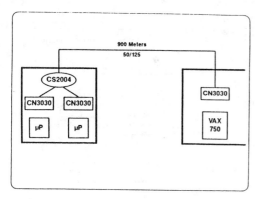

Figure 12. Fiber Optic Network Implementation for Steel Plant Application

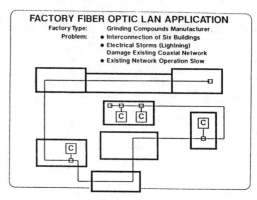

Figure 15. Ethernet Coaxial Cable Implementation for Grinding Compounds Manufacturing Application

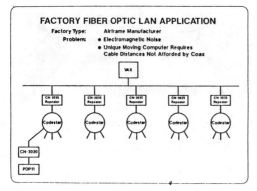

Figure 13. Fiber Optic Network Implementation for Airframe Manufacturing Application

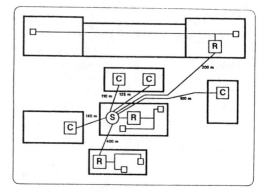

Figure 16. Fiber Optic Network Implementation for Grinding Compounds Manufacturing Applications

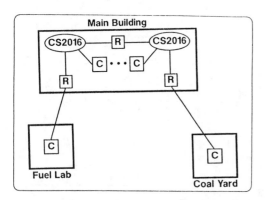

Figure 14. Fiber Optic Network Implementation for Power Generating Station

Fiber Optic LANs for the Manufacturing Environment

W.B. Hatfield
M.H. Coden
F.W. Scholl

The Physical Layer requirements for Local Area Networks (LANs) which satisfy the set of environmental requirements unique to factory automation applications are discussed. In many cases these requirements are best met with fiber optic LANs compatible with the IEEE 802.3 (CSMA/CD) or 802.4 (Token Bus) network standards. Implementation of a passive fiber optic star coupler based LAN is presented in terms of two simple fiber optic system design rules. A number of applications of the passive star coupler technology LAN are briefly described in order to highlight the benefits of fiber optics as a data communication medium.

Each application environment for a data communication or Local Area Network (LAN) has a certain set of characteristics which serve to define the media access technology and associated physical data transmission media which are best suited for that particular application. The factors most often mentioned for the manufacturing or factory environment [1] are large physical extent, harsh physical environment (particularly as regards electromagnetic interference) and the presence of different types of industrial systems, many having real time requirements.

Until recently, the choice of electromagnetic data transmission media has been limited to copper base conductors, primarily coaxial cable and twisted pair. With these as the options, it can be argued [2] that broadband transmission on coaxial cable is superior to baseband transmission on either coax or twisted pair as regards both distance capability and electromagnetic noise immunity. For just this reason the Manufacturing Automation Protocol (MAP) standard [3] was developed with broadband transmission on a coaxial cable data bus as the primary data transmission media. A token passing access method was specified to satisfy the perceived need for a deterministic system.

Although broadband coaxial networks do offer superior distance capability and noise immunity, they also contain tuned circuitry which might be expected to make them more expensive to implement and more difficult to maintain in the LAN environment than baseband systems of equivalent data capacity. A data transmission technology which could combine the positive attributes of broadband transmission with the relative simplicity of a baseband system should be of great interest for application in manufacturing environments.

Fiber optic transmission systems are totally immune to electromagnetic radiation as well as any ground potential differences which may exist between two separated locations. At the operating data rate of most current LANs (1-10 Mbs), optical fiber allows greater transmission distances than are possible with copper media. Furthermore, the wide inherent bandwidth of optical fibers allows migration to higher data rate LANs without recabling. This is a very important aspect of fiber optics since historically it has not been the case with copper based LANs where each succeeding standard has required a new cable plant (see Fig. 1).

For several years fiber optics has been the transmission medium of choice for new long-haul telecommunication system installations. This has been due to a combination of factors. The first is that, depending upon construction, optical fibers can offer transmission rates greater than 1 Gbs [4]. In long haul systems, end terminal costs are not as important as signal regeneration (repeater) costs. The extremely low attenuation achievable in optical fibers (less than one dB per kilometer for single mode fibers) [5] allows very large repeater spacings. As a result, long haul fiber optic systems can be installed at a lower cost per bandwidth than transmission systems employing competing technology [6]. The results of the research and development effort required to realize the very extensive fiber optic telecommunications network now in place in this country are now available to the Local Area Networking community in the form of highly reliable, low cost fiber optic data transmission components including, in addition to the optical fiber itself, optical emitters, detectors and connectors.

NETWORK	DATA RATE	CABLING
ASCII	19.2 Kbps	Twisted Pair
IBM 5251	1.0 Mbps	Twin-Ax
PC NETWORK	2.0 Mbps	75 ohm Coax
IBM 3278/9	2.35 Mbps	92 ohm Coax
TOKEN RING	4.0 Mbps	Dual Twisted Pair Shielded Data Cable
IEEE 802.4 MAP (G.M.)	5/10Mbps	75 ohm Coax or Fiber Optic
IEEE 802.3 ETHERNET	10 Mbps	50 ohm Coax
ADVANCED TOKEN RING	16 Mbps	Fiber Optic
IEEE - 802.6	50 Mbps	Fiber Optic
ANSI - FDDI	100 Mbps	Fiber Optic

Fig. 1. Evolution of Local Area Network Cable Specifications.

Having asserted that fiber optics is now a competitive technology which specifically addresses many of the problems characteristic of the manufacturing environment, we shall describe in the remainder of this paper a fiber optic LAN implementation which is applicable to two of the more popular LAN standards in use today.

LAN Standards

There has been considerable activity in recent years [7] directed towards developing standards for the interconnection of computers and other electronically intelligent equipment into local area networks. All existing or developing LAN standards today are based upon the structure of the International Standards Organization (ISO) Open System Interconnection Reference Model for data transmission shown in Fig. 2. This model provides a structure for the protocols needed to perform reliable data communications and provide interworking between users' systems. This paper will focus on the lowest layer in the ISO model, the Physical Layer, which consists (see Fig. 2) of the actual data transmission Medium, the Medium Dependent Interface (MDI), the Medium Attachment Unit (MAU), the Attachment Unit Interface (AUI), and the Physical Signaling (PLS) technique. It is the definition of these elements that is of concern in developing fiber optic LANs. Since most LAN Standards allow a multiplicity of Physical Layer implementations, one requirement on this layer is that it be transparent to all higher ISO layers.

The Manufacturing Automation Protocol (MAP) standard was developed specifically to meet the requirements of the factory environment. MAP incorporates a token passing protocol to satisfy the perceived need for a deterministic system to control factory equipment and a broadband transmission medium for noise immunity and greater transmission distances. The MAP standard includes all seven Layers of the ISO model. As Layers 1 and 2, however, it incorporates the IEEE 802.4 Token Passing Bus standard [8]. For various reasons, MAP has not yet evolved into a stable standard. This has caused many potential users of factory LANs to consider alternative networking standards. A frequently considered alternative because of its relative maturity as a standard and its large installed base is the Ethernet or IEEE 802.3 standard [9]. A major argument against the use of Ethernet in the factory environment is that rather than a deterministic token passing protocol, Ethernet utilizes the Carrier Sense Multiple Access with Collision Detect (CSMA/CD) access method. Under this non-deterministic media access method, any station on a common bus can transmit whenever it detects a quiet period on the medium (Carrier Sense). If two stations attempt to transmit simultaneously (Multiple Access) both must stop (Collision Detect), back off and re-transmit at a later time. Ethernet was developed primarily for the office environment where strict determinism was not regarded a necessity. Under many traffic conditions it does, however, allow more rapid access to a network than does a token passing protocol. The applicability of Ethernet to a factory environment [10] will depend upon the specific equipment involved; however, it should be noted that the growth of distributed processing has tended to reduce the need for determinism in a factory LAN.

At the present time neither the MAP nor the IEEE 802.3 standards provide for a fiber optic implementation of the Physical Layer. A revision to the IEEE 802.4 Token Bus standard which does include a fiber optic implementation is now being considered and fiber optic networking subsystem components conforming to this draft standard have been announced. The demand for longer distances, greater noise immunity and greater reliability than is inherent in existing copper based networks has led to the development of IEEE 802.3 compatible LANs which utilize optical fiber as the transmission medium [11-14]. The need for a fiber optic based IEEE 802.3 LAN standard has been recognized and a study group established within IEEE 802.3 to begin the standard drafting process.

Fiber Optic LAN Implementation

The IEEE 802.3 and IEEE 802.4 (MAP) standards share the common attribute of being broadcast type networks. That is, the function of the tapped coaxial cable bus, is to distribute a signal from any transmitting node in the network simultaneously to all other network nodes as indicated in Fig. 3. To implement this function in an identical fashion in fiber optics, that is, a tapped optical fiber, is virtually impossible without incurring excessive power losses. There is, however, a very elegant means of achieving the equivalent broadcast function in fiber optics utilizing a different topology.

The fiber optic star coupler is a well developed fiber optic network component having some number of INPUT ports and an equal number of OUTPUT ports as indicated schematically in Fig. 4. The basic principle of the star coupler is that an optical signal coupled into any of the INPUT ports is divided approximately equally among all the OUTPUT ports. The optical star coupler is, therefore,

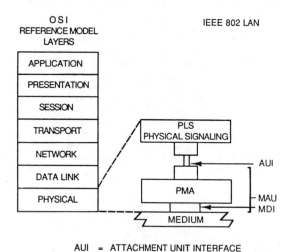

Fig. 2. *Open System Interconnection Reference Model for Data Transmission.*

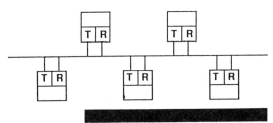

Fig. 3. *Tapped Coaxial Cable Bus LAN Topology.*

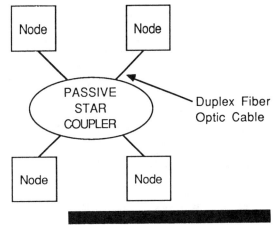

Fig. 5. *Passive Fiber Optic Star Coupler LAN Topology.*

functionally equivalent to the tapped coaxial cable. A fiber optic LAN can then be configured by connecting the Media Attachment Unit (MAU) associated with each network node to a pair of INPUT/OUTPUT ports on the star coupler, as indicated in Fig. 5, using a duplex fiber optic cable. A Fiber Optic MAU (FOMAU) performs the same functions as an electronic MAU, including Collision Detection in IEEE 802.3 LANs. It will also contain the circuitry and components to perform the necessary electro-optic transducing functions.

The passive fiber optic star coupler based LAN is favored over other possible implementations [13] for a number of reasons. Passive stars should be more reliable than stars containing active electrical components. They can be placed without regard to the power and environmental constraints of active devices. This is an especially important consideration in many factory environments where active electrical devices constitute safety hazards. The star configuration itself provides additional benefits. While a problem at any point in a tapped bus will tend to affect operation of every node on the network, in a fiber optic star the failure of any node or its associated cable will affect only that node, leaving the remainder of the network undisturbed [15]. In other words, the passive fiber optic star topology does not exhibit a single point of failure. This same characteristic extends to network reconfiguration. Adding or removing nodes from a passive fiber optic star network will not affect overall network operation. Also, the centralized wiring topology implicit in the star configuration is flexible and well suited to the design of an optical LAN for the factory environment [16].

Fiber Optic LAN Design

For reasons of cost and reliability, fiber optic LANs normally incorporate LEDs with a center wavelength of approximately 830 nanometers as the light emitting element. PIN-PD (p-type, intrinsic, n-type photodiode) based receivers are used to detect the optical signal. The amount of power which can be launched into optical fibers will depend upon both the LED design and the dimensions of the optical fiber. It has been shown [17] that with edge emitting LEDs specifically designed for fiber optic application, -3 to -7 dBm can be easily launched into the popular sized 100 micron core multimode optical fiber. Product literature from several vendors indicates that fiber optic receivers exhibiting sensitivity values in excess of -35 dBm at the 10 Mbs data rate characteristic of IEEE 802.3 and 802.4 networks are representative of current industry practice. Extensive reliability studies [17] for edge emitting LEDs have demonstrated room temperature operational lifetimes in excess of 10 million hours. Therefore, the performance benefits of fiber optic LANs include highly reliable system operation.

All fiber optic data communications systems must satisfy two criteria in order to assure reliable optical performance.

Criteria #1—Sensitivity Limit

The optical losses (attenuation) due to the passive star coupler, optical fiber attenuation, any patch panel connectors or splices, together with a designated operating margin should not cause the signal launched into a fiber (at the transmitter) to be attenuated below the level at which it can be detected by a receiver. Noting that since each FOMAU should detect its own "reflected" signal, this condition becomes

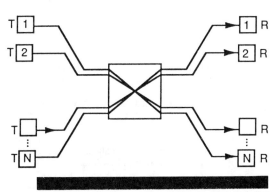

Fig. 4. *Schematic Representation of a Passive Fiber Optic Star Coupler.*

Launch power (dBm) −
Receiver Sensitivity (dBm) > Coupler Loss (dB)
+ Attenuation of two times the maximum coupler to node fiber length (dB)
+ Patch panel connector loss (dB)
+ Operating Margin (dB)

Criteria #2—Dynamic Range Limit

Optical receivers have an upper limit to their detection range above which they saturate. The optical circuit must contain sufficient loss to insure that the optical power reaching the receiver does not exceed the dynamic range of the receiver. That is,

Launch power (dBm) −
Receiver Sensitivity (dBm) − Dynamic Range (dB) < Coupler Loss (dB)
+ Attenuation of twice the minimum coupler to node fiber length (dB)
+ Patch panel connector loss (dB)

The passive fiber optic star coupler is a major contributor to the system optical loss calculation. The relationship between the input and output powers through any two ports on the star coupler is given by

$$\text{Coupler Loss} = 2C + E + 10 \log N$$

where: C is the optical connector loss (typically 1.5 dB)
E is the star coupler "excess loss" (typically 1 to 4 dB)
N is the number of optical ports.

There are currently a number of manufacturers of passive fiber optic star couplers with up to 32 ports. Based upon the above analysis, a reasonable maximum loss for a 32 port coupler is 21 dB. For design purposes, 5dB/km is a conservative value for the attenuation of multimode optical fiber.

The launch power and optical receiver sensitivity values discussed above are consistent with a nominal system gain of 30dB. A network configured about a 32 port star coupler, having no patch panel connectors could then have a maximum coupler-to-node separation of 500 meters and operate with an entirely adequate 4 dB operating margin. Such a network would cover an area of 0.79 square kilometers. Smaller couplers would allow correspondingly greater coupler-to-node spacings. For instance, an 8 port coupler might have a maximum loss of 13 dB. The coupler-to-node distance could then be in excess of 1 km and still maintain a 4 dB operating margin.

The optical receivers in the hypothetical network described above would need to have 17 dB dynamic range to assure adherence to the dynamic range limit. Dynamic range values in excess of 20 dB are more typical of current industry practice.

Because of the token passing protocol, there will be only one signal present on the network at a time in an IEEE 802.4 compatible passive star coupler based fiber optic LAN. In an IEEE 802.3 compatible network, on the other hand, an attempt by two nodes to transmit simultaneously will result in two optical signals present on a receive fiber at the same time. It is the function of the FOMAU to detect this collision condition. A number of techniques for detecting the simultaneous presence of two fixed data rate optical signals have been proposed and analyzed. These include optical power sensing [18], code violations based on pulse-width detection [18], code violations based on partial response [18,19], time delay violations [18] and sequence weight violation [16].

In all these techniques, a limiting factor is the magnitude of the optical power inbalance (the difference in amplitude of the optical power from the two transmitting nodes as seen by a single receiver) that the collision detection technique can tolerate. An issue in fiber optic LAN design then becomes the matter of trading off the cost of optical fiber network balancing against collision detection circuit complexity.

Fiber Optic LAN Applications

Passive star coupler based fiber optic LANs have been commercially available for over four years. Hundreds of networks encompassing thousands of nodes have been installed during this time in factories as well as offices. A sampling of just a few of these installations serves to highlight the benefits that fiber optics can provide as the Physical Layer Medium for LANs in the manufacturing environment.

1) There was a requirement in a large electric power generating facility to network equipment in three buildings spaced approximately 2 kilometers apart. In this harsh electromagnetic environment, a fiber optic Ethernet LAN provided the noise immunity and distance capability not achievable in a coaxial cable implementation.

2) A grinding compound manufacturer occupied a complex of six buildings which were interconnected with a coaxial cable LAN. This network was very susceptible to electrical storms which on some occasions actually damaged terminal equipment. Reliable data transmission under all weather conditions was achieved when the coaxial cable was replaced by a fiber optic Ethernet LAN.

3) Failure in a single network node caused total collapse of the trading floor coaxial LAN at a Wall Street based brokerage firm. The resulting 14 hour loss in trading activity meant the loss of hundreds of millions of dollars in business. The coaxial cable network was replaced by a passive star fiber optic network to eliminate the potential of another single point of failure crashing the entire system.

4) A government owned research facility in Europe wished to network 25 buildings spread over a 3 Km distance. The attendant ground loops resulting from the long extent of this facility made reliable data transmission virtually impossible until a fiber optic network was installed.

5) It was desired to network several thousand work stations in a new high-rise corporate headquarters. Unfortunately the architect had not provided

sufficient duct space in the building plan for copper conductors. The goal was attained by installing a space conserving fiber optic network to the workstation level throughout the building.

Conclusion

The brief sketches of actual LAN operating experiences presented above serve to emphasize the conclusion that fiber optic based Ethernet or IEEE 802.3 compatible LANs for the manufacturing environment are not only technically feasible but have been shown, over several years time and in many real installations, to offer such significant performance benefits that they should be considered in all potential network designs which require high data integrity and system reliability, large area coverage, or efficient utilization of space.

References

[1] G. Kim, "What is MAP?" a guide to manufacturing automation protocol," *LAN Magazine,* pp. 32-36, Nov. 1986.
[2] G. T. Hopkins and N. B. Meisner, "Choosing between broadband and baseband local networks," *Advances in Local Area Networks,* K. Kummerle, J. O. Limb and F. A. Tobagi, editors, pp. 62-66, IEEE 1987.
[3] General Motors Corporation, "MAP Specifications-Version 3.0, implementation release subject to errata changes," General Motors 1987.
[4] T. Li, "Advances in lightwave systems research," *AT&T Technical Journal,* vol. 66, no. 1, pp. 5-18, Jan./Feb. 1987.
[5] Corning Glass Works, Guidelines, vol. 1, no. 4, p. 3, Corning, NY, 1985.
[6] R. J. Sanferrare, "Terrestrial lightwave systems", *AT&T Technical Journal,* vol. 66, no. 1, pp. 95-107, Jan./Feb. 1987.
[7] R. W. Kilgore, "Status of standards for fiber optic LANs today," *Proc. Of The SPIE Conference On Fiber Optic Networks,* San Diego, CA, Aug. 1987.
[8] IEEE Computer Society, "Token-passing bus access method and physical layer specifications," ANSI/IEEE 802.4 (ISO/DIS 8802/4), IEEE 1985.
[9] IEEE Computer Society, "Carrier sense multiple access with collision detect (CSMA/CD) access method and physical layer specifications," ANSI/IEEE 802.3 (ISO/DIS 8802/3), IEEE 1985.
[10] W. R. Johnson, "Keynote address," MAP/TOP Users Group Summary, vol. 2, no. 2, Pittsburgh, PA, May 1987.
[11] J. R. Jones, J. S. Kennedy, and F. W. Scholl, "A prototype CSMA/CD local network using fiber optics," *Local Area Networks '82,* Los Angeles, CA., Sept. 1982.
[12] M. H. Coden and F. W. Scholl, "Implementation of a fiber optic ethernet local area network," *SPIE Technical Symposium East,* Arlington, VA, April 1983.
[13] G. O. Thompson, "Comparison of methods of implementing a fiber optic IEEE 802.3 ethernet," *Proc. Of The SPIE Conference On Fiber Optic Networks,* San Diego, CA, Aug. 1987.
[14] W. B. Hatfield, F. W. Scholl, M. H. Coden, R. W. Kilgore and J. H. Helbers, "Fiber optic ethernet in the factory automation environment," *SPIE Conference On Fiber Optic Networks,* San Diego, CA, Aug. 1987.
[15] M. H. Coden, F. W. Scholl and W. B. Hatfield, "Reliability of fiber optic LANs," *SPIE Conference On Reliability Considerations In Fiber Optic Applications,* Cambridge, MA, Sept. 1986.
[16] M. Kavehrad and C. W. Sundberg, "A passive star-configured optical local area network using carrier sense multiple access with a novel collision detector," *Journal of Lightwave Technology,* vol. LT-5, no. 11, pp. 1549-1563, Nov. 1987.
[17] F. W. Scholl, M. H. Coden, S. J. Anderson, B. Dutt, J. Racette and W. B. Hatfield, "Reliability of components for use in fiber optic LANs," *SPIE Conference On Reliability Considerations In Fiber Optic Applications,* Cambridge, MA, Sept. 1986.
[18] J. W. Reedy and J. R. Jones, "Methods of collision detection in fiber optic CSMA/CD networks," *IEEE Journal On Selected Areas in Communications,* vol. SAC-3, no. 6, pp. 890-896, Nov. 1985.
[19] N. Yokota, K. Ozawa and A. Hirata, "ISO 8802/3 compatible CSMA/CD fiber optic communications and local area network," *International Fiber Optic Communications and Local Area Networks Exposition,* Anaheim, CA, Oct. 1987.

Dr. Walter B. Hatfield received his PhD in Physics from Stanford University. Prior to joining Codenoll Technology in 1982 as Vice President of Subsystem Engineering, he had twenty years' Industrial R & D and product development experience at Bell Laboratories, Singer Co. and AM International. Areas of interest included plasma discharge and liquid crystal display technologies, thin film magnetic devices, magnetographic printing and advanced electronic packaging concepts. Currently, as Vice President, Standards and Reliability, Dr. Hatfield is concerned with the development of standards for fiber optic based Local Area Networks.

Mr. Michael H. Coden has been Chairman of the Board, President and Chief Executive Officer of Codenoll Technology Corporation, a vertically integrated manufacturer of fiber optic components and data communications systems, since its incorporation in 1980. From 1977 to 1979, he was the Manager of Exxon Enterprises Optical Information Systems, a manufacturer of semiconductor laser and related products, and, from 1975 to 1977, he was a Program Manager at Exxon Enterprises Inc. with responsibility for evaluating venture capital investments. Prior to 1975, Mr. Coden held management positions at Digital Equipment Corporation, Hewlett-Packard Company and Maher Terminals, Inc. Mr. Coden has a Bachelor's Degree in Electrical Engineering from the Massachusetts Institute of Technology, a Master's Degree in Business Administration from Columbia University and a Master's Degree in Mathematics from the Courant Institute of Mathematical Sciences at New York University. He has taught computer science at Columbia University Graduate School of Business (1974-1975) and semiconductor physics at Fairleigh-Dickinson University Electrical Engineering Department (1979).

Dr. Frederick W. Scholl received his PhD in 1974 from Cornell University, where he developed and characterized ternary non-linear optical materials. From 1974 to 1977 with Rockwell International Science Center, he developed new GaAsSb avalanche photodiodes and AlGaAs photodiodes for study of picosecond phenomena. From 1977 to 1979 at Optical Information System Division of Exxon Enterprises, Dr. Scholl was responsible for semiconductor laser research and production. In 1980 Dr. Scholl co-founded Codenoll Technology Corporation. In addition to his responsibilities as Senior Vice President and Director of Codenoll, Dr. Scholl has been adjunct associate professor at Polytechnic Institute of New York. Dr. Scholl has eighteen publications and eight patents.

IV

INVESTMENT BANK TRADES UP TO FIBER OPTIC ETHERNET

Stephen J. Anderson and Frederick W. Scholl

INTRODUCTION

On multiple floors of an office building in the financial district of New York City, millions of dollars are traded each day in transactions as fleeting as bursts of light on a fiber optic local area network. An innovative system developed by a major investment bank connects traders and salesmen serving the financial needs of corporations and institutional investors with a fiber network designed to meet the strictest requirements for reliable information transfer. The sometimes hectic pace of the financial trading desk is moderated by the ability to accurately service the needs of corporate customers seeking the most cost effective funding of their short term cash needs. Through the network, they make contact with investors looking for the best return on their money in an environment where a mistake or even a lost moment can have severe financial consequences. Experience has taught the investment banking group that fiber optic local area network implementation demonstrates greater reliability than any other cabling method.

The process by which the network was developed is described in this article. Hopefully it will give other potential network users an idea of the design, installation, and operating experiences which they can expect to encounter. The well documented benefits of a fiber system will also become evident in the description of the important day to day capabilities which the network provides to customers and clients.

The decision to network the financial trading desk was made by the investment bank in 1985. It was clear that an automated system that would allow traders to enter and display information on terminals would greatly reduce the chance of transaction errors compared to the existing system involving verbal communication, hand written entries, and slips of paper carried by clerks. While the need for an interconnection of information was clear, it was less clear which type of network solution would be the best one for this application.

Optical fiber quickly became the media of choice for several reasons. First, terminated thin coax sections that were left unused were more prone to damage from equipment and furniture being rearranged than were the fiber connectors. Secondly, the availability of a totally passive fiber hub was appealing from a reliability point of view. In addition, one of the goals of the corporation was to remain ahead of the competition through the adoption of the fastest and most powerful information handling technology available. That meant that within a few years a Fiberoptic Distributed Data Interface (FDDI) network, running at 125 Mbps, would be the system of choice. In order to prepare for that technology, it was recognized that running an Ethernet system on fiber optics would allow experience to be gained in dealing with fiber and at the same time, any investment in fiber cable plant could be maintained in the switchover to higher speed fiber networks as they became available.

The system has been in routine use for more than a year and a half, allowing a doubling to tripling of sales volume and an increase in client service with the same staff. Extensive proprietary software was developed which made the network even more valuable. Since many individuals have simultaneous access to the status of trades, a system of checks and balances has developed to eliminate errors that might be caused by trading on incorrect information. Traders enter sales quantity and salesmen complete the trade with detailed customer information. At each stage the trades are checked to match the amounts and prices agreed to by the traders, salesmen, and corporate clients. Also, the efficiency of each individual is greatly increased because someone with slack time can assist an account of an otherwise occupied trader or salesman.

INSTALLATION

The system consists of about 20 SUN workstations and 40 Compaq 386 terminals, each with 6-10 Megabytes of memory, connected to Codenoll 3037A repeaters through Codenoll 32 port passive star couplers. The fiber connection at the SUN workstations and at the PC's is made through Codenoll 3030C fiber optic Ethernet transceivers. A central Codenoll 16 port passive star coupler ties the computer segments to two SUN servers. The central star also provides connection through two T1 lines to the banks' Boston office and to another Manhattan location. The network configuration is shown in the attached diagram.

The network has evolved over a period of time, but the network manager estimates that a system of this size could be installed in one to two weeks, including field termination of the fiber optic cables. The electricians are responsible for all cabling in the building and have become skilled in fiber optic cable termination. In the view of the banking group, it is absolutely necessary to have an experienced fiber optic cable installer take the overall responsibility for training, support and warranty of the cable installation. The network manager believes that vendors of fiber optic equipment are an excellent source of information on fiber optic cable installers or system integrators. In addition, he suggests that the best way to make the final selection is to talk with the installers' customers.

In some other installations, preterminated cable has been used, but it does require accurate knowledge of the exact distance between connections which is usually not available in the environment at the bank. Recent introduction of connectors with ceramic ferrules has greatly improved the success rate of field terminations because it is very easy to avoid overpolishing the connector, as sometimes happens with metal ferruled connectors.

BUILDING WIRING

The installation involves both horizontal and vertical cabling runs which intersect at a wiring closet provided on each floor of the building during its construction. The wiring closet is a secured area with access by a limited number of people. It is the hub for both telephone and network installations. As a result, the wiring closet is quite full. The much smaller size of the fiber cables compared to coaxial copper cables is a big advantage. To do the same network in copper would be nearly unworkable in the space available.

All of the network servers, fiber optic repeaters and passive star couplers shown in the attached network layout are located in the wiring closet. From there, horizontal fiber optic cables run underneath the floor to the SUN workstations and Compaq 386 personal computers at the trading desk. In this case, a raised floor provides a cabling distribution plenum that gives access to any location on the floor, allowing complete flexibility in the layout of the work area.

Two vertical runs are made from the wiring closet as well. In one case, fiber runs in ductwork to the floor below where the T1 connection to the "cage" is located. The cage is a highly secure office at another location in Manhattan where the notes are stored. Even though the transactions of the financial trading desk are carried out electronically, the various financial instruments of value are still pieces of paper which are very carefully controlled. When trades are made, the cage must have an accurate record so that the notes can be received from the issuer's bank and delivered that same day to their new holder. Deadlines are involved for both the notes coming into the cage and for their arrival at similar locations in banks and brokerage houses throughout Manhattan. Any errors will result in additional costs, at least for an overnight period. It is very costly to cover such errors, but the fiber optic network in place now has eliminated the kinds of problems inherent in manual processing.

A second vertical run goes to an upper floor where there is a connection, via a T1 telephone line, to the bank's Boston office. In this case, fiber optics really came into its own. The vertical duct system in the building which interconnects wiring closets only runs part of the way up. The building begins to taper at this point so that straight vertical runs from the outer portions of the building to upper floors do not exist. To reach the required floor, short horizontal runs had to be combined with vertical runs through an elevator shaft. Without the excellent flexibility and noise immunity of the fiber optic cable, this installation would have been very much more difficult. In fact, the small size of the eight-fiber cable allowed the use of this convenient path, which probably would not have been posssible at all with a copper coaxial wiring system.

OPERATION

One of the complexities faced by the financial trading desk is that in reality there are about ten different types of financial instruments that are traded. The greatest effort goes into trading of commercial paper, a source of financing for the short term cash flow needs of corporations. Bank holding company paper, Bankers' Acceptances, and Medium Term Notes are also traded. The more familiar CD's, and the financing of the bank's money market inventory are transacted on the network as well. To keep track of this large and varied base of information and to assure that each of the traders stays current with the changes which occur continuously, a network is essential.

The network has met all of the investment bank's expectations in operation. The passive star fiber configuration has proven to be less prone to failure than thin coax linear bus arrangements where the cable itself is a single point of failure. A coax cable break can bring down multiple nodes on the network. In the fiber network, the layout of nodes can be distributed over a number of repeaters (as shown in the attached network diagram) so that a problem with the cable can at most cause the failure of only a portion of the network, not the entire network as in the coax case. The bank also observed that a single intermittent transceiver on a thin coax segment caused problems throughout the leg on which it was installed for three weeks before it was isolated as the problem. In contrast, on the fiber network, a problem with a single transceiver affects only that node.

In this application, 99% of the work involves viewing the data and only 1% in changing the data as trades are completed. As a result, each node on the network has a very large memory to store the entire data base and software locally. In this way, only the changes to the data need be sent over the network so that the throughput of information is maximized.

FUTURE PLANS

The investment banking group plans to continue the expansion of its network with fiber optics. Although the current fiber optic system has shown a reliability beyond that of coax systems, the importance of the financial data on the network and the high dollar value involved means that systems with even greater fault tolerance are desired as the technology becomes available. In addition, the size of the network and the amount of information will be continually increasing, with additional remote sites being tied into the network. An upgrade of all computers to Sun workstations is planned, to allow a greater array of information to be displayed on the screen at one time.

The availability of fiber optic Ethernet has been instrumental in allowing the bank to supply an accurate and timely service to its clients. The learning process for installation was easily manageable in a few weeks time. Fiber optics is scheduled to support future network requirements. In all, the experience with fiber optics has been a very positive one.

FIBER OPTIC NETWORK LAYOUT HORIZONTAL BUILDING WIRING INVESTMENT BANK

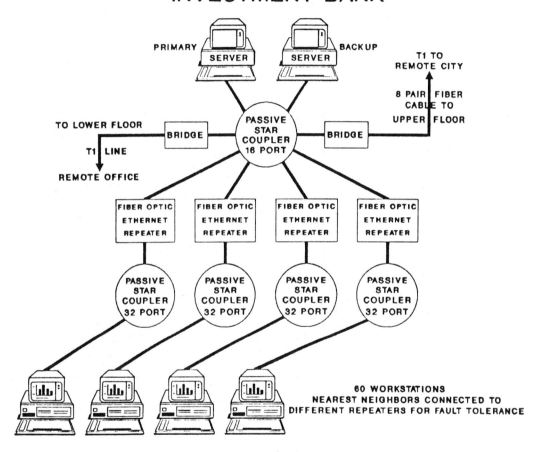

HIGH-RISE OPTICAL COMMUNICATIONS NETWORKS

David E. Stein

Southwestern Bell Telephone Company
St. Louis, Missouri USA

One Bell Center is the tallest building in the state of Missouri. It is 608 feet tall and has 1,500,000 square feet of floor space. It is 44 stories tall and has two basement levels.

Figure 1

Floors 1 thru 6, 43 and 44, and the B1 and B2 levels are building service floors.

1	- Lobby and Retail Areas
2	- Kitchens and Building Service Offices
3	- Cafeteria (1,000 Seat Capacity)
4	- Conference Floor
5	- Conference Floor and Building Mechanical
6	- Television Production and Mechanical
7 thru 41	- General Purpose Office Floors
42	- Executive Floor
43 and 44	- Building Mechanical Floors

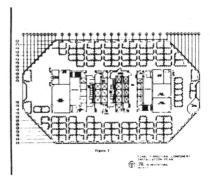

Figure 2

The building core area provides elevators, rest rooms, mail handling, copy services, vending machines, and mechanical and telephone equipment room areas. Three elevator banks are provided:

1) Low-Rise   (Floors 1 - 18)

2) Medium-Rise   (Floors 18 - 30)

3) High-Rise   (Floors 30 - 43)

The rest rooms are provided in the unused elevator lobby spaces of the low-, medium-, and high-rise elevator banks. The general office space is provided on the outer perimeter of each floor and is approximately 24,500 square feet of usable space per floor (7 thru 41). The maximum occupancy per floor is 185 work stations with the average being 130 to 140 work stations.

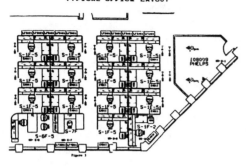

ONE BELL CENTER
TYPICAL OFFICE LAYOUT

Each of the wire closets is served by a vertical riser shaft from the B2 thru 42nd floor levels. Floor levels 7 thru 41, which are the general office space floors, are equipped with an underfloor duct system.

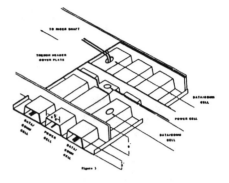

No floor-to-ceiling walls are permitted in the office areas on the north and south ends of the building. All floor-to-ceiling offices, libraries, terminal rooms, et cetera, are provided on the east and west ends of the building.

The building was not designed to provide any floor space for on-premise switching systems (PBXs) or 1A2-type key telephone equipment. Centrex lines equipped with single line telephone sets or electronic key telephone systems will serve the voice requirements of the building occupants.

Each floor is equipped with two 8' x 9' telephone closets, one on the NE end of the core area and one on the SW end of the core area. These closets also serve as AC electrical distribution closets.

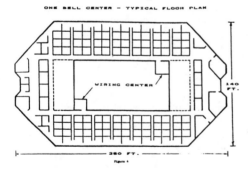

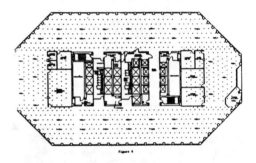

Three different types of wire closets have been provided:

1) Single door with most of the three wall surfaces available for equipment mounting.

2) Double door entry (every other floor).

3) Single door entry with 3' x 3' x 5' electrical transformer in the closet (every 5th floor).

Trench headers radiate from the two wire closets, crossing and giving access to the duct system which repeats every five feet. Each duct system consists of three cells, one for data, one for AC electrical power, and one for telephone circuits. The cross section of each duct cell is approximately 15 square inches.

In January 1983, the project of equipping One Bell Center with the most modern cost-effective communications system for the distribution of voice, data, and video began. The voice needs would be filled by single line telephone sets and electronic key telephone sets. Data transmission requirements were an unknown. It was anticipated that by 1990, over 90% of the building population would be equipped with some sort of an information terminal as well as a telephone set for voice communication.

The number of host computers and terminals had to be determined. We also had to ascertain the types of interfaces, bit rates, and protocol types which the transmission system would have to accept and deliver to the end users.

One Bell Center Data Communications Physical Inventory Summary   1

1. **EQUIPMENT SUMMARIES**

1.1 HOSTS

```
                Z  T        S  S     S     S  S     U  T
             5  1  O  M  3  Y  Y  4  W  Y  P  E     8  N  O
             5  L  W  V  B  S  S  3  A  K  C  E  H  V  3  I  T
             2  O  E  T  2  B  3  3  6  N  E  T  R  /  A  H  8  S  A
FL DEPT.     5  G  R  T  8  2  6  8  1  G  S  M  Y  1  X  P  8  C  L
42 SBC....   .  .  .  .  .  .  .  .  .  .  .  .  .  .  .  .  .  .  .
41 SBC....   .  .  .  .  .  .  .  .  .  .  .  .  .  .  .  .  .  .  .
40 SBC/LEG   .  .  .  .  .  .  .  .  .  .  .  .  .  .  .  .  .  .  .
39 SBC....   .  .  .  .  .  .  .  .  .  .  .  .  .  .  .  .  .  .  .
38 SBC....   .  .  .  .  .  .  .  .  .  .  1  .  .  .  .  .  .  .  1
37 RPA/PRL   .  .  .  .  .  .  .  .  .  .  .  .  .  .  1  1  .  .  2
36 RPA....   .  .  .  .  .  .  .  .  .  .  .  .  .  .  .  .  .  .  .
35 RPA....   .  .  .  .  .  .  .  .  .  .  3  .  .  .  .  .  .  .  3
34 RPA....   .  .  .  6  .  .  .  .  .  .  .  .  .  .  .  .  .  .  6
33 STF/RPA   .  .  .  .  .  1  .  .  .  .  .  .  .  .  .  .  .  .  1
32 STF....   .  .  .  .  .  .  .  .  .  .  .  .  .  .  .  .  .  .  .
31 PER/STF   .  .  .  .  .  .  .  .  .  .  .  .  .  .  .  .  .  .  .
30 PER....   .  .  .  .  .  .  .  .  .  .  .  .  .  .  .  .  .  .  .
29 C&T....   .  .  .  .  .  .  1  .  .  .  .  .  .  .  .  .  .  .  1
28 C&T....   .  .  .  .  .  .  .  .  .  .  .  .  .  .  .  .  .  .  .
27 C&T....   .  .  .  .  .  .  .  .  .  .  .  .  .  .  .  .  .  .  .
26 C&T....   .  .  .  .  .  .  .  .  .  .  .  .  .  .  .  .  .  .  .
25 ISO....   .  .  .  .  .  .  .  .  .  .  .  .  .  .  .  .  .  .  .
24 ISO....   .  .  .  .  .  .  .  .  .  .  .  .  .  .  .  .  .  .  .
23 ISO....   .  .  .  .  .  .  .  .  .  .  .  .  .  .  .  .  .  .  .
22 ISO....   .  .  .  .  .  .  .  .  5  4  .  .  .  .  .  .  .  .  9
21 ISO....   .  .  .  .  .  .  .  .  .  .  .  .  1  .  .  .  .  .  1
20 ISO....   .  .  .  .  .  .  .  .  .  .  .  .  .  .  .  .  .  .  .
19 ISO....   .  .  .  .  .  .  .  .  .  .  .  .  .  .  .  .  .  .  .
18 ISO/STF   1  .  .  .  .  .  .  .  .  .  .  1  .  .  .  .  .  .  2
17 ISO....   .  .  .  .  .  1  1  .  .  .  .  .  .  .  .  .  .  .  2
16 ISO....   .  .  .  .  1  .  .  .  .  .  .  .  .  .  .  .  .  .  1
15 STF....   .  .  .  .  .  2  .  .  .  .  .  .  .  .  .  .  .  .  2
14 NTW....   .  2  .  1  .  .  .  .  .  2  .  .  .  1  .  .  .  .  6
13 NTW....   .  2  .  .  .  .  .  .  .  .  .  .  .  .  .  .  .  .  2
12 NTW....   1  .  .  .  .  .  .  .  .  .  .  .  .  .  .  .  .  .  1
11 MKT/NTW   .  2  .  .  .  .  .  .  .  .  .  .  .  .  .  .  .  .  2
10 MKT....   .  .  .  .  .  .  .  .  .  .  .  .  .  .  .  .  .  .  .
09 MKT....   .  .  .  .  .  .  .  .  .  .  .  .  .  .  1  .  .  .  1
08 MKT....   2  .  .  .  .  .  .  .  .  .  .  .  .  .  .  .  .  .  2
07 MKT/STF   .  .  .  .  .  .  .  .  .  .  .  .  .  .  .  .  .  .  .
06 STF....   .  .  .  .  .  .  .  .  .  .  .  .  .  .  .  .  .  .  .
05 STF....   .  .  .  .  .  .  .  .  .  .  .  .  .  .  .  .  .  .  .
04 STF....   .  .  .  .  .  .  .  .  .  .  .  .  .  .  .  .  .  .  .
03 STF....   .  .  .  .  .  .  .  .  .  .  .  .  .  .  .  .  .  .  .
02 STF/PER  .  .  .  .  .  .  .  .  .  .  .  .  .  .  1  .  .  .  1
01 STF....   .  .  .  .  .  .  .  .  .  .  .  .  .  .  .  .  2  .  2

TOTAL        4  2  4  1  6  3  2  2  5  4  2  4  1  1  1  2  1  2  48
```

Figure 7

June 8, 1985   dcpi-s

As of today, 48 computers have been identified—a total of 18 different types.

One Bell Center Data Communications Physical Inventory Summary   2

1.2 OVERALL TERMINAL SUMMARY

FL DEPT.	DATA GENL	DEC	IBM	TELE TYPE	WANG	OTHER	TOTAL
42 SBC....	.	.	.	.	.	2	2
41 SBC....	.	.	.	.	.	2	2
40 SBC/LEG	.	.	.	.	.	2	2
39 SBC....	.	10	.	.	.	2	12
38 SBC....	.	.	.	.	.	10	10
37 RPA/PRL	.	54	.	.	.	2	56
36 RPA....	.	.	.	.	.	2	2
35 RPA....	.	.	.	.	.	2	2
34 RPA....	.	.	.	.	.	2	2
33 STF/RPA	.	1	5	5	.	4	15
32 STF....	.	.	.	.	.	2	2
31 PER/STF	.	.	.	.	.	3	3
30 PER....	.	.	.	.	.	2	2
29 C&T....	.	.	.	.	.	22	22
28 C&T....	.	.	.	.	.	2	2
27 C&T....	.	.	.	.	.	2	2
26 C&T....	.	.	.	.	.	2	2
25 ISO....	25	.	98	.	12	2	137
24 ISO....	.	.	15	6	1	23	45
23 ISO....	.	1	89	4	17	42	153
22 ISO....	.	.	75	.	4	17	96
21 ISO....	.	1	67	12	14	43	137
20 ISO....	.	.	165	.	2	2	169
19 ISO....	.	.	88	2	1	2	93
18 ISO....	.	1	68	2	27	35	133
17 ISO....	.	.	63	.	1	48	112
16 ISO....	.	4	98	8	8	42	152
15 STF....	.	4	26	16	.	113	159
14 NTW....	78	6	4	19	.	95	202
13 NTW....	.	4	5	20	.	61	90
12 NTW....	.	.	5	46	1	7	59
11 MKT/NTW	1	4	7	17	.	25	54
10 MKT....	.	10	13	21	.	13	57
9 MKT....	.	5	49	51	.	12	117
8 MKT....	.	1	65	13	.	23	102
7 MKT/STF	.	.	1	.	17	.	18
6 PRL....	.	10	.	.	.	.	10
5	.	.	.	.	.	.	.
4	.	.	.	.	.	.	.
3	.	.	.	.	.	.	.
2 STF/PER	.	.	.	.	.	15	15
1 STF....	.	.	.	.	.	1	1
TOTAL	104	122	1038	214	87	684	2249
	5%	5%	46%	10%	4%	30%	

Figure 8

June 8, 1985   dcpi-s

The present terminal population is 2,249 and grows every day as decisions to provide mini and micro computer systems are made concurrent with a department's move into One Bell Center.

It was decided that two separate communications systems should be provided:

1) A copper-based system for voice, DC, and alarm system requirements, and

2) A fiber-optic-based transmission system for data and video requirements.

Fiber as the backbone data transmission system made the most sense considering:

1) Fiber is lightweight and small in diameter. The size of the underfloor duct system and wire closets posed a severe space limitation problem in using conventional coaxial, twin-axial, and shielded copper-pair transmission systems. It also precluded pulling most pre-connectorized cables from the host computers to their associated terminals.

2) Fiber could be reused when the existing host computer and terminals were changed to a different system. Only the type of electro-optical converter would have to be changed. No new cable or connector work would be required. This would prevent duct congestion and later abandonment caused by additions,

rearrangements, and disconnection of computer systems.

3) New systems can be added with ease. Since the building is "pre-fibered," only the addition of the correct type of electro-opto converter is required. No new underfloor cabling is necessary.

4) Wide bandwidth of the fiber would allow future wave division multiplexing as this type of technology becomes available. No new fiber runs would be required.

5) Electrical isolation would prevent ground loop, EMI, RFI, and EMP problems.

In January 1984, Aetna Telecommunications consultants were engaged to provide a feasibility and fiber design study. Their recommendation was to provide a fiber transmission system that served the work stations on the floors as well as a vertical backbone high-speed system.

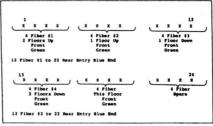

Figure 10

In addition, an extra 12-fiber cable in each of the shafts was installed from 22 to the 16th thru 25th floors, to provide for the extra data needs of the Information Systems Department. To serve the remaining floors, a 4-fiber cable was run to each of the two floors above and below the group patch panel floors and terminated in a patch panel. This gives us eight fibers from 22 to floors 1 thru 42; i.e., four fibers in each riser shaft.

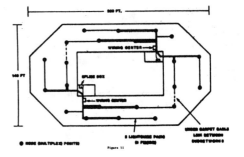

Figure 9

Refer questions to Steve Matthew 247-8059   10/30/84

A fiber-optic distribution room and a local area network equipment room were provided on the 22nd floor. A fiber patch panel was provided on 22 with two 12-fiber cables in each of the east and west riser shafts to the 4th, 9th, 14th, 19th, 25th, 30th, 35th, and 40th floors which terminated in group patch panels (splice centers) on each of these floors.

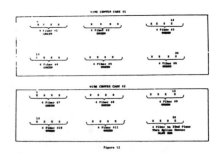

Figure 12

A horizontal fiber distribution system has been provided on floors 7 thru 41, which consists of a 4-fiber cable from the patch panel in the wire closets to 11 nodes served by that wire closet. These are the floors served by the underfloor duct system.

This gives us a total of 22 fiber nodes per floor or 88 fibers total.

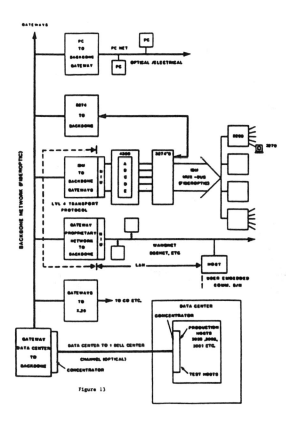

Figure 13

The ultimate plan is to interconnect all the different operating systems through a high-speed fiber backbone which will have access to the X.25 packet network, digital Centrex, et cetera.

Fiber: The fiber cable specification was for 85 micron inner conductor, 125 micron outer diameter multimode, graded index with a minimum numerical aperture of 29, optical loss of 3.7 DB/KM at 850 NM and 116 DB/KM at 1300 NM.

Fiber-Optic System Statistics:

FIBERS	TOTAL	ALLOCATED	FEET	MILES
Vertical	864 :	180 21%	244,448 /5280 =	46.3
Horizontal.2,678 :		459 17%	241,780+/5280 =	45.8+
TOTAL......3,542 :		639 18%	486,228+/5280 =	92.1+

Connectors:

The connectors chosen are all mechanical crimp-type connectors to the fiber with no fusion or elastomeric splicing required. The loss of the connectors was specified to be 1.5 DB or less and has been testing in the 0.4 to 1.3 DB range. The cables were all pre-measured and connectorized on both ends at the fac-

tory. No "on the job" splicing or connectorization has been required.

Patch Panels and Boxes:

These were all custom-designed and fabricated to meet our specifications.

Underfloor Mounting Brackets:

These were also custom-designed and fabricated to allow mounting below the floor. Patch cords are used to connect the electro-optical converters.

Figure 14

This figure shows a schematic design for a circuit from the main computer room on the 22nd floor to an associated terminal or other data communication device on any other floor in the building. The loss budget is 9.5 DB maximum.

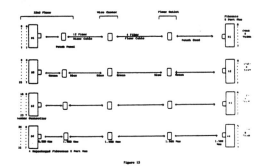

Figure 15

This figure shows a schematic design using the special cables direct from the 22nd floor to floors 16 thru 25.

I believe One Bell Center is unique in the United States. Southwestern Bell Telephone Company is pioneering an advanced communications system that will serve our information dissemination needs far beyond the year 2000.

IV

A FEW FIBER OPTICS APPLICATIONS

When Southwestern Bell planned the construction of its St. Louis headquarters—at 44 stories, the tallest building in Missouri—it also planned for a modern, cost-effective communications system for the distribution of voice, data and video signals. Selecting a twisted-pair system for voice, the company decided, for a variety of reasons, to base its data-transmission system on fiber optics. Space limitations and concerns about the ability to pull cable were overcome because the fiber was lightweight and small in diameter; at the same time, fiber optics' wide bandwidth accommodates future higher-speed network upgrades (when available) without the need to run new fiber. Additional benefits achieved by using glass transmission media include the elimination of ground loops and a total absence of electromechanical and radio frequency interference problems.

Employing products from the Bridge Communications Division of 3Com Corporation and Codenoll Technology Corporation, Southwestern Bell's entire building was cabled with 92 miles of fiber, using three floor ducts 5-3/8 inches wide by 3 inches high (dimensions that would be impossible for use with half-inch diameter standard Ethernet coax). Each floor was provided two wiring centers with fiber cable running from the wire centers to eleven clusters of eight offices each. Currently, the building contains well over 4,000 computer terminals and workstations, 224 modems and more than 60 host computers. By 1990, over 90% of the building's workers will be equipped with some sort of information terminal.

While Southwestern Bell had its eye on future expansion, Wall Street's Goldman, Sachs brokerage firm switched to a fiber-optic Ethernet last year to ensure reliability. Because fiber cable does not transmit electrical signals, there is no possibility of a short circuit or an electrical connection failure. And because it transmits light impulses, fewer parts are involved, thereby reducing the possibility of component failure.

With copper-based systems, a single electrical component failure can bring down an entire network. Such a problem had already hit the brokerage firm, causing a 14-hour trading lapse and the loss of hundreds of million dollars in transactions. While the network was under repair, Goldman, Sachs began the process of ensuring that another such failure would not collapse its entire system again.

To take advantage of fiber optics' reliability and network availability the firm supports 14 PC AT workstations with coaxial-cable Ethernet, linked via asynchronous gateways to a host through which trading is conducted via fiber facilities. The workstations, also linked through the network to Quotron Systems' financial information services, run a program that integrates a variety of trading functions in each microcomputer, thus eliminating the need for multiple terminals on traders' desktops.

As a further backup measure, Goldman, Sachs plans to run individual fibers from a Codenoll fiber-optic star coupler to each trader's desk; essentially, then, there will be two LANs—one in a star configuration and the other a ring fiber net. Use of the star wiring topology assures that failure of the media will affect only the single station connected to the spoke of the star.

In yet another case, Apple Computer employed fiber-optic technology for its own connectivity capabilities. After Apple acquired a Cray X-MP/48 computer in March, 1987, it required high-speed communications between the supercomputer and various CPUs, engineering workstations and Macintosh computers—all operating in two buildings on two different Ethernets three blocks apart from each other at the company's Cupertino, California, headquarters.

Pacific Bell, the local telephone company that designed and installed the network, proposed a configuration of fiber-optic media utilizing Codenoll equipment to tie the three Ethernets together—a "wide area" Ethernet with fully redundant high-speed links. After stringing 5,600 feet of fiber-optic cable and installing the requisite equipment, Pacific Bell engineers installed the Cray and linked it to the nodes in the two remote buildings in 15 working days.

Another company faced with linking three separate facilities was GTE Communications Systems. Each of the facilities at its Northlake, Illinois, manufacturing center—devoted to computer-aided design, manufacturing and engineering (CAD, CAM and CAE)—was linked in 1985 over a fiber-optic Ethernet. Today, design and graphics information is transmitted to a wide variety of computers and workstations (including VAXs and PDP-11s from Digital and Sun Microsystems workstations) at 10 million bits per second.

Incorporating Codenoll's star wiring concept (also known as FOCUS—Fiber Optic Cable Universal System) allows GTE to easily and efficiently change or adjust its network configuration, and to operate without the noise problems—such as those that can be caused by fluorescent lights—which can adversely affect the copper connections common in coaxial installations. Additionally, the absence of radiated emissions—one of fiber's many appealing elements—secures the network from unauthorized eavesdropping.

The need for communications security led the United States Air Force to install a Codelink-2000M system, a specialized version of Codenoll's Codelink-2000 system used in a variety of government and military fiber-optic point-to-point applications. Housed in a 19-inch chassis that can hold up to four fiber-optic channels per shelf, each shelf of the system features two modular power supplies and a status/alarm card that alerts the station to any optical failures to the shelf. All electrical connections to the shelf are surge-protected and all optical modules are completely compatible and interchangeable with all interface modules.

IV

244

Passive star based optical network for automotive applications

L. K. Di Liello, G. D. Miller, and R. E. Steele

Packard Electric Division, General Motors Corporation
P.O. Box 431, Warren, Ohio 44486

ABSTRACT

Fiber optic networks in the telecommunications and Local Area Network (LAN) applications are typically active networks (point to point fiber optic links interconnected using active devices). Short distance automotive networks have a sufficient flux budget to allow passive optical distribution. This paper reviews the differences between Telecommunications, LAN, and automotive network applications. The benefits of a passive star architecture will be described and a passive star implementation, based on large diameter plastic fibers, will be described.

1. INTRODUCTION

As the vehicle electronics content continues to increase through the addition of new electronic features and the replacement of mechanical controls with electronic control, the use of serial data links and networks will proliferate. The data rates, which started out low, will increase as the applications change from diagnostics and data sharing to control applications. As they increase, the conductor-based data distribution systems will become much more difficult to fabricate. As more uniform transmission structures are required, fiber optics will become a viable alternative. On-board fiber optic network architecture will be determined by not only functional requirements such as fault tolerance, power moding, and latency but also other considerations such as total cost and elasticity. Total cost includes piece cost and vehicle build and service costs. Elasticity is an indication of how easily the network can be expanded and contracted.

This paper will look at automotive data communications trends, compare telecommunications, LAN, and automotive network applications, examine the advantages of fiber optics with the automotive application in mind, define vehicle considerations, and present an evaluation of a passive star developed for automotive network applications.

2. AUTOMOTIVE DATA COMMUNICATIONS TRENDS

Automotive applications of serial data communications started to appear in the early 80's. Point to point, function specific applications were introduced followed by networks connecting the various vehicle computer systems. Projected growth out to the 1990's, shows the potential need for data rates in the Mega bit/sec. region. As data rates grow, transmission medium complexity increases. Today, most applications are implemented with single random laid wires. Fiber optics or any other uniform transmission medium cannot compete with a single wire on the basis of cost alone. This situation will change as data rates increase to the point where single random laid wire does not function acceptably. At higher data rates, twisted pair or twisted shielded pair will be required depending on data rate and vehicle EMC requirements. As more stringent vehicle requirements evolve, fiber optics will become a viable alternative.

3. TELECOMMUNICATIONS VS LAN VS AUTOMOTIVE

Telecommunications, LAN, and automotive applications have very distinct differences. The main reason fiber optics is used in telecommunications applications is the high bandwidth distance product which allows the long distances and high data rates to be achieved. Fiber optic links minimize the requirements for repeaters which are a high cost portion of the total system. They represent not only an initial cost but also an ongoing repair cost, especially when located at the bottom of the ocean. Telecommunication networks consist of long point to point links interconnected electronically. Passive optical distribution is not an option since the optical flux budget is consumed in the long fiber runs.

LANs are significantly shorter in length than telecommunications links, but are still quite long compared to an automotive network. A LAN can have node to node distances of up to several kilometers. Passive distribution could be an option depending on distances involved and the available optical flux budget.

Automotive applications can be classified as very short distance LANs. Node to node distances in excess of 10 meters would be difficult to comprehend. With these short distances, fiber losses are negligible even with plastic fiber. This means the majority of the optical flux budget is available for passive distribution. Passive distribution has a number of added benefits which will be discussed later.

4. ADVANTAGES OF FIBER OPTICS

The typical advantages of fiber optics are weight, size, electromagnetic compatibility (EMC), dielectric nature, and bandwidth. When applied to the automobile these advantages must be reassessed using the automotive application as the criterion.

4.1 Weight

A fiber replacement for a cable containing 3200 twisted pair is a clear winner, but a single copper conductor or a twisted shielded pair are not heavier when compared to an automotive grade glass or plastic fiber optic cable.

4.2 Size

Individual fibers are quite small, but by the time a fiber is strengthened to the point it can survive the total automotive environment, it will be comparable to the data wires in the vehicle today. This condition will change as the data rate increases and the need for a twisted or twisted shielded or coaxial transmission medium becomes necessary.

4.3 EMC

Due to the many electronic systems, including a sensitive radio receiver, being packaged in the close confines of the automobile, EMC is a very significant factor. With fiber optics, the fiber can be routed anywhere without causing a problem in another system or having another system interfere with its operation.

4.4 Dielectric

One of the common faults in the automobile is a short to ground. Since a fiber cannot be electrically shorted, it is possible for the majority of the bus to continue to function. This will depend heavily on the network architecture.

4.5 Bandwidth

Data communications in the automobile are still very slow compared to telecommunications and LANs. Automotive communications data rates today are less than 10K bits/sec. Projections of past growth would put this at 1M bit/sec. in the not too distant future. These data rates are not sufficiently high to justify conversion to fiber on the basis of bandwidth alone.

The typical advantages of fiber optics do not all apply to the automotive application. Automotive functional requirements are not as severe as those in the Telecomm or LAN applications. As data rate requirements for automotive applications increase, EMC becomes an increasingly significant issue. The main issue is cost effectively attaining the required level of electromagnetic compatibility.

5. VEHICLE NETWORK ARCHITECTURE CONSIDERATIONS

There are a number of possible architectures that could be implemented in an automotive network: the single and double ring, the active and passive star, and the linear tapped bus. The issues to be used in the decision process are: cost, complexity, power moding, fault tolerance, expandability, serviceability, and latency.

5.1 Cost

Total system cost (not just initial piece cost) will determine which network architecture will prevail. Initial cost will be the combination of the fiber ends, receiver-transmitter pairs, and fiber length. Total cost will include assembly, warranty, and service costs. A relative comparison is made here to determine system advantage or disadvantage. (The lowest cost system will have the fewest components, one receiver-transmitter pair per node, and one fiber per node.)

Cost factors that can be readily quantified are fiber ends, receiver-transmitter pairs, fiber length, and any other significant system component differences.

5.1.1 Fiber ends. Due to the large number of fiber ends to finish in an Automotive network, the number of fiber ends will be proportional to a significant part of the system cost and be inversely proportional to reliability (the more mated ends, the more opportunities for a fault to occur).

5.1.2 Receiver-transmitter pairs. Active devices are costly components. Minimizing receiver-transmitter pairs will minimize cost and also result in reduced number of failure modes.

5.1.3 Fiber length. Although the amount of fiber per vehicle will be small, it still represents a cost factor that is easily quantified.

5.1.4 Distribution components. Additional passive or active components required by the system could have a significant cost impact.

5.2 Physical Complexity.

Physical complexity is an issue that will impact the vehicle assembly plant operation. A physically complex system will be difficult to assemble reliably. This could add to assembly costs, dealer warranty cost, or owner maintenance cost. (The least complex system will avoid concentrations of complexity and will have the fewest fibers, nodes, and receiver-transmitter pairs)

5.3 Power Moding

The electronic systems on the vehicle are not all powered up at all times. Some systems may be purposely powered down during a portion of the time while some systems will be maintained on for a time after the vehicle is turned off. Times such as ignition off, crank, or accessory operation require only a portion of the vehicle system to be operational to conserve battery charge. The network should allow for this reduced level of network operation. (Ideally the network should operate with any combination of nodes powered.)

5.4 Fault Tolerance

Continued network operation under a fault condition is a desirable feature. The tolerance of a system to faults could impact the type of functions that could be included on the network. The more tolerant the network is to faults, the more critical the functions that can be implemented on it. (Ideally the network will continue to function with any single point fault)

5.5 Expandability

Expandability for vehicle variability, future vehicle improvements, and dealer installed equipment must be allowed for in the initial design. Easy network expansion and contraction is a desirable feature. (Ideally the network will be expandable without cost penalty to the lesser system)

5.6 Serviceability

Second only to initial functionality, serviceability is a very important issue. The vehicle must be serviceable within the existing service network. This will include local service stations. (Ideally the signal distribution network will permit full access to diagnostic information.)

5.7 Latency

The time between data updates will impact the type of function performed on the network. Control applications will require rapid data exchange updates. (The best system will have minimum transmission latency)

6. Architecture Evaluation

The transmissive star approach has the most benefits of the networks considered and the fewest negative factors. The benefits of the transmissive star are: low cost, portions of the network can function independently, the star is not susceptible to faults and a fault in the remainder of the network will only affect one node, it has the lowest latency, and it can be designed with spare expansion ports. The negative factors of the transmissive star are: the added star element and the complexity of the interconnecting fibers at the star.

Although the single ring is simpler and least costly, its longer latency and lack of fault tolerance make it undesirable. Even though the double ring has the advantage of complete redundancy, cost, complexity and latency rule it out. The active star and the linear tapped bus are both expensive and add a significant level of complexity to the system in the form of added components in addition to long latency.

There are a number of methods to build a passive transmissive star, twisted biconic, slab, mixing rod, etc. The remainder of this paper will concentrate on the evaluation of a plastic fiber star using the mixing rod approach.

7. TRANSMISSIVE STAR REQUIREMENTS

Components and systems that are designed for the automotive applications must satisfy a number of requirements. In addition to meeting the functional requirements of distributing optical energy among the system nodes, the star must also be dimensionally compatible with the vehicle physical constraints, manufacturable, assemblable, installable, environmentally compatible, and serviceable all at a very low cost.

7.1 Functional

Using low cost devices currently on the market, the available flux budget is about 18 db. Allowing for fiber loss (2 db), LED life (3 db), routing losses (1.5 db), the flux budget available for splitting is 11.5 db. In an ideal 7 node star the splitting loss is 1/7 or 8.5 db. the remaining 3.0 db is available for excess loss in the splitting element. In addition the star must divide the light evenly, that is insertion loss must not vary by more than 1 db.

7.2 Dimensionally Compatible

Many locations in the vehicle are already tightly cramped. The size of the device size will impact the potential vehicle packaging locations which could in turn impact system cost. If the device is too large, it could preclude its use. A package larger than 210 mm X 130 mm X 40 mm would most likely be unpackageable in any area of the vehicle except the trunk.

7.3 Manufacturable

Automotive volumes require components that can be manufactured at very high rates. Component designs must be capable of being manufactured at low cost, with excellent quality, and high yield. Tweaking and sorting are undesirable.

7.4 Assemblable

The star must be designed so that it can be integrated into the existing power and signal distribution system. The star could be integrated in one of two fashions. First, as a separate component which could be built into the vehicle and connected to the distribution fibers or, as a part of the power and signal distribution system and shipped as a part of the vehicle wiring assembly. The second scenario is desirable because it would allow the optical distribution system to be fully tested prior to installation into the vehicle.

7.5 Installable

The star must be robust enough to withstand the rigors of the assembly line installation. It should also be configured such that when assembled into the vehicle it does not add significant routing or packaging complexity.

7.6 Environmentally Compatible

The automotive passenger compartment environment consists of temperatures extremes from $-40^{\circ}C$ to $+85^{\circ}C$, humidity, shock and vibration, and dust and dirt. The final star assembly must operate over the life of the vehicle under all expected conditions.

7.7 Serviceable

The star design must be serviceable within the existing service environment. Service operations should not require an excessive amount of equipment or be beyond the expected capabilities of trained service personnel.

8. TRANSMISSIVE STAR IMPLEMENTATION

There are several recognized approaches to making passive transmissive stars. These include twisted biconic, waveguide structures created using photolithographic techniques, and mixing elements created by molding or drawing. A low startup cost and easily producible method was desired for this implementation.

8.1 Selection of Approach

The twisted biconic approach involves a complicated, low yield process. Special care must be taken throughout the process to ensure uniform splitting of the light. These difficulties are not encountered with the waveguide and mixing element approaches.

The waveguide structure approach involves a high yield, highly repeatable process. This process, however, requires complicated equipment and technology not needed for the mixing element approach. A difficulty also exists in assuring uniform cladding of the waveguide. This is not a problem when using a mixing element produced by conventional techniques. For these reasons a mixing element approach was chosen for this implementation. A transmissive star was designed and 12 prototypes manufactured using this approach.

8.2 Selection of the Mixing Element

The star uses a 3mm diameter plastic fiber as the mixing element. This element was chosen because the fiber is readily available and mates nicely with a circular bundle of seven 1mm diameter fibers as shown in figure 1, thus creating a 7x7 star. Since the cladding thickness on both the fibers and the mixing element is only 0.02mm, it does not cause excessive loss. Thus it is not necessary to remove the cladding as it is when using glass fibers.

The mixing rod used was 60mm in length. It is desirable to keep the mixing rod as short as possible to minimize attenuation and overall physical size of the star. At the same time making the mixing rod too short will not provide sufficient mixing of the light. Computer modeling using a ray tracing approach was used to determine the optimum length. Figure 2 shows a graph of mixing rod length versus minimum, average, and maximum insertion loss. Insertion loss is the attenuation between the input port and any one of the output ports. Beyond a length of 60mm the curves have reached a steady state, that is a longer length does not provide better mixing. A peculiarity of using a circular mixing rod is that the minimum insertion loss deviates much farther from the average than does the maximum insertion loss. This can be seen in figure 2. This minimum insertion loss always occurs when light is input at the center fiber and received by the opposite center fiber.

8.3 Assembly

Figure 3 shows the star in various stages of assembly. A brass tube was placed over the length of mixing rod as shown in figure 3 to protect it and maintain axial alignment. The 14 fibers were cut to the necessary length and the jacket stripped back several millimeters. Molded plastic ferrules were used to fix the ends of the mixing rod and the two 7 fiber bundles. The ferrules were epoxied on, then the end faces were polished. Brass compression fittings were used to hold the mixing rod and fiber bundle together.

8.4 Packaging

Once assembled, the star was packaged inside a protective box. It was arranged inside the box so that the input and output fibers all exit from the same opening. This makes routing of the fibers away from the star much neater. Overall packaged size of the star was 200 mm X 76 mm X 25 mm.

9. TRANSMISSIVE STAR TEST DATA

9.1 Test Method.

Each of the 12 prototype stars was tested as shown in figure 4. Input optical power P_o is measured with a one foot reference fiber connected directly between the LED and the optical power meter. One of the seven input fibers is then connected to the LED and optical power output is measured at each of the output fibers. The star is also measured in the opposite direction, that is with the input and output fibers reversed. Insertion and excess losses are then calculated as shown in figure 4. These values actually represent the node-to-node loss including loss due to attenuation in the fiber (approximately 0.2 db per meter). Loss due solely to the star is found by subtracting out fiber loss. For example if the node-to-star distance is 3 meters, 0.2 db / m * 6 m = 1.2 db is subtracted from the measured values.

9.2 Insertion Loss

Figure 5 shows the minimum, average, and maximum insertion loss for the 12 prototype stars. The average insertion loss for all 12 stars is 11.4db which meets the performance requirement. However the variation between the minimum and maximum insertion loss is much greater than the desired 1db. In each case the minimum insertion loss occurs when light enters through the center fiber and is measured at the center fiber. This value is much less than all other insertion loss measurements as was the case with the computer model discussed earlier. In addition the maximum insertion loss occurs when the center fiber and one of the outside fibers is involved. Thus while mixing uniformity is very good between the 6 outside fibers, it is not as good when the center fiber is included.

9.3 Excess Loss

Excess loss is a measure of how much of the input light actually reaches one of the output fibers. The greatest amount of light is lost due to packing fraction loss. The area of the end face of the 3mm diameter mixing rod is 9*(PI/4) while the combined area of the seven fibers is 7*(PI/4). Thus if light is uniformly distributed across the face of the mixing rod, 7/9 of it will be received by the seven fibers. This is equivalent to an excess loss of 1.09db. Other sources of loss include Fresnel reflection loss at the interfaces and attenuation in the mixing rod core. Figure 6 shows the minimum, average, and maximum excess loss for the 12 prototype stars. The average excess loss for all 12 stars is 2.89 db which meets the performance requirement.

9.4 Environmental Testing

One of the stars was also measured at $-40^{\circ}C$ and $+85^{\circ}C$. Figure 7 shows the excess loss measurements at all three temperatures. In general there is more loss at $-40^{\circ}C$ than at room temperature and less loss at $+85^{\circ}C$ than at room temperature. However the difference, about 0.5db, is not significant because the devices compensate for this loss over temperature.

10. CONCLUSIONS

Automotive data communication applications are expected to increase in number and complexity. As data rates increase to Mega bits/sec, where twisted or twisted-shielded wires are required in order to meet vehicle EMC requirements, fiber optics will become a viable alternative.

Fiber optics are used in telecommunications applications primarily because of the high bandwidth-distance product. These applications cover very long distances and most of the flux budget is consumed by fiber losses. LANs, while shorter than telecommunications networks, are still much longer than automotive networks. In the automotive networks, distances are on the order of 10 meters. At these lengths the flux budget is not consumed by fiber losses and is available for passive distribution.

The typical advantages of fiber optics, such as weight, size, bandwidth, and EMC, do not all apply in an automotive application. The main advantage which will make fiber a viable alternative is the capability of cost effectively attaining the required level of electromagnetic compatibility.

When choosing a network architecture for an automotive application, several factors are taken into consideration. These include cost, physical complexity, power moding, fault tolerance, expandability, serviceablity and latency. Based on these factors the transmissive star architecture holds the most advantages. To be useful in an automotive application, the transmissive star must meet certain requirements in the areas of functionality, size, manufacturablilty, assemblability, installability, and environmental compatibility.

A prototype transmissive star was designed and built using a 3 mm diameter mixing rod. It met most of the original requirements. It functioned within the available flux budget over the temperature range and was well within the dimensional constraints. The components used were low in cost and, based on the prototypes built, the yield should be very high in production.

There are, however, several issues that need to be addressed. The manufacture and build of the pigtailed assembly in a high volume situation would be challenging. In addition, increasing the node count beyond seven with a circular mixing rod will adversely affect the excess loss and increase the splitting loss, requiring a larger flux budget.

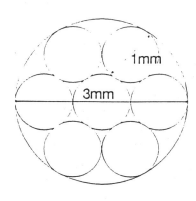

Figure 1. Arrangement of seven 1 mm fibers with 3 mm diameter mixing rod.

Figure 2. Insertion loss versus mixing rod length.

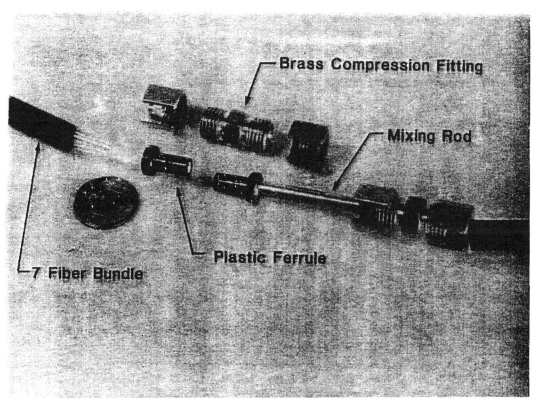

Figure 3. Transmissive star during assembly.

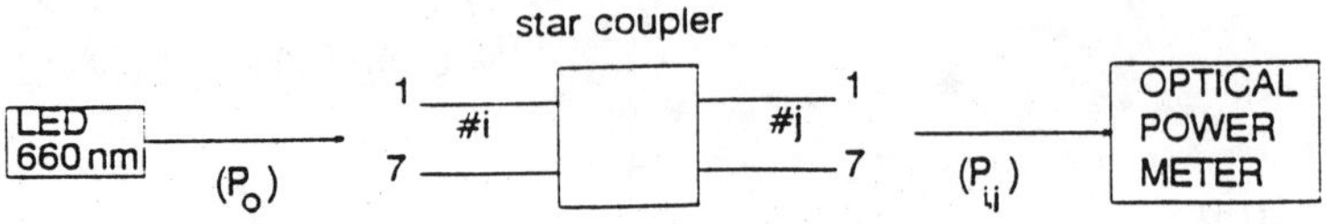

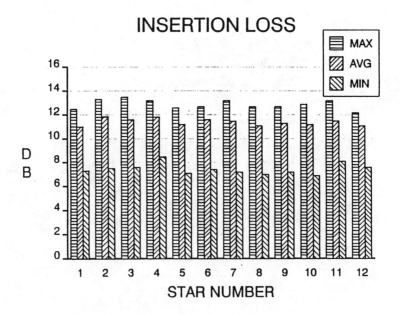

$$(\text{Excess Loss}) = -10 \log \left(\sum_{j=1}^{7} P_{i,j} / P_o \right) \qquad 1 \leq i < 7$$

$$(\text{Insertion Loss}) = -10 \log (P_{i,j} / P_o) \qquad \begin{array}{l} 1 \leq i \leq 7 \\ 1 \leq j \leq 7 \end{array}$$

Figure 4. Transmissive star test method.

Figure 5. Insertion loss of 12 prototype stars.

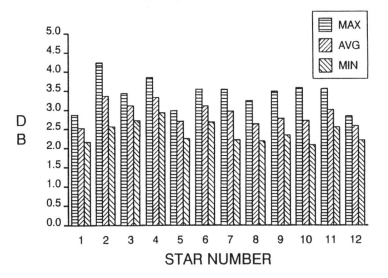

Figure 6. Excess loss of 1a prototype stars.

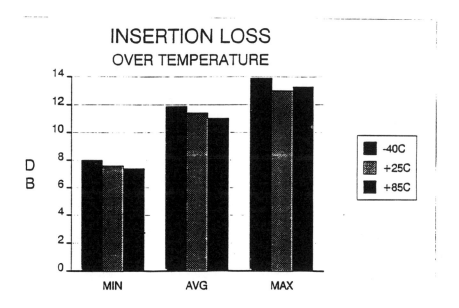

Figure 7. Insertion loss over temperature.

IV
258

INTRODUCTION TO CSMA/CD NETWORK DESIGN

Dr. Walter B. Hatfield

ABSTRACT

This paper describes the factors which determine the physical extent of an IEEE 802.3 CSMA/CD (Ethernet) Local Area Network (LAN). Two simple techniques, one tabular and the other graphical, for verifying the conformance of a LAN to the IEEE 802.3 standard are demonstrated for the two important cases of a fiber optic passive star building LAN and a fiber optic passive star campus backbone. Data is presented which allows these design techniques to be easily extended to any fiber optic or mixed media LAN.

1. INTRODUCTION

1.1. CSMA/CD Media Access Method

A CSMA/CD Network is based on the following media access method:
When any node (commonly referred to as a DTE) on a common data bus wants to transmit a packet, it first listens to determine if anyone else is currently transmitting. If the network has been quiet for at least an interframe spacing time, the node transmits immediately. If not, the node defers until the network has been quiet for an interframe spacing time and then begins transmitting. It is possible that more than one node will start transmitting when the network is quiet. Therefore, transmitting nodes are required to monitor the media to determine if a collision has occurred. (A collision is defined as two or more nodes simultaneously transmitting.) When a transmitting node detects a collision, it enforces the collision by continuing to transmit to ensure that all other transmitting nodes will detect the collision, and then stops transmitting. The node then waits for a random amount of time chosen in accordance with the backoff algorithm, defers if necessary, and again attempts to transmit. A collision is thus a naturally occurring event in a CSMA/CD network; however in most operational networks the probability of a collision taking place is fairly small.

Actually, it is not sufficient for a node to wait until the network is quiet and then begin to transmit immediately. There must be some time between packets on the network so receiving nodes can distinguish that the prior packet has ended and prepare to receive a new packet. The IEEE 802.3 standard requires that a node wait at least the interpacket gap time of 9.6 microseconds after the network becomes quiet before beginning to transmit.

1.2 Frame Format

In the IEEE 802.3 CSMA/CD standard, the following distinction is made between frames and packets:

Frame: Consists of the Destination Address, Source Address, Length Field, LLC Data, Pad, and Frame Check Sequence.

Packet: Consists of a frame, as defined above, preceded by the Preamble and the Start Frame Delimiter.

The format for an IEEE 802.3 packet or frame is shown in Figure 1. In order of transmission, it consists first of a Preamble (7 octets*) and a Start Frame Delimiter (1 octet). These are followed by Destination and Source Addresses (6 octets each) and an indication of the Frame Length (2 octets). Finally there is the Data Field followed by the Frame Check Sequence (4 octets), which contains a Cyclic Redundancy Check (CRC) value for the frame.

In general, the link layer is only concerned with frames. The physical layer is usually not aware of frames and deals with data in packet form. It is important to realize that data is handled and viewed differently depending upon what layer is involved and whether the data is transmitted or received for a particular DTE. That is, a transmitting DTE perceives data differently than a receiving DTE.

With that in mind, let us discuss two parameters that determine the constraints on network timing; Network Acquisition Time and Minimum Frame Length.

1.3 Network Aquisition Time

As discussed above, we have seen that not all transmissions are made without suffering a collision. Collisions may occur within a window of time beginning with the first Preamble bit transmitted by a DTE. The duration of this window is called the Network Acquisition Time. This concept has also been referred to as the Network Round Trip Delay. It is important to realize the the Net Acquisition Time is not the same as the Minimum Frame Length. Minimum Frame Length will be discussed in the next section.

What is the significance of Network Acquisition Time?

This time window is important to a transmitting DTE. The window represents the period of time in which a collision may occur as signaled by the MAU of the transmitting DTE. If the DTE transmits for a period exceeding the Network Acquisition Time window without having a collision, then the DTE is said to have acquired the network. The DTE will send the remainder of its frame without the possibility of having a collision. By this time, all DTEs in the network have detected network activity and are deferring to it.

* 1 octet = 8 bits
  1 bit   = 100 nanoseconds = 0.1 microseconds

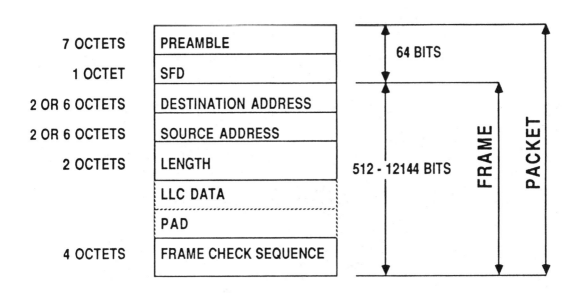

FIGURE 1. ETHERNET PACKET FORMAT

Network Acquisition Time is determined by the worst case round trip delay of the network. Small networks (in terms of time) have a small Network Acquisition Time. If a collision is going to occur, it will occur relatively quickly. For example, the Network Acquisition Time on a single segment 500 meter coaxial network is approximately 10.8 microseconds.

Consider a situation where a DTE suffers a collision at the very last instant before the Acquisition time window expires. What does the transmitting DTE do? In accordance with the IEEE 802.3 procedures, the DTE will jam the network for a short period of time and quit sending. The jam period has been set by IEEE 802.3 to be 32 bits (3.2 microseconds), provided the total transmission time for the DTE exceeds 96 bits. In any event, it is clear that the transmitting DTE will send for a period of time exceeding the Network Acquisition Time.

In the next section we will consider how the receiving DTE perceives the situation described above.

1.4 Minimum Frame Length

A receiving DTE examines the validity of a packet by first discarding the Preamble and Source Frame Delimiter as containing no further useful information. It then verifies that the resulting frame (1) contains an integral number of octets; (2) is between the min Frame Size (measured from the Destination Address) of 64 octets (512 bits) and the max Frame Size of 1518 octets (12144 bits), see Fig. 1; and (3) generates a CRC value identical to the one received. Frame reception ends when Carrier Sense indicates that there is no longer activity on the line. Because there are two (or more) transmitters on the network at the same time during a collision, it is difficult to predict the bit stream a receiving DTE will decode from the collision. Another thing to consider is that the number of bits received by the DTE is dependent upon the combination of the bit streams from each transmitting DTE. As IEEE 802.3 does not require the receiving DTE to monitor the Control In (Collision) line during frame reception, the DTE can not determine via the MAU whether there has been a collision. Instead, the receiving DTE uses the received frame length to check if the received frame had a collision. If the frame length is less than the Minimum Frame Length, then the receiving DTE knows the frame was involved in a collision and discards the data as invalid.

The Minimum Frame Length has been set by IEEE 802.3 at 512 bits. This number applies to both transmitted and received frames. All valid frames transmitted must be at least 512 bits in length. For a receive frame to be valid, it must contain at least 512 bits. The maximum length of an IEEE 802.3 network is based upon maximum collision fragment length. In order for the IEEE 802.3 network to work as intended, the network size must be limited to insure that no collision causes a received frame to be in excess of 511 bits in length.

1.5 Repeater Rationale

This section describes why a repeater is necessary to expand an IEEE 802.3 network to its maximum allowed length in time.

We have seen that the overall length of the network is constrained by a time delay budget. Different components, performing different functions within the physical layer, each contribute to the network delay. Unfortunately, time is not the only constraint on the size and topology of the network.

All the copper media approved by IEEE 802.3 have constraints which make it impossible to construct a single segment maximum length network. The coaxial cable, for instance, is not an ideal transmission line. It causes the signals transmitted over it to be attenuated. This causes the data to be more susceptible to corruption because of noise. DC losses in the coaxial cable cause the reliability of collision detection to decrease. The low pass characteristics of the cable also cause timing distortion in the data which makes the data more difficult to decode at the receiving DTE. Similar restrictions apply in varying degree to Twisted Pair Media.

In the case of fiber optics, the distance a signal may propagate before noise effects become important is related to the difference (in dB) between the optical power coupled into a fiber at the transmitter and the sensitivity of the fiber optic receiver. With current device technology, this difference can easily be as much as 30 dB. Given that the attenuation of an optical fiber might typically be 4dB/Km, it can be seen that transmission distances of several kilometers are technically feasible, even when other loss factors such as optical connectors and couplers are considered. In principle then, a single fiber optic LAN segment could cover a distance comparable to or even exceeding the maximum extent allowed by the timing considerations contained in the IEEE 802.3 standard. An understanding of these considerations therefore becomes quite important in the design of fiber optic based LANS.

These constraints have been considered by the IEEE 802.3 committee in designing the network. For instance, it has been determined that the maximum length of any one segment of coaxial cable is 500 meters. Two fiber optic LAN standards are currently (1989) being developed by the IEEE 802.3 committee. Both are based upon a star topology. In one case the star is a passive, power splitting device while in the second the star is active. That is, the incident optical signals are regenerated in the star. In the passive implementation, the committee is considering separations between DTE's of up to 1000 meters. For active stars, where the signal is regenerated, the maximum spacing between DTE's might be as great as 4000 meters depending upon timing considerations.

As stated earlier, the Network Acquisition Time for a single segment 500 meter coaxial network is only 108 bit times. In order to expand a network out to its maximum allowed length in time, a repeater may be required to compensate for the restrictions imposed by the physical properties of the medium. From this discussion, it is clear that the repeater must be capable of at least two functions:

1. Receive and decode data under worst case noise and signal amplitude conditions.

2. Transmit the received data with re-established amplitude and timing.

It is not possible to design repeater sets (a repeater and its associated MAU's and AUI cables) which are completely transparent to the data. Even if a single DTE were to transmit a uniform stream of packets separated by the IEEE 802.3 mandated minimum inter-packet gap of 96 bits, the inherent variability and non-instantaneous response times of the repeater sets would give rise to a reduction, or shrinkage, in the inter-packet gap spacing. To assure that this shrinkage does not extend to a point where a DTE is unable to distinguish between valid packets, the IEEE 802.3 standard limits the maximum number of repeater sets in a network to four.

2. NETWORK DESIGN

2.1 Example Network

Consider the CSMA/CD Network of Fig. 2. The network is assumed to incorporate optical fiber as the data transmission media and consists of a Codestar 20XX Passive Fiber Optic Star Coupler (PFOSC) connected to a second layer of PFOSC through a Codenet 4300 Multiport Repeater containing Codenet 3331 Passive Star Modules. In other, more descriptive, terms individual work stations are connected through the "horizontal" building wiring to workgroup hubs (PFOSC1, PFOSC3). These in turn are connected to a central hub (PFOSC2) through the building's "vertical" wiring system. In an idealized situation, one might envision a single central hub per building with a workgroup hub located in a wiring closet on each floor. If implemented entirely with 32 port PFOSC (CS2032), this network could accommodate up to 992 nodes or DTEs. Here we will focus on three DTE's. DTE 1 and 2 are located on the same PFOSC. DTE 1 and 3 are assumed to have the maximum separation allowed by the topology. They are further assumed to be connected to the network through Codenet 3030 stand-alone MAUs and AUI cables while DTE 2 is assumed to contain an integrated Codenet 3051 MAU. There is no significance to the equipment choice in this example, they are merely illustrative. At this point the fiber optic cable lengths, L1-L7, are unspecified.

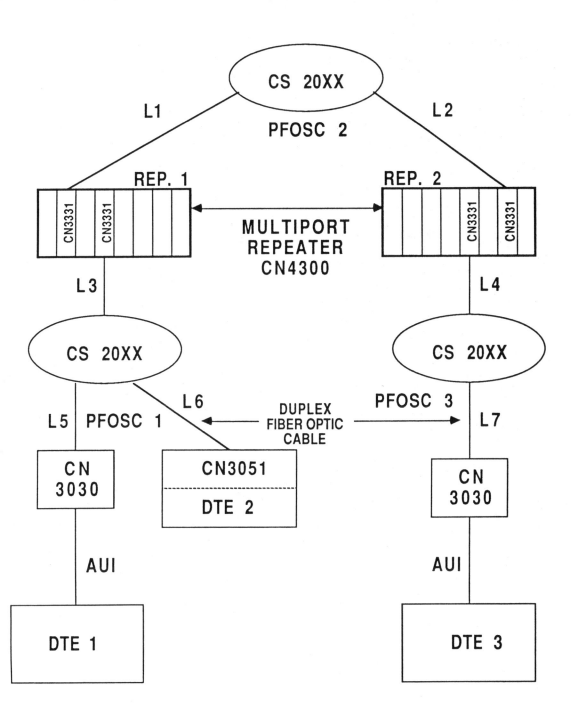

FIGURE 2. EXAMPLE PASSIVE STAR FIBER OPTIC CSMA/CD NETWORK

Consider the following scenario:

(1) DTE 1 transmits to DTE 2, also on PFOSC 1.

(2) A transmission from DTE 3 collides with the transmission from DTE 1.

(3) DTE 3 is assumed to be the worst case distance from DTE 1 and its transmission just misses deferring to the DTE 1 message.

(4) When a DTE detects the presence of a collision, under the IEEE 802.3 standard it will continue to send a collision enforcing JAM signal for a minimum of 32 bits or until its total transmission exceeds 96 bits. The 96 bit collision fragment from DTE3 then travels back down the network to inform DTE 1 that a collision has occurred on its message.

The maximum network extent is not determined by a single parameter, but rather is constrained by two conditions, one applied to participants in a collision (transmitting stations) and the other to innocent bystanders (non-transmitting stations).

In order for DTE 1 to detect the presence of a collision, it must still be transmitting when the collision fragment arrives from DTE 3. With reference to Figure 1, this says that the round-trip signal propagation time must be less than the minimum packet length (i.e. < 576 bits). When DTE 1 detects a collision, it does not stop transmitting immediately. Rather it will transmit a packet whose length is given by

DTE 1 Packet Length = Round Trip Signal Propagation Time

+ Delay Between SQE Present and Start of JAM

+ JAM Signal

The second term is identified in TABLE I as D11 and is equal to 10 bits. The JAM signal is given above as 32 bits.

The message reaching the listening station, DTE 2 (from either DTE 1 or DTE 3 through REP1), must be short enough that DTE 2 can ignore it on the basis of its being too short. Again, from Figure 1, this says that

Message Length at DTE 2 - Preamble - SFD < 512 bits.

The two preceding paragraphs yield two conditions which determine the maximum permissible network extent:

Condition I: The maximum round-trip signal propagation time must be < 534 bits.

Condition II: The maximum length message reading the addressed station, from whatever source, must satisfy the requirement that

Message Length - Preamble - SFD < 512 bits

Having determined that network extent is limited by signal propagation time, we will next consider the contributors to signal propagation time and tabulate them for various Medium Attachment Unit and Baseband Medium Specifications. Following that, a calculational procedure which results in a straight forward evaluation of Conditions (I) and (II) will be described.

In principle every CSMA/CD network should be verified for conformance to Conditions (I) and (II). For the original coaxial (10 BASE5) Ethernet, a generic maximum extent network was defined -- the well-known 5 segment/4 repeater network. Any 10BASE5 network which falls within the parameters of this generic maximum extent network will satisfy Conditions (I) and (II). Unfortunately, generation of generic maximum extent networks has not kept pace with the proliferation of CSMA/CD media and mixed media networks. Verification of Conditions (I) and (II) now rests largely with the designer of such networks. In this Application Note, a generic maximum extent passive star fiber optic network will be defined. We will also explore the bounds of a network application of particular interest - that of a fiber optic campus backbone.

2.2 Delay Times

In this section, the delay times associated with the various IEEE 802.3 or Ethernet system components are tabulated (Table I). In addition to passive (10BASE-FP) and active (10BASE-FA) fiber optic star LANs, delay times are also tabulated in Table I for standard coax (10BASE5), thin coax (10BASE2), twisted pair (10BASE-T) LAN implementations as well as the fiber optic Inter Repeater Link (FOIRL). The 10BASE-FP and 10BASE-FA fiber optic standards are currently under development in IEEE 802.3. Consequently, the delay time values for these LANs are tentative. The delay times for Codenoll's current passive star LAN product (CN3030C) are also given in Table I. In Table I, all the delays are expressed as "bit times" (1 bit time = 100 nanoseconds).

TABLE I

MEDIUM

MEDIUM	SYMBOL	PROPAGATION VELOCITY	MAXIMUM SEGMENT LENGTH (METERS)	START UP DELAY (BITS)	LAST IN TO LAST OUT DELAY (BITS)
10BASE5	C1	.77C	500M	0.0	21.65
10BASE2	C2	.65C	185M	0.0	9.50
10BASET	T1	.59C	100M	0.0	5.61
FOIRL	F1	.65C	1000M	0.0	50.00
10BASEF	F1	.67C	1000M	0.0	50.00
AUI	A1	.67C	50M	0.0	2.57

SYMBOL	DESCRIPTION	START UP DELAY (BITS)	LAST IN TO LAST OUT DELAY (BITS)
	STATION EQUIPMENT (DTE)		
D2	Output-->DataOut Assert	3	
D7	SQE Assert-->Signal Status=Error	3	
D8	SQE Deassert-->Signal Status=No Error	5	
D10	Commit-to-Send to DataOut	10	
D11	Signal Status=Error-->Jam Output	10	
D12	Jam Output	32	
	MEDIUM ACCESS UNIT (MAU)		
M1	DataIn-->DI Assert		
	10BASE-FP	5.5	3.5
	10BASE-FA	2.0	
	FOIRL	3.5	
	10BASE5	6.0	0.5
	10BASE2	3.0	0.5
	10BASE-T	6.0	0.5
	CODENET 3030C	3.5	0.5
M2	DO In-->DataOut Assert		
	10BASE-FP	6.5	3.5
	10BASE-FA	10.0	7.0
	FOIRL	3.5	
	10BASE5	3.0	0.5
	10BASE2	6.0	0.5
	10BASE-T	3.0	0.5
	CODENET 3030C	3.5	0.5

SYMBOL	DESCRIPTION	START UP DELAY (BITS)	LAST IN TO LAST OUT DELAY (BITS)
M3	DataInCollision-->SQE Assert		
	10BASE-FP		
	Electronic (ORX Before DO)	4	
	Non-Synchronous Packets	13	
	Synchronous Packets	45	
	10BASE-FA		
	FOIRL	3.5	
	10BASE5	17.0	
	10BASE2	9.0	
	10BASE-T	9.0	
	CODENET 3030C		
	Electronic (ORX Before DO)	4.0	
	Non-Synchronous Packets	9.0	
	Synchronous Packets	---	
M4	Collision Deassert-->SQE Deassert		
	10BASE-FP		
	Transmit	35.0	
	Receive	35.0	
	10BASE-FA		
	FOIRL	7.0	
	10BASE5	20.0	
	10BASE2	20.0	
	10BASE-T	9.0	
	CODENET 3030C		
	Transmit	4.0 (from DO deassert)	
	Receive	160.0 (from CI assert)	

10BASEF - FIBER OPTIC HUB

PASSIVE	HP1	Optical DataIn-->Optical DataOut	0	0
ACTIVE	HA1	Optical DataIn-->Optical DataOut	9	

REPEATER

R1	Input 1,2-->Output 2,1	8.0	
R2	Input Idle 1,2-->Output Idle 2,1		12.5
R4	SQE-->JAM	6.5	
R5	Jam Output (Fragment Extension)--> Output Idle	96	

3. NETWORK CALCULATIONS

There are two techniques for making network extent calculations. In the tabular technique, tables are generated which list and sum the transit delays seen by signals traveling certain critical paths. The same result can be obtained by a graphical or space/time diagram method. Inspection of these space/time diagrams often brings out aspects of network performance not otherwise apparent.

Many find it most effective to combine the two techniques. In the following sections we will apply the tabular and graphical techniques to two cases: a maximum extent all fiber optic passive star network (3.1) and a passive star fiber optic LAN backbone (3.2).

3.1 Maximum Extent Fiber Optic Passive Star LAN

The maximum extent fiber optic passive star network currently proposed in the IEEE 802.3 10BASE-F Task Force is the passive star fiber optic network of Figure 2 with all fiber optic cable lengths L1-L7 taken to be 500m. The AUI cables are taken to their maximum length of 50m. Such a network supposes an optical receiver sensitivity limited passive star segment consisting of a passive star hub of up to 21 dB attenuation (approximately 32 ports), two patch panel connectors and up to 500 meters of 62.5/125 duplex fiber optic cable between each MAU and the HUB.

We first generate tables which list and sum the transit delays seen by signals traveling certain critical paths. For the case of Figure 2, we will need to calculate the Forward path from DTE1 to DTE3 to determine the time signals first arrive at DTE's 2 and 3. We assume that DTE3 just misses deferring to the signal arriving from DTE1. DTE3 will, therefore, generate a 96 bit fragment before ceasing transmission. We need to calculate the reverse path for both the beginning and end of this fragment, the former to verify that DTE1 is transmitting long enough to be aware of the collision (Condition I) and the latter to determine that the packet seen by DTE2, the listening DTE, is too short (Condition II). Once DTE1 becomes aware of the collision, it will continue to transmit for 32 bits before ceasing. This path also enters into the evaluation of Condition II. The propagation delays described above are calculated in Tables II A-F, based upon delay times appropriate to Codenoll's current product.

The graphic technique is illustrated in Figure 3 for the Passive Star Network of Figure 2 with all fiber optic cable lengths (L1-L7) = 500m. The network of Figure 2 is laid out on the horizontal axis of Figure 3 with the positions of all network equipment indicated. Although it may be convenient to do so, it is not necessary that all distances on the horizontal axis be to the same scale. The vertical scale in Figure 3 is time. Beginning with the initial activity at DTE1, the propagation of signals with time is plotted on Figure 3. Start-up delays within a piece of equipment appear as vertical lines (proceeding in time at fixed position) while signal transmission through the various media appears as a line whose slope is equal to the signal propagation velocity.

TABLE II. Propagation Delays for Passive Star Fiber Network (Figure 2).

TABLE IIA

FORWARD PATH DTE1 TO DTE3

SYMBOL	PROP. DELAY (BITS)	TOTAL DELAY THIS SECTION (BITS)	TOTAL DELAY FROM ZERO (BITS)
	0	0	
D2	3	3	
A1	2.57	5.57	
M2	3.5	9.07	
L5	25.00	34.07	
HP1	0	34.07	
L3	25.00	59.07	
M1	3.5	62.57	
R1	8.0	70.57	
M2	3.5	74.07	
L1	25.00	99.07	
HP1	0	99.07	
L2	25.00	124.07	
M1	3.5	127.57	
R1	8.0	135.57	
M2	3.5	139.07	
L4	25.00	164.07	
HP1	0	164.07	
L7	25.00	189.07	
M1	3.5	192.57	
A1	2.57	195.14	195.14
DTE3 sees first activity			195.14

TABLE IIC

RETURN PATH DTE3 START OF FRAGMENT TO DTE1

SYMBOL	PROP. DELAY (BITS)	TOTAL DELAY THIS SECTION (BITS)	TOTAL DELAY FROM ZERO (BITS)
			195.14
D10	10		205.14
A1	2.57		207.71
M2	3.5		211.21
L7	25.00		236.21
HP1	0		236.21
L4	25.00		261.21
M3	9.0		270.21
R4	6.5		276.71
M2	3.5		280.21
L2	25.00		305.21
HP1	0		305.21
L1	25.00		330.21
M3	9.0		339.21
R4	6.5		345.71
M2	3.5		349.21
L3	25.00		374.21
HP1	0		374.21
L5	25.00		399.21
M3	9.0		408.21
A1	2.57		410.78
D7	3		413.78
DTE1 sees Collision			413.78

TABLE IIB

FORWARD PATH DTE1 TO DTE2

DTE1 Signal Reaches			
HP1		34.07	
L6	25.00	59.07	
M1	3.5	62.57	62.57
DTE2 sees first activity			62.57

TABLE IID

RETURN PATH DTE3 END OF FRAGMENT

SYMBOL	PROP. DELAY (BITS)	TOTAL DELAY THIS SECTION (BITS)	TOTAL DELAY FROM ZERO (BITS)
			195.14
D10	10.0		205.14
Send 96 bits	96		301.14
A1	2.57		303.71
M2	0.5		304.21
L7	25.00		329.21
HP1	0		329.21
L4	25.00		354.21
M4	160		430.21
R2	12.5		442.71
M2	0.5		443.21
L2	25.00		468.21
HP1	0		468.21
L1	25.00		493.21
M4	160		499.21
R2	12.5		511.71
M2	0.5		512.21
L3	25.00		537.21
HP1	0		537.21
L5	25.00		562.21
M1	0.5		562.71
A1	2.57		565.28
End of DTE3 Transmission reaches DTE1			565.28

TABLE IIE

LAST ACTIVITY AT DTE2 FROM DTE1

SYMBOL	PROP. DELAY (BITS)	TOTAL DELAY THIS SECTION (BITS)	TOTAL DELAY FROM ZERO (BITS)
			413.78
D11	10	10	
D12	32	42	
send activity ends at DTE1			
D2	3	45	
A1	2.57	47.57	
L5	25.00	72.57	
HP1	0	72.57	
L6	25.00	97.57	
M1	0.5	98.07	511.85
last activity at DTE2 from DTE1			511.85

TABLE IIF

Last Activity At DTE2 From DTE3			537.21
L6	25.00	25.00	
M1	0.5	25.50	562.71
last activity at DTE2 from repeater			562.71

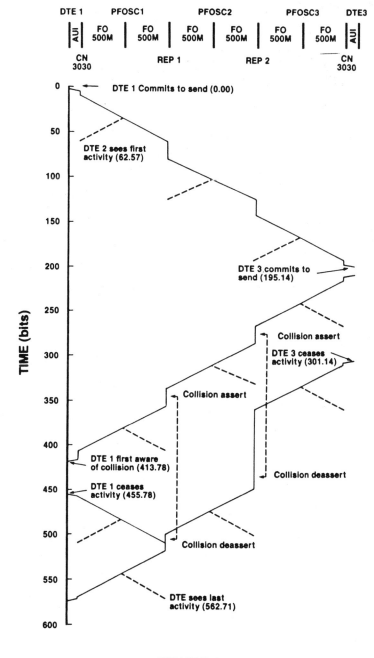

FIGURE 3
SPACE/TIME DIAGRAM FOR MAXIMUM EXTENT PASSIVE STAR, FIBER OPTIC NETWORK (FIGURE 2) -BASED UPON CURRENT PRODUCT DELAY TIMES

The Maximum Extent Network Conditions I and II can be summarized from either Table II or Figure 3 as:

CONDITION I

Round-trip signal propagation time < 534 Bits

From Table IIC, the round-trip time DTE1 to DTE3

and return is <u>413.78 Bits < 534 Bits</u>.

CONDITION II

Message Length at Listener - Preamble - SFD < 512 Bits

DTE2 first sees a signal at 62.57 Bit times. It last sees

a signal (from DTE 3) at 562.71 Bit times.

$(562.71 - 62.57) - 56 - 8 = $ <u>436.14 Bits < 512 Bits</u>

Therefore, since conditions I and II are satisfied, the Passive Star Network of Figure 2 falls within the size limitations imposed by IEEE 802.3.

3.2 Fiber Optic Passive Star Backbone

Figure 4 depicts a fiber optic passive star "campus" backbone. Each of the network nodes can be considered to be located in a separate building of the campus. Each node in Figure 4 is shown terminated in a bridge rather than a DTE. Each of these bridges would be connected to the building LAN (such as that shown in Figure 2). Bridges serve two functions. From a timing point of view, they de-couple the two networks to which they are connected. The timing constraints of IEEE 802.3 are then applicable to each LAN individually rather than to the two together as would be the case if they were connected by repeaters. In addition, bridges provide a filtering function passing on only those packets destined for a remote LAN. This then serves to minimize the traffic on the backbone. All the bridges (designated B1, B2, B3) are assumed to be located a distance L from the hub. The question addressed in this example is, based solely upon timing constraints, what is the largest possible fiber optic passive star backbone LAN. It should be understood that optical design considerations (power and loss budgets) form, a second, entirely independent set of system constraints.

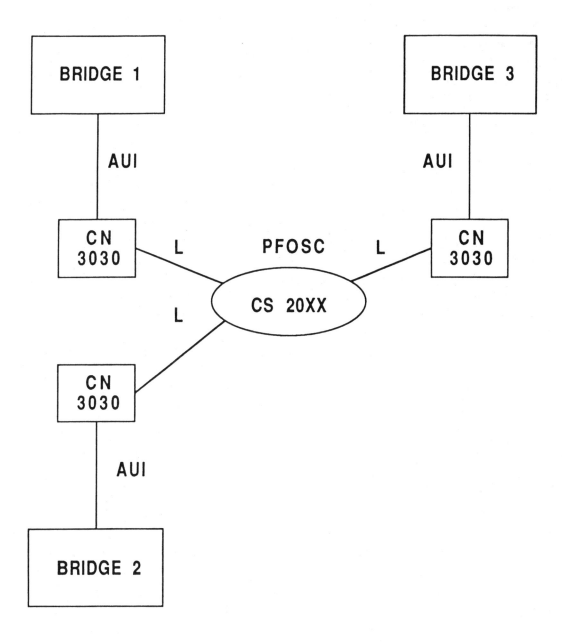

FIGURE 4. PASSIVE FIBER OPTIC BACKBONE LAN

The calculation here is similar to that in the preceding section. First, we need to calculate the time it takes a signal directed from B1 to B2 to reach B3, and then assuming B3 begins to transmit at the last possible moment the time for the return signal to reach B1. This round trip transit time will determine for what values of L Condition I is satisfied. In order to satisfy Condition II, the last signal reaching B2 (from either B1 or B3) must be determined.

These various signal propagation times are tabulated in Table III. Applying Conditions I and II to Table III, we find

CONDITION I:

$42.78 + 4L < 534$

CONDITION II:

$(127.28 + 4L) - (15.14 + 2L) - 64 < 512$

or

$(93.42 + 6L) - (15.14 + 2L) - 64 < 512$

The maximum allowable value of L will be the smallest value resulting from the above conditions

or L_{MAX} = 122 BITS = 12,200ns

Assuming a propagation velocity of 5 ns/m

$$L_{MAX} = 2440 \text{ m}$$

Thus, in principle, a fiber optic passive star backbone LAN could span an area of 2440 m radius or

LAN Area = 18 square Km

As noted before, an independent set of optical design constraints would also have to be met and, in practice, these will tend to limit the backbone LAN to somewhat smaller radii.

4. SUMMARY

In this note we have discussed the factors which limit the size of any CSMA/CD network and more particularly those conforming to the IEEE 802.3 standard. Two techniques, tabular and graphical, were demonstrated which allow the conformance of any network configuration, either single or multi-media, to the requirements of the standard to be verified.

It was shown that any three segment, two repeater set fiber optic passive star network in which none of the optical fiber lengths exceed 500 meters will conform to the requirements of IEEE 802.3 as well as the emerging 10BASE-F fiber optic CSMA/CD standard.

TABLE III. Propagation Delays for Passive Star Fiber Optic Backbone Shown in Figure 4.

TABLE IIIA
FORWARD PATH B1 TO B3

SYMBOL	PROP. DELAY (BITS)	TOTAL DELAY THIS SECTION (BITS)	TOTAL DELAY FROM ZERO (BITS)
		0	0
D2	3.0	3.0	
A1	2.57	5.77	
M2	3.5	9.07	
L1	L	9.07+L	
HP	0	9.07+L	
L3	L	9.07+2L	
M1	3.5	12.57+2L	
A1	2.57	15.14+2L	
B3 sees first activity		15.14+2L	

TABLE IIIB
FORWARD PATH B1 TO B2

		0	0
Signal reaches HP		9.07 + L	
L2	L	9.07+2L	
M1	3.5	12.57+2L	
A1	2.57	15.14+2L	15.14+2L
B2 sees first activity			15.14+2L

TABLE IIIC
RETURNED PATH B3 START OF FRAGMENT TO B1

SYMBOL	PROP. DELAY (BITS)	TOTAL DELAY THIS SECTION (BITS)	TOTAL DELAY FROM ZERO (BITS)
		0	15.14+2L
D10	10	10	
A1	2.57	12.57	
M2	3.5	16.07	
L3	L	16.07+L	
HP	0	16.07+L	
L1	L	16.07+2L	
M3	9.0	25.07+2L	
A1	2.57	27.64+2L	42.78+4L
B1 sees collision			42.78+4L

TABLE IIID
RETURN PATH B3 END OF FRAGMENT

		0	15.14+2L
D10	10	10	
96 bits	96	106	
A1	2.57	108.57	
M2	.5	109.07	
L3	L	109.07+L	
HP	0	109.07+L	
L1	L	109.07+2L	
M1	.5	109.57+2L	
A1	2.57	112.14+2L	127.28+4L
Last activity at B1 and B2 from B3			127.28+4L

TABLE IIIE
LAST ACTIVITY FROM B1

		0	42.78+4L
D11	10	10	
D12	32	42	
D2	3	45	
A1	2.57	47.57	
L1	L	47.57+L	
HP	0	47.57+L	
L2	L	47.57+2L	
M1	.5	48.07+2L	
A1	2.57	50.64+2L	93.42+6L
Last activity at B2 from B1			93.42+6L

It was also shown that a single segment, fiber optic passive star backbone LAN can span a distance as great as 4.88 kilometers without violating any of the timing constraints of IEEE 802.3.

5. REFERENCES

Carrier Sense Multiple Access with Collision Detection (CSMA/CD) Access Method and Physical Layer Specifications.
ISO 8802-3:1989 (E), ANSI/IEEE Std 802.3 - 1988
The Institute of Electrical and Electronics Engineers, Inc.
345 East 47th Street
New York, NY 10017

Supplements to Carrier Sense Multiple Access with Collision Detection.
ANSI/IEEE Std. 802.3 a, b, c and e - 1988.
The Institute of Electrical and Electronics Engineers, Inc.
345 East 47th Street
New York, NY 10017.

A MIGRATION STRATEGY TO FDDI FOR THE FIBER OPTIC LAN CABLE PLANT

Walter B. Hatfield and Michael H. Coden

INTRODUCTION

The increasingly rapid introduction of ever more powerful work stations and desktop computers, full screen graphics and high speed I/O peripherals has created a need to share and exchange information at ever increasing speeds. This need is reflected in the establishment of standards for local area networks (LANs).

The primary standards being implemented today are those of the IEEE 802 family - Carrier Sense Multiple Access with Collision Detect (CSMA/CD) or Ethernet (IEEE 802.3); Token Bus (IEEE 802.4) and Token Ring (IEEE 802.5). These standards define LANs which will operate at maximum data rates of 10, 20 and 16mbs respectively. Already it is perceived that these data rates will soon become limiting, if they are not already so, for many LAN applications. This explains then the very keen interest in the Fiber Distributed Data Interface (FDDI) standard currently under development by ANSI.

FDDI is a 100Mbs data rate, token-ring LAN with an optical transmission rate of 125 Megabaud. It will accommodate as many as 500 nodes spaced up to 2km apart. Typical applications which will utilize the high bandwidth available in FDDI include communication between main frames, as a backbone to interconnect lower-speed (eg. IEEE 802) LANs through file servers, bridges, gateways, and a means of linking work-stations and high-end PCs especially in CAD/CAM/CAE, financial, advertising and other image processing environments.

Because of its very high speed, the FDDI standard has been written around optical fiber as the transmission media (F stands for Fiber). As the advantages of fiber optics over twisted pair or coaxial cable become increasingly evident - higher bandwidth, smaller size and weight, immunity to EMI- optical fiber is more and more seen as the ideal cabling for office building, campus and industrial environments. Recognizing this fact, all three IEEE 802 standards have either developed or are in the process of writing standards for fiber optic media. The inherent advantages of fiber notwithstanding, for the lower speed (IEEE 802) LANs where copper conductors can provide an adequate solution there is a very spirited cost based competition between the various transmission media.

Given that the labor cost associated with installing a building cabling system often exceeds the cost of the materials and that a system once installed should be expected to have a 10-20 year useful operating life, it would seem that rather than allowing the cabling system to evolve with the supported LAN, it would be better to install a single cabling system which could support any LAN evolution. As it is evident that this evolution will almost certainly include higher data rates, fiber optics is the only media which can support this strategic approach.

The key to installing a fiber optic cabling system is to select a wiring topology and termination scheme that will accommodate any future choice of LAN protocol and to choose fiber and connector types for long term hardware compatibility. The cabling strategy described here addresses both of these major issues - wiring topology and optical fiber specification - and is applicable to all current and future LAN implementation scenarios.

CABLE PLANT TOPOLOGY

The Electronic Industries Association (EIA) TR-41.8.1 Working Group has for some time been developing a Commercial and Industrial Building Wiring Standard. Although the work of this group is not yet complete, it does serve as a very good example of the recommended cable plant wiring topology.

Basically a three level star wiring topology such as that illustrated in Figure 1 is suggested. The maximum distances shown in Figure 1 are those currently recommended by EIA TR-41.8 for optical fiber. That portion of the star wiring from the work area (wall outlet) to the wiring closet is referred to as the Horizontal Wiring from the fact that most of the wiring on this level will be in the horizontal plane of the building. The wiring from the Wiring Closet to the first Cross-Connect and between Cross-Connects is referred to as the Vertical Wiring or Building and Campus Backbones. These terms in all cases are meant to be descriptive. The Wiring Closet and Cross Connects are sometimes referred to as Hubs (Work Group, Building and Campus).

The objective of the star wiring topology, particularly the Backbone sections, is to enable installation of the wiring to proceed without prior detailed knowledge of the communications service which will be supplied. The limitation to three levels of hubs will adequately serve virtually all network applications while remaining simple to administer. On smaller sites a single cross-connect system may be desirable.

In many cases the user has requirements for both voice and data equipment at a work location. Although the focus here is on fiber optics for LAN applications, the considerations apply equally well to voice requirements and in many instances, particularly the horizontal wiring, cable or cables containing more than one transmission media (say, fiber optics and unshielded twisted pair) will be installed.

It should be noted that the wiring strategy being proposed here differs from the developing EIA standard in at least one important aspect. In its present (incompleted) state EIA - TR41.8.1 does not recognize fiber optics as an option for the horizontal wiring. This is very short-sighted. As noted in EIA - TR41.8.1 after construction of a building, the horizontal wiring is usually much _less_ accessible than the vertical wiring. The time to make changes can be extremely high with attendant disruption to the occupants in the area and their work.

These factors make the choice of horizontal cable types and their layout very important. The strategic arguments for fiber optics apply

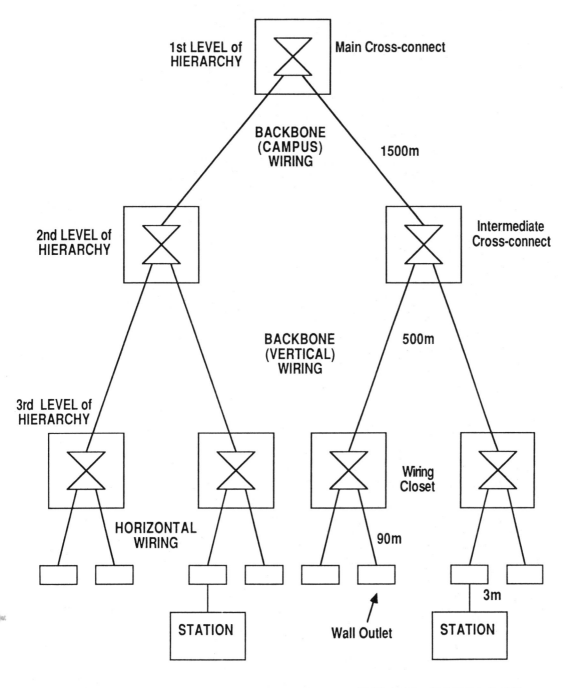

FIGURE 1. RECOMMENDED STAR WIRING TOPOLOGY

equally well to the horizontal and vertical wiring. Consideration of a wide range of user applications must be given in order to reduce or eliminate the probability of requiring changes to the horizontal wiring in the face of unforeseen user requirements.

The functionality of a Hub, be it wiring closet or cross-connect, is illustrated in Figure 2. Generally speaking, all incoming and outgoing transmission lines, fiber optics or other, are terminated in patch panels. To minimize current costs, some users prefer to leave unused fibers un-terminated. Also present in the Hub will be signal distribution and regeneration equipment. This will be specific to the particular LAN protocol being implemented. Use of the star wiring and patch panels allows any desired LAN topology and protocol to be implemented now and changed to meet any future, presently unknown, requirements making changes only in the electronics and interconnections without changing the basic cable plant. Figures 3-5 illustrate this point, showing how the basic star wiring topology can be configured as a ring (Figure 3) passive or active star (Figure 4), or linear bus (Figure 5). Figure 6 illustrates the important case where the upper level of the star wiring is configured as a ring (FDDI) while the lower levels are configured for either rings (IEEE 802.5) or passive stars (IEEE 802.3).

One of the questions to be addressed when planning a fiber optic cable plant is just how many fibers to install. What makes this question so difficult is that once in place any cable plant, because of its installation expense, should have a 20 year life. Users thus need to consider the nature of their network requirements over this period. These should include compatibility with emerging standards, evolving capacity requirements, the need for multiple sub-nets, redundancy, etc. Also to be considered is the desirability of other services, such as color video at the workstation.

Multimode optical fiber has more than enough bandwidth to meet all of today's and most of tomorrow's LAN applications. Its use, rather than higher bandwidth single mode fiber, results in lower overall system costs. However, in looking to the future, some network planners, in anticipation of increasing bandwidth requirements, specify cables containing one or more single mode fibers, in addition to multimode fibers.

The point is that if a cable plant should contain enough fibers for both current and future requirements, the cost of increasing capacity becomes incremental - just terminate the fibers and install the appropriate electronics.

All of the IEEE 802 standards require a single duplex fiber optic cable to connect a node (one fiber for the transmit signal and one for the receive signal). FDDI, on the other hand, provides two wiring options. Stations may be attached to either a _dual_ counter-rotating ring or through concentrators in lower cost _single_ ring sub-networks. These 100 Mbs sub rings are connected to each other and to the main dual ring through concentrators. Which of the two implementations a user will choose depends upon a number of factors such as existing cable plant, reliability requirements, etc.

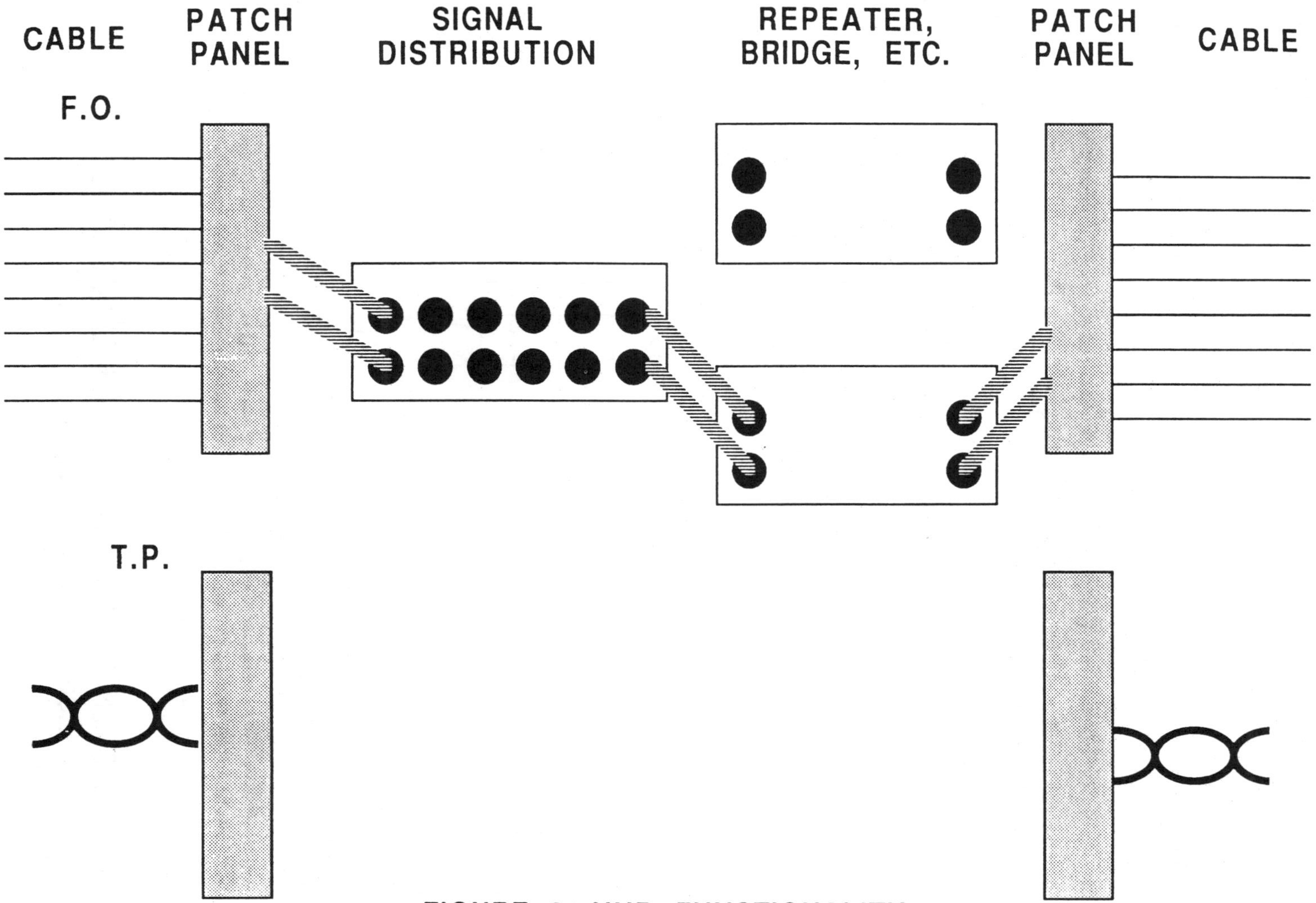

FIGURE 2. HUB FUNCTIONALITY

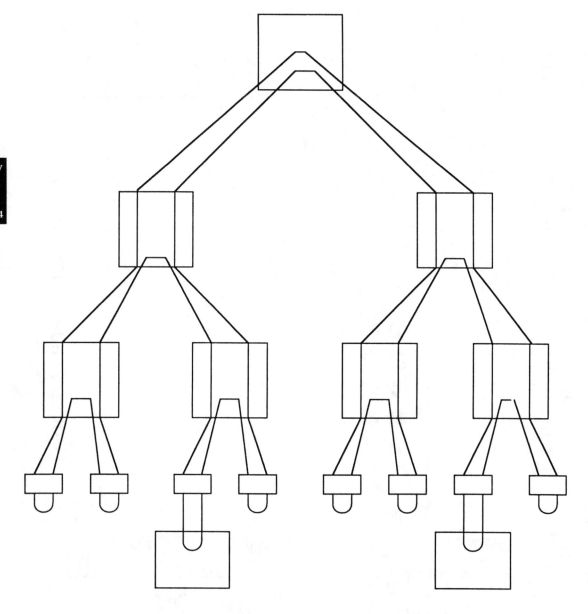

FIGURE 3. RING TOPOLOGY IMPLEMENTED USING ALL PURPOSE PHYSICAL STAR WIRING

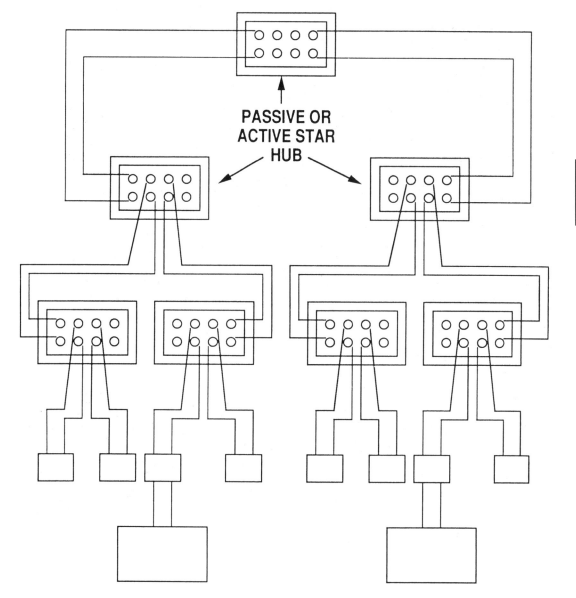

FIGURE 4. PASSIVE OR ACTIVE STAR TOPOLOGY IMPLEMENTED USING ALL PURPOSE PHYSICAL STAR WIRING

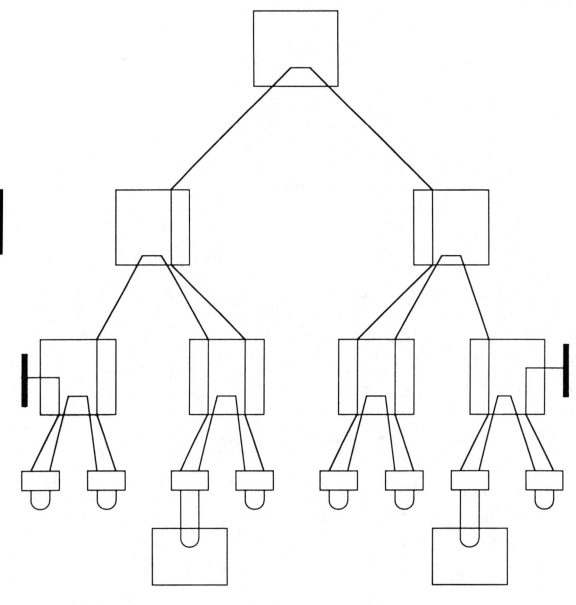

FIGURE 5. LINEAR BUS TOPOLOGY IMPLEMENTED USING ALL PURPOSE PHYSICAL STAR WIRING

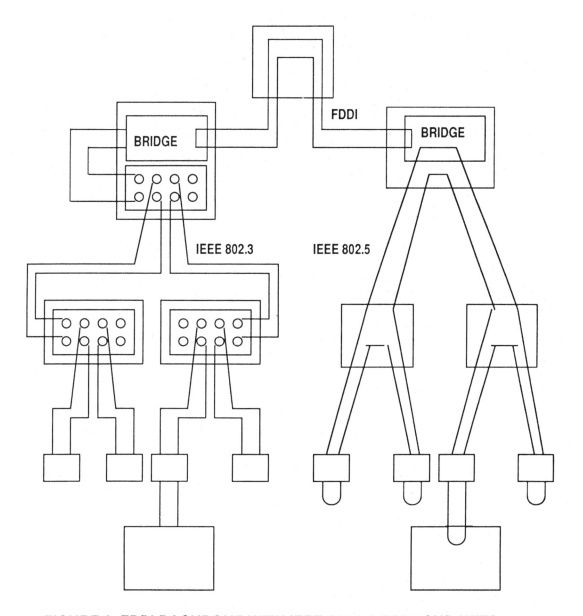

FIGURE 6. FDDI BACKBONE WITH IEEE 802.3 & 802.5 SUB-NETS IMPLEMENTED USING SAME ALL PURPOSE PHYSICAL STAR WIRING

At the lowest or horizontal level of the three layer wiring hierarchy, a single ring sub-network will usually best serve the user's requirement. Therefore, the two fiber cables initially installed for an IEEE 802 network will map directly into a single ring FDDI sub-network. It is only in the vertical wiring, and very likely only at the highest layer, that one might wish to make provision for a dual FDDI ring. This is demonstrated in Figure 7 which illustrates how, with proper planning, a fiber optic Ethernet (or IEEE 802.3) star topology segment can be easily migrated to a dual, counter rotating FDDI ring. At the time the Ethernet network is installed, a second duplex fiber optic cable is run from the Hub to every network station. The cost of installing this initially spare fiber cable is minimal if it is done at the same time as the primary installation. When it is time to upgrade the network to FDDI, the passive star coupler is simply disconnected and jumpers on the patch panel re-configure the cable plant into a ring. What were initially spare fiber optic cables now become the second of the dual counter-rotating rings. The major cost of this conversion lies not in the cable plant but rather in replacing the Ethernet Station Interface with the one appropriate to FDDI.

Unless one has the prescience to determine beforehand which nodes on some future FDDI network will be on the dual main ring and which will be on some single ring subnet, it is probably best to plan on running at least four fibers to each potential FDDI node on the horizontal wiring and certainly at least four fibers on the vertical or backbone wiring.

OPTICAL FIBER CHARACTERISTICS

The fiber optic cable properties currently specified in the various IEEE 802 and ANSI FDDI standards are summarized in Table I and Figure 8. It should be noted that while the media portion of all these standards are well established and stable, in many cases the standards themselves still have Draft status. This means that however unlikely, some important changes are still possible.

The discussion in this section is from the point of view of a user who wishes to install and use fiber now with an IEEE 802 type LAN with the provision that in the future some or all of the fiber optic cable plant would be upgraded to FDDI use. It was shown in the preceding section how star wiring allows for this topologically. The purpose of this section is to assure that the fiber installed now will have the proper characteristics for use in both IEEE 802 and FDDI LANs.

The first point to be kept in mind is that the IEEE 802 LANs have been specified for operation at a nominal optical wavelength of 850nm while FDDI is intended to operate at 1300nm. Any fiber intended for dual use must, therefore, exhibit the proper characteristics at both wavelengths. Most glass fiber sold today with core diameters of 50um, 62.5um, 85um and 100um operate satisfactorily with both wavelengths. You should, however, check the manufacturers specifications.

All of the LAN standards define a 62.5/125 micon reference fiber for purposes of testing and verifying various fiber optic parameters. All of the standards, however, explicitly include the use of other

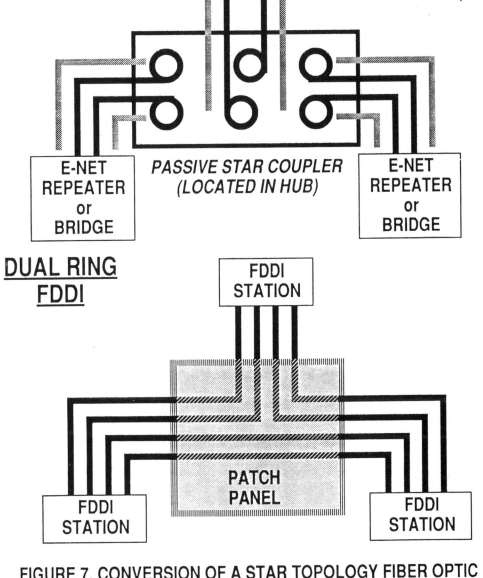

FIGURE 7. CONVERSION OF A STAR TOPOLOGY FIBER OPTIC ETHERNET SEGMENT INTO AN FDDI RING

TABLE I - Summary of Fiber Optic Cable Characteristics

	802.3 FOIRL	802.31 10BASE-F[1]	802.4	802.5[1]	FDDI[1]
Optical Fiber (Core/Cladding-NA) Reference	62.5/125-.275	62.5/125-.275	62.5/125-.275	62.5/125	62.5/125-.275
Other Allowed	50/125 85/125 100/140	50/125-.23 85/125-.26 100/140-.275	50/125 85/125 100/140	50/125 100/140	50/125-.20 50/125-.22 85/125-.26 100/140-.29
Nominal Wavelength (nm)	850	850	850	850	1300
Maximum Attenuation Optical Fiber (dB/Km)	8	3.75		4.0	2.5
Total Cable Plant (dB)		FA-12[2] FP-26[3]	27	12	11
Minimum Modal Bandwidth (MHz - KM)	150	150		150	500
Chromatic Dispersion Slope (ps/nm^2-km)				.093	See Fig 7
Zero Dispersion Wavelength(nm)				1365	See Fig 7
Maximum Fiber Length(km)	2.0	FA-2.0[2] FP-0.5[3]			2.0
No. Fibers Pairs per Attachment	1	1	1	1	1 or 2
Connectors Simplex (one fiber per connection)	SMA	ST			
Duplex (two fibers per connection)			FDDI	FDDI	FDDI

Notes: (1) These standards exist in draft form at the present time
(2) Active Star Network
(3) Passive Star Network; Loss = Star Coupler + 2x (Loss of Longest F.O. Link) + Connector Loss

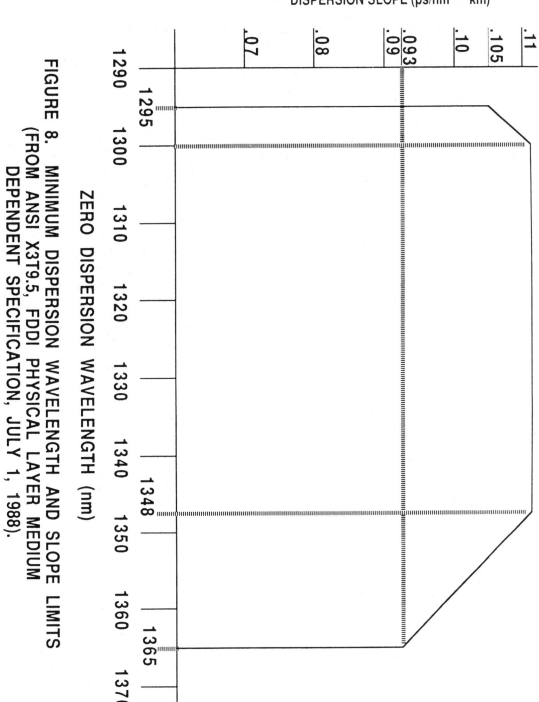

FIGURE 8. MINIMUM DISPERSION WAVELENGTH AND SLOPE LIMITS (FROM ANSI X3T9.5, FDDI PHYSICAL LAYER MEDIUM DEPENDENT SPECIFICATION, JULY 1, 1988).

multimode fiber sizes. This is both in recognition of a large installed base of fiber other than 62.5/125 and the expected continuation of the popularity of other fiber sizes in certain areas. In any event, all multi-mode fibers are included in the various standards with the understanding that use of fiber other than 62.5/125 may change some parameters, such as coupled optical power or optical fiber attenuation, which in turn might affect other parameters such as maximum achievable link length. The effect, if any, can be determined by relative simple arithmetic calculation.

The fiber dispersion characteristic is typically presented in the form of a modal bandwidth. For the purposes of specification, fiber bandwidth is approximated as an inverse linear function of length. That is, a 500 MHz-Km (FDDI) fiber will present approximately 250 MHz of bandwidth (four times the optical line rate of 125 M baud = 62.5 MHz) at the maximum FDDI distance of 2km. It is not necessary that all fiber optic cable intended for use in FDDI LANs exhibit a 500MHz-km modal bandwidth. If the known maximum node separation is $L_{max} < 2km$, then it is only necessary that

$$\text{Modal Bandwidth} > 250 \, L_{max} \quad \text{(MHz-km)}$$

The chromatic dispersion parameters provide a measure of the speed at which different colors (wavelengths) of light travel through a multimode optical fiber. Since all of the existing and developing fiber optic LAN standards employ wide spectral width LEDs as the optical source, chromatic dispersion will contribute, together with modal dispersion, to the overall system bandwidth. Chromatic dispersion has relatively little influence upon the performance of lower data rate IEEE 802 LANs, but becomes more important when upgrading a system to the higher data rates of FDDI if cable lengths are long.

The Appendix contains manufacturer's data on fiber optic cables. The cables in this data sheet will conform to the specifications of both the 850nm IEEE 802 standards and the 1300nm FDDI standard.

With the exception of IEEE 802.3, all of the IEEE 802 standards are deferring to FDDI as regards the fiber optic connector specification. The duplex connector specified by FDDI, however, is not yet in widespread use. It should be noted that the various standards define only those connectors which form part of the media dependent interface to the equipment being defined by the standard. Other connectors, particularly those at Hub patch panels and the wall panels, are user's choice. It is necessary only that their normal optical loss, together with the fiber attenuation and other normal optical losses present in the transmission path (such as a passive star coupler) satisfy the total cable plant loss requirement specified in the standard. If the fiber optic cable plant is terminated at the wall outlet as indicated in Figure 1, with connectors of the user's choice, then a change of LAN protocol will at most require a change in the short (1 to 3 meter) equipment to wall plate hook-up cable.

SUMMARY

This paper has presented a fiber optic LAN wiring strategy based upon two basic concepts. The first is the three tier star wiring topology. From this basic topology any desired LAN topology - ring, star or bus - can be derived. The three tier structure provides adequate flexibility for most applications including the very important case where a hierarchy of 802 LANs interconnected by an FDDI backbone is required.

The second premise concerns the media itself. The media portions of what can be expected to be, over the next ten years or so, the primary, standard LANs are today well enough defined to permit their installation with little fear that they will be changed. Moreover, with little effort, a single fiber optic media can be chosen which will not only satisfy the current requirements of IEEE 802 LANs operating at 4-20 Mbs, but will equally well serve the 100 Mbs requirements of FDDI in the years to come.

Data Sheet
-PRELIMINARY-
(July, 1989)

Codenoll Multimode LAN Fiber Optic Cable

Features

- **Conforms to Standards**
- **Low Loss**
- **High Bandwidth**
- **Wide Range of Core Diameters**
- **Variety of Options Available**

Codenoll
TECHNOLOGY CORPORATION

Description

Codenoll's Multimode LAN fiber optic cable(s) conform to the requirements of all IEEE 802.3, IEEE 802.4, IEEE 802.5 and FDDI Standards. This makes them ideal for local area networks, premises networks, and data processing networks and systems. Codenoll LAN cables are assembled using high quality Corning optical fiber, certified to meet the Specifications contained in this data sheet, and provide the ideal strategic solution to all facility wiring requirements.

Installed now for use with 4-15Mbs IEEE 802 networks operating at 850nm, these cables will also meet the requirements of evolving 100Mbs, 1300nm FDDI Standards. A single fiber optic cable plant will meet all foreseeable network requirements.

Codenoll Multimode LAN fiber optic cable is available with up to 12 fibers per cable in a variety of cable configurations -- interior/exterior cable, armored cable, heavy duty breakout -- and jacketing materials. Cable can also be provided cut to length(s) and terminated.

Please contact your dealer or authorized Codenoll reseller for pricing and other details.

Specifications

	CFOCLAN 50/125	CFOCLAN 62.5/125	CFOCLAN 100/140
Core Diam	50.0 ±3.0µm	62.5 ±3.0µm	100.0 ±4.0µm
Numerical Aperture	0.200 ±0.015	0.275 ±0.015	0.290 ±0.015
Maximum Attenuation (dB/Km) at 830nm at 1300nm	3.75 2.0	3.75 2.0	3.75 2.0
Minimum Modal Dispersion (MHz-Km) at 830nm at 1300nm	160 500	160 500	160 500
Chromatic Dispersion Zero Dispersion Wavelength (nm) Zero Dispersion Slope (ps/nm^2 - Km)	1297-1316 <0.101	1332-1354 <0.097	1332-1361 <0.100

Codenoll Technology Corporation 1086 North Broadway Yonkers, N.Y 10701. (914) 965-6300

IV

295

IV

296

Plastic Optical Fiber

Plastic Optical Fibers Take Aim at LANs, p. 299
  F. W. Scholl

Applications of Plastic Optical Fiber to Local Area Networks, p. 303
  F. W. Scholl, M. H. Coden, S. J. Anderson, B. V. Dutt

Communicating in New Ways, p. 309
  U. von Alpen, M. H. Coden

Implementation of a Passive Star Based Fiber Optic Network for Full Vehicle Control, p. 317
  R. H. Lefkowitz, M. H. Coden, F. W. Scholl, U. von Alpen

High Speed Polymer Optical Fiber Networks, p. 329
  D. V. Bulusu, T. E. Zack, F. W. Scholl, M. H. Coden, R. E. Steele, G. D. Miller, M. A. Lynn

V
298

Electronic Engineering TIMES

Issue 474 — The Industry Newspaper For Engineers And Technical Management — Monday, February 22, 1988

Plastic Optical Fibers Take Aim At LANs

POF systems can meet one of the major driving forces in the marketplace: low cost. Plastic fiber is less expensive than competing silica-based fiber and LEDs.

By Frederick W. Scholl
*Senior Vice President
Codenoll Technology Corp.
Yonkers, N.Y.*

Recent improvements in the properties of plastic optical fiber (POF) and increasing supply of optical components designed for it suggest that POF-based local-area networks will represent a greater part of installed optical-fiber systems. POF systems can meet one of the major driving forces in the marketplace: low cost. Plastic fiber is less expensive than competing silica-based fiber and other interconnect components like LEDs. Of course, POF retains the advantages of glass, like noise immunity, high bandwidth and ease of cabling.

The technologies for plastic-fiber LANs are all moving along rapidly.

Plastic Optical Fiber

The key advantage of plastic optical fiber is the large core that is technically and economically feasible with polymer materials. The core size allows the use of low-cost optical sources and detectors, as well as low-cost connectors. To put POF in perspective, a comparison of the properties of glass and plastic reveals significant information about both fiber properties and the system as well.

The core size POF results in superior LED-to-fiber coupling. The core dimensions of POF permit the use of plastic molding technology for both fiber-to-fiber connectors and source- and detector-to-fiber adapters. Various POF sizes are available, with 500 and 1,000 microns being common. Note that the cladding thickness of 5 to 20 microns for POF is generally less than that for glass. The core diameter for POF is generally sufficient for mechanical strength; a cladding thickness of 5 microns is all that is required for good optical properties. POF is protected from abrasion by an additional non-optical jacketing.

The fiber cost per meter scales approximately with fiber cross-sectional area, while launch power also scales approximately with area. The numerical aperture (NA) of POF is typically larger than glass. In addition, POF is designed with a step-index construction between core and cladding, while glass-core fiber has a graded index. POF offers good LED-fiber coupling efficiency (efficiency proportional to NA^2) resulting from the large numerical aperture.

Fiber attenuation is being studied at various industrial laboratories. The current minimum of 160 dB/km at 660 nm is influenced by intrinsic fiber parameters (such as scattering) and extrinsic parameters (such as impurities). Theoretical estimates predict that POF loss can be reduced to less than 20 dB/km. These attenuation specifications allow calculation of loss-limited point-to-point links. The maximum length is given by Lmax = Flux Budget/Fiber Attenuation, where the flux budget is established by the optical transmitter and receiver utilized. For a DC to 10 Mbps system, which would allow operation of several current LAN protocols, a flux budget of 30 dB is easily achieved. Link length is then determined to be Lmax = 188 meters (619 feet) for present fibers and Lmax = 1.5 km (4,950 feet) for future material with 20-dB/km attenuation.

Optical bandwidth for POF is determined by the numerical aperture. For applications involving LEDs, chromatic dispersion must, in principle, also be included. The chromatic dispersion-limited bandwidth of POF is approximately 60 MHz/km, so modal dispersion (determined by fiber NA) is actually the limiting factor. Recently, larger bandwidths have been measured for fibers made by Hoechst A.G. A 24-meter length of POF was characterized to have a bandwidth of 23 MHz/km.

A 3-cm typical minimum bend radius is specified for both plastic and glass fiber. In the case of glass, this is determined by the residual long-term strain that the material can support. For POF, the polymer core and cladding can sustain considerable strain, but light loss begins to increase for bend radii less than 3 cm. The thermal properties of POF and glass fiber are remarkably similar due to the fact that the glass fiber requires a polymer coating for protection from abrasion.

As noted, POF research is an active field now and advances are expected in several areas. The POF manufacturing process of coextrusion is expected to remain constant in principle. With polymethyl methacrylate (PMMA) core, copolymer cladding materials can be used to give a wide variety of numerical aperture, from approximately 0.5 down to 0.23. Attenuation will be reduced toward the 20-dB/km level by reducing extrinsic and intrinsic losses. As these are reduced to below 50 dB/km, the presence of absorbed water molecules in the fiber core becomes a critical, additional loss. Such water vapor can permeate the fiber core from the surrounding atmosphere. Two possible routes are being pursued to eliminate this problem. One is to use core materials that have extremely low saturated water vapor levels. With them, water molecules will penetrate the core, but the net effect on loss will be small. A second approach is to use an impenetrable sheath material to prevent access of water vapor to the POF cladding surface. Other polymeric materials are being investigated for higher temperature operation. Such materials will offer work up to 130°C to 150°C. As indicated, glass-fiber cables also have high temperature limitations resulting from the polymer buffer coating used. The same comments apply to solvent resistance, since both glass and POF materials use polymer materials.

Light-emitting diodes and PIN detectors are the most common optoelectronic components for glass, as well as polymer-fiber LANs. A significant difference is that the POF systems optimally operate at 660 nm (red), while 830 nm (infrared) is optimum for glass. The 660 nm wavelength offers a good combination of low fiber attenuation and availability of efficient sources and detectors. A second difference between components for glass and plastic systems is in the packaging. The larger-core POF permits plastic active device receptacles, while the glass systems usually require precision machined components for best active component to fiber-coupling efficiency.

Gallium arsenide LED technology is preferred for both plastic and glass. The active area of the LED is actually fabricated from an alloy of aluminum arsenide and gallium arsenide; the bandgap and, hence, emission wavelength can be centered on either 660 nm or 830 nm (or values in between).

The choice of center wavelength 660 nm or 850 nm reflects the minimum attenuation of plastic and glass fiber. The source spectral width is important for establishing an 830-nm system bandwidth but is less important for typical POF systems. A critical LED parameter is launched power, P_{fiber}. LEDs are packaged in active-device-mount receptacles for direct connection to fiber connectors. The coupled power figures include the following parameters: 70-mA device current and 1,000-micron fiber diameter for 850 nm. Although the 660-nm device launches more power, in fact, these components are internally less efficient than 850-nm LEDs. The latter have nearly 100 percent internal conversion efficiency of electrons to photons. However, the 660-nm LED has competing direct transitions (photon emission) and indirect transitions (non-radiative process), leading to reduced overall output power.

Several factors contribute to the reduced bandwidth of 660-nm LEDs. First, most applications to date have been for systems operating at under 1 MHz. Thus, manufacturers have optimized devices for power output and not bandwidth. Second, even for optimized emitters, 830-nm devices will be faster since the slower non-radiative emission processes are not present. Future technology development should result in a selection of 660-nm devices optimized for both speed or power output.

Considerations for optical receivers are more straightforward since silicon photodetectors are preferred for both POF and glass. Responsivity, measured in amperes per watt, is lower at 660 nm than at 850 nm, simply because there are fewer photons per watt at the short wavelength. For devices optimized for 600-nm to 850-nm operation, this results in a 1.1-dB receiver sensitivity degradation going from 850 nm to 660 nm. Detectors optimized for 850 nm may show 2-dB sensitivity degradation going from 850 to 660 nm.

Three basic types of optical-fiber connector technologies exist at present. One is the precision machined metal or metal/ceramic connector for glass fibers. Modifications to these designs for POF have appeared using lower-

KEY PARAMETERS OF POF AND GLASS FIBERS FOR LAN USAGE

	Typical Core Size	Typical Fiber Size	Numerical Aperture	Minimum Attenuation	Bandwidth	Minimum Bend Radius	Maximum Operating Temperature	Approximate Simplex Cable Cost
Plastic	460 Micron 960 Micron	500 Micron 1,000 Micron	0.47	160 dB/km (660 nm)	6 MHz-km	3cm	90°C (500 Micron)	$0.22/m
Glass	50 Micron 100 micron	125 Micron 140 Micron	0.29	5 dB/km (830 nm)	400 MHz-km	3cm	70°C	$0.60/m (100 Micron)

KEY PARAMETERS OF LEDs FOR LAN SYSTEMS

	(nm)	(nm)	P_{Fiber} (Microwatts)	Analog Bandwidth	Cost	Packaging
LED for POF	660	30	140[1]	7 MHz	$2.93 (1 K = Q)	Truncated Plastic Dome
LED for Glass	850	70	80[2]	150 MHz	$24.75 (1 K = Q)	Metal Can

(1) Device Current = 70 mA; Fiber Diameter = 1,000 Microns
(2) Device Current = 70 mA; Fiber Diameter = 1,000 Microns

Technology Update: INTERCONNECTIONS

cost plastic inserts and ferrules. Some cost reductions have been achieved with these designs. A third generation of molded connectors has been designed specifically for POF. The true cost reductions available with POF connection hardware has been achieved with these designs.

Each in-line termination consists of two connectors and one bulkhead or receptacle. In-line connection using glass fibers and SMA connectors would cost $2 \times \$5.60 + \$2.75 = \$13.95$. In-line connection using POF would cost $3 \times \$0.40 = \$1.20$. These numbers illustrate the excitement generated by this new interconnect technology.

The above prices refer only to connector hardware costs. Further significant reductions results from lower termination costs associated with POF. The plastic molded-plug termination does not require polishing—the plastic fiber is "terminated" using a hot-knife process. This also makes connector termination easy for field applications.

The standard SMA achieves 1 dB using a 100-micron-core glass fiber. Using similar precision SMA technology, significantly lower loss is achieved using large-core 1,000-micron POF. The molded plastic connector technology trades off cost and attenuation, resulting in 2-dB maximum attenuation per in-line connection.

Systems Applications

Systems applications for POF that have been identified to date include LANs for future automobiles, workstation interconnects for office LANs, LANs for home consumer applications and backplane interconnects for advanced computers. The driving forces in these applications are low interconnect cost (including cable, sources, detectors and connectors) and ease and simplicity of installation. Applications requiring large numbers of connections per unit length of fiber are often candidates for large-core POF.

Undoubtedly, additional applications will be discovered as the technology matures.

Probably the most discussed application for POF is the automotive data bus. Trends in automotive data communication show that by the 1990s, operations in the megabit range will be needed. The short distances (<20 feet) and high connector density make POF a strong contender for the future.

Another potential market for POF is LANs for office automation. Such LANs fall into several types established by international standards committees. At present, Ethernet CSMA/CD LANs are most popular; also of importance are token-ring and token-bus standards. Fiber-optic versions of each of these are commercially available using glass-fiber cable.

In order for POF to penetrate the LAN market, significant reductions in fiber attenuation must occur. Present optical transceivers will accommodate 30 dB of optical loss over the network at 10 MHz. For a 16-port passive star, this results in 15-dB loss available for the cable plant. A star radius of 100 meters (200 meters round-trip between hub and node) covers most workstation hookups. This implies a fiber attenuation of less than $15/0.2 = 75$ dB/km. This number is clearly achievable, but requires further improvements from the 160 dB/km presently available.

Two other POF applications of note are "home-bus" consumer networks and advanced computer backplane hookups. As single-mode fiber service to individual homes begins in the early 1990s (following the trials occurring now) the need will arise for LANs within the home to interconnect television, personal computers, security service, etc., to the external information channel. POF technology is extremely attractive because of the low cost of connectors and the ease of field termination of fiber-optic connectors. Again, ease of interconnection makes POF attractive for large computer backplane interconnects. Current aggregate serial bandwidths are in the 0.5 to 5 Gbps range. POF arrays of four or eight channels would allow reliable transmission of this data between printed-circuit boards or between subassemblies within a larger computer cabinet. ∎

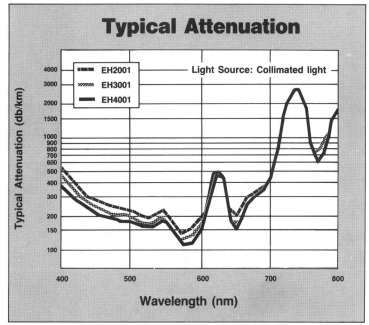

Attenuation of commercial plastic optical fiber. Theoretical estimates predict that POF loss can be reduced to less than 20 dB/km.

APPLICATIONS OF PLASTIC OPTICAL FIBER TO LOCAL AREA NETWORKS

Frederick W. Scholl, Michael H. Coden,
Stephen J. Anderson, Bulusu V. Dutt
Codenoll Technology Corporation
1086 North Broadway
Yonkers, New York 10701

I. Introduction:

Recent improvements in plastic optical fiber properties and increasing availability of optical components designed for this fiber suggest that POF based local area network (LAN) systems may represent an increasing fraction of installed optical fiber systems. POF based systems can satisfy one of the major driving forces in the marketplace: low cost. The plastic fiber is lower in cost than competing silica based optical fiber and most coaxial or twisted pair media options. Optical interconnect components such as LEDs, detectors, and optical connectors can be similarly low in cost for POF based systems. Of course, POF retains the advantages of glass fiber cabling systems such as noise immunity, high bandwidth and ease of cabling.

System experiments reported in this paper illustrate how current standards based LANs may be implemented in the new fiber. At present no standard product offering in the LAN marketplace makes use of POF. This situation is likely to change as improvements in the fiber, optoelectronic component and connector technology occur.

In the first section of this paper we give a review of each of the communications systems components. Following this we will describe the LAN applications in which POF seems most likely to contribute. Since costs are important, comparison will be made between present-day glass fiber LANs and projected costs for POF systems. Our view for future systems is that POF will dominate in the most cost sensitive areas, predominantly short distance applications, whereas glass fiber systems will be favored for applications requiring longer distance between DTEs.

II. Plastic Optical Fiber

The key advantage of plastic optical fiber is the large core size that is technically and economically feasible with polymer materials. The large core size allows the use of low cost optical sources and detectors, as well as low cost connectors. To put POF in perspective, refer to Table I which gives a comparison of properties of glass fiber and plastic fiber. In this Table, the POF is a commercially available PMMA core, fluoropolymer clad fiber.

Examination of the parameters in Table I reveals significant information about not only the fiber properties but also systems use as well.

The large core size for POF results in superior LED to fiber coupling, as well as lower cost optical connectors. The core dimensions of POF permit the use of plastic molding technology for both fiber-to-fiber connectors and source and detector to fiber adapters. Various POF sizes are available -- 500 micron and 1000 micron are common. Note that the cladding thickness of 5-20 microns for POF is generally less than for glass fiber. The core diameter for POF is generally sufficient for mechanical strength; a cladding thickness of 5 microns is all that is required for good optical properties. POF is protected from abrasion by an additional non-optical jacketing material.

The fiber cost per meter scales approximately with fiber cross-sectional area, while launch power also scales approximately with area. The Table also shows that the numerical aperture (NA) of POF is typically larger than glass fiber. In addition, POF is designed with a step-index construction between core and cladding, while glass core fiber has a graded index. POF offers good LED-fiber coupling efficiency (efficiency proportional to NA^2) resulting from the large numerical aperture.

Fiber attenuation is a key property being studied at various industrial laboratories. A typical attenuation spectrum for a commercial fiber is shown in Figure 1. The most practical transmission window is located at 660 nm where the current minimum attenuation is 160 dB/km. This attenuation results from intrinsic fiber parameters such as scattering and molecular vibration as well as extrinsic parameters such as impurities. Theoretical estimates predict that POF loss can be reduced to less than 20 dB/km.[1] These attenuation specifications allow calculation of loss-limited point-to-point link distances. The maximum link length is given by

$$L_{max} = \frac{\text{Flux Budget}}{\text{Fiber Attenuation}} \quad (1)$$

where the flux budget is established by the optical transmitter and receiver utilized. For a DC to 10 Mbps system which would allow operation of several current LAN protocols, a flux budget of 30dB is easily achieved. Link length is then determined to be L_{max} = 188 meters (619 feet) for present fibers and L_{max} = 1.5 km (4950 feet) for future material with 20dB/km attenuation.

The optical bandwidth listed in Table I for POF is determined by the numerical aperture. For a step index fiber the pulse spreading at the end of a fiber of length L and core index n_c is:

$$\Delta t = \frac{L\,(NA)^2}{c\,2n_c} \quad (2)$$

	Typical Core Size	Typical Fiber Size	Numerical Aperture	Minimum Attenuation	Bandwidth	Minimum Bend Radius	Maximum Operating Temperature	Approximate Simplex Cable Cost
Plastic	460 micron, 960 micron	500 micron 1000 micron	0.47	160dB/km (660nm)	6MHz-km	3cm	90°C	$0.22/m (500 micron) 0.40/m (1000 micron)
Glass	50 micron, 100 micron	125 micron 140 micron	0.29	5dB/km (830nm)	400MHz-km	3cm	70°C	$0.60/m (100 micron)

KEY PARAMETERS OF POF AND GLASS FIBERS FOR LAN USAGE.

TABLE I

The corresponding 3dB optical bandwidth is

$$f_{3dB} = \frac{0.6}{\Delta t} \quad (3)$$

Real POF samples show measured bandwidth greater than the above due to selective excitation of lower order modes and selective attenuation of higher order modes. Figure 2 shows experimental optical bandwidths of POF and comparison with the theory above. For applications involving LED light sources chromatic dispersion must, in principle, also be included. The chromatic dispersion limited bandwidth for POF is approximately 60MHZ-km, so modal dispersion is generally the limiting factor.

Table I specifies 3cm typical minimum bend radius for both plastic and glass fiber. In the case of glass fiber, this is determined by the residual long-term strain that the material can support. For POF, the polymer core and cladding can sustain considerable strain, but light loss begins to increase for bend radii less than 3cm. The thermal properties of POF and glass fiber are remarkably similar due to the fact that the glass fiber requires a polymer coating for protection from abrasion. The final column in Table I gives a typical current pricing for light duty cable structures typical for office installation.

As noted above, POF research is an active field now and advances are expected in several areas. The POF manufacturing process of coextrusion is expected to remain constant in principle. With PMMA core, copolymer cladding materials can be used to give a wide variety of numerical aperture, from approximately 0.5 down to 0.23. Attenuation will be reduced toward the 20dB/km level by reduction of extrinsic and intrinsic losses. As these losses are reduced to below 50dB/km, then the presence of absorbed water molecules in the fiber core becomes a critical, additional loss. Such water vapor can permeate the fiber core from the surrounding atmosphere. Two possible routes are being pursued to eliminate this problem. One is to use core materials that have extremely low saturated water vapor levels. With these materials, water molecules will penetrate the core, but the net effect on loss will be small. A second approach is to use an impenetrable sheath material to prevent access of water vapor to the POF cladding surface. Other polymeric materials are being investigated for higher temperature operation. Such materials will

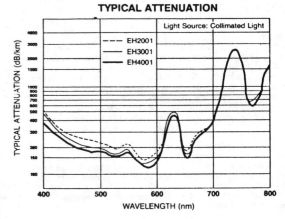

FIG. 1

offer operation up to 130-150°C. As noted above, glass fiber cables also have high temperature limitations resulting from the polymer buffer coating used on these fibers. The same comments apply to solvent resistance, since both glass and POF fibers use polymer materials.

III. Sources & Detectors

Light emitting diodes and PIN detectors are the most common optoelectronic components for glass, as well as polymer fiber LAN systems. A significant difference is that the POF systems optimally operate at 660nm (red), while 830nm (infrared) is optimum for glass fiber transmission. The attenuation of a commercial POF is shown in Figure 1. The 660nm wavelength offers a good combination of low fiber attenuation and availability of efficient sources and detectors. A second difference between components for glass and plastic systems is in the packaging. The large core of POF permits plastic active device receptacles, while the glass systems usually require precision machined components for best active component to fiber coupling efficiency.

Gallium arsenide LED component technology is presently the preferred emitter technology for both plastic and glass systems. The active area of the LED is actually fabricated from an alloy of aluminum arsenide and gallium arsenide; the bandgap and hence, emission wavelength can be centered on either 660nm or 830nm (or values in between). Some important properties of red and infrared emitters are shown in Table II.

The choice of center wavelength 660 or 850 reflects the attenuation minima for plastic and glass fiber. The source spectral width is important for establishing 830nm system bandwidth, but, as pointed out above, is less important for typical POF systems. A critical LED parameter is launched power, P_{fiber}. Values in Table II reflect high performance commercially available components; both LEDs are packaged in active device mount receptacles for direct connection to fiber connectors. The coupled power figures include the following parameters: device current 70mA and fiber diameter 1000 microns for 660nm, 100 microns for 850nm. Although the 660nm device launches more power, in fact these components are <u>internally</u> less efficient than 850nm LEDs. These latter LEDs have nearly 100% internal conversion efficiency of electrons to photons. However, the 660nm LED has competing direct transitions (photon emission) and indirect transitions (non-radiative process), leading to reduced overall output power.

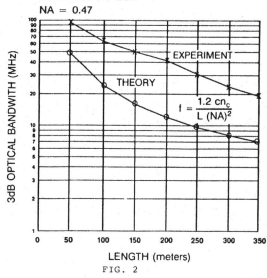

FIG. 2

	λ_o (nm)	$\Delta\lambda$ (nm)	P_{fiber} (microwatts)	Analog Bandwidth	Cost	Packaging
LED for POF	660	30	140(1)	7MHz	$2.93 (1K = Q)	Truncated Plastic Dome
LED for Glass	850	70	80(2)	150MHz	$24.75 (1K = Q)	Metal Can

KEY PARAMETERS OF LEDS FOR LAN SYSTEMS.

(1) Device Current = 70mA, fiber diameter = 1000 microns.

(2) Device Current = 70mA, fiber core diameter = 100 microns.

TABLE II

Several factors contribute to the reduced bandwidth of commercial 660nm LEDs. Firstly, most applications to date have been for systems operating at under 1MHz. Thus, manufacturers have optimized devices for power output and not bandwidth. Second, even for optimized emitters, 830nm devices will be faster since the slower non-radiative emission processes are not present in these devices. Future technology development should result in a selection of 660nm devices optimized for both speed or power output.

Recent developments[2] we have carried out with $Ga_{.5}In_{.5}P$ edge emitting LEDs show that this material system should be considered for future POF LAN systems. Devices were made using organometallic vapor epitaxy with an active layer of $Ga_{.5}In_{.5}P$ and confining layers of $(Al_xGa_{1-x})_{.5}In_{.5}P$. Risetimes of 5 ns and coupled power of 100 microwatts into 500 micron core POF were observed.

Emission wavelength was centered at 665 nm with full width at half maximum intensity of 20 nm. These devices are the fastest reported red emitters and are suitable for Ethernet (10MHz) and token ring (16MHz) LAN applications.

Considerations for optical receivers are more straight forward since silicon photodetectors are preferred for both POF and glass systems. Responsivity, measured in amperes per watt, is lower at 660nm than at 850nm, simply because there are fewer photons per watt at the short wavelength. For devices optimized for 600-850nm operation, this results in a 1.1dB receiver sensitivity degradation in going from 850nm to 660nm operation. Detectors optimized for 850nm may show a 2dB sensitivity degradation in going from 850nm to 660nm operation.

IV. Connectors

Three basic types of optical fiber connector technologies exist at present. The first is the precision machined metal or metal/ceramic connector designed for glass fiber interconnection. Modifications to these designs for POF have appeared using lower cost plastic inserts and ferrules. Some cost reductions have been achieved with these designs. A third generation of molded connectors has been designed specifically for POF. The true cost reductions available with POF connection hardware has been achieved with these designs. Representative prices for these commercial designs, referred to as SMA, modified SMA and plastic molded are shown in Table III.

Each inline termination consists of two connectors and one bulkhead or receptacle. In-line connection using glass fibers and SMA connectors would cost 2 x 5.60 + 2.75 = $13.95. In-line connection using POF would cost 3 x .4 = $1.20. These numbers illustrate the excitement generated by this new interconnect technology.

The above prices refer only to connector hardware costs. Further very significant reduction results from lower termination costs associated with POF. The plastic molded plug termination does not require polishing -- the plastic fiber is "terminated" using a hot knife process. This also makes connector termination easy for field applications.

	Price	Attenuation per Connection
SMA Connector	$5.60	<1dB (100 micron fiber)
Modified SMA	$3.00	<0.5dB (1000 micron fiber)
Bulkhead	$2.75	
Plastic Molded Plug	$ .40	<2dB (1000 micron fiber)
Plastic Moded Receptacle	$ .40	

REPRESENTATIVE CONNECTOR PRICES (Q = 10,000) AND ATTENUATION INFORMATION (8/88)

TABLE III

Typical connector losses are also shown in Table III. The standard SMA achieves 1dB using 100 micron core glass fiber. Using similar precision SMA technology, significantly lower loss is achieved using large core 1000 micron POF. The molded plastic connector technology trades off cost and attenuation resulting in 2dB maximum attenuation per in-line connection.

V. Systems Applications

Systems applications for POF that have been identified to date include LANs for future automobiles, workstation interconnects for office LANs, LANs for home consumer applications and backplane interconnects for advanced computers. The driving forces in these applications are low interconnect cost (including cable, sources, detectors and connectors) and ease and simplicity of installation. Applications requiring large number of connections per unit length of fiber are often candidates for large core POF.

Applications involving office or factory LANs will probably be the first to make use of POF. Such LANs are configured around backbones which interconnect departments in different buildings and work group LANs which interconnect a particular office area. Most of the interconnections within a work group LAN fall within a 100 meter radius. The trend today is to make use of standards based LANs whenever possible. Table IV outlines the key features of several of these systems.

At present all of these networks are commercially available with metallic media physical layer hardware as well as fiber optic media. The FDDI standard has been developed exclusively as an optical fiber network. The five other LAN systems originated as coaxial or twisted pair systems, but are now commercially available as glass optical fiber systems. The installation of optical fiber systems has become sufficiently widespread that standards activities in IEEE committees 802.3, 802.4, and 802.5 are all incorporating or planning to incorporate optical fiber.

The utilization of POF therefore depends on the economics and performance of glass fiber systems and also copper media systems. Table V shows the comparative media costs for systems utilizing twisted pair, coax, thin coax, glass fiber and plastic fiber.

The first demonstration of a POF Ethernet LAN system was presented by us at Productronica '87 in Munich. This was a passive optical star system using POF and 830 nm LEDs. Due to high absorption losses at 830 nm distances were limited in this demonstration. Recent developments of high speed red InGaP LEDs noted above have allowed us to demonstrate a more practical Ethernet system using POF. The test system is shown in Figure 3.

An 80 meter point-to-point data link using 500 micron core 0.48 NA polymer optical fiber (POF) was demonstrated with 10 Mb/s Ethernet Transceivers. The transmitters were modified to incorporate AlGaInP/GaAs edge emitting LEDs operating at 665 nm red wavelength matching the low attenuation window in the fiber transmission spectrum. The rise and fall times of these LEDs were 5 ns. The link was configured as shown in Figure 3 and consists of two data generators (Model 4972A LAN Protocol Analyzer manufactured by HP), two Fiberoptic Multiport transceiver units (Codenoll Model 3038), two Internetwork Bridges (Bridge Communications Inc. Model ESPL), two Fiberoptic Ethernet Transceivers (Codenoll Model 3310) and two 80 meter long POFs terminated with the standard SMA connectors. The connectorized fibers showed a loss of 21 to 22 dB at 665 nm. The system operated collision-free in both directions with no framing or CRC errors. The available flux budget was estimated to be approximately 23 dB.

The point-to-point capability demonstrated here can be extended to active hub type Ethernet systems. With a flux budget of 23 dB and a 3 dB system margin, 20 dB remains for fiber loss. A spectrally weighted average loss of <200 dB/km would allow a network radius of 100 meters. This seems attainable with commercial PMMA core fibers.

SUMMARY OF INFORMATION DESCRIBING LAN NETWORKS

Network	Access Protocol	Line Rate	Maximum No. Stations	Maximum Network Size	Optical Line Coding
802.3 (Ethernet)	CSMA/CD	10 MHZ	1024	4.5km	Manchester
802.4	Token Bus	10 MHZ			Manchester
802.5	Token Ring	4/16 MHZ	260/segment		
ARCNET	Token Bus	2.5 MHZ	255	6.4km	Miller
PC Net	CSMA/CD	10 MHZ	1000	4.8km	Manchester
FDDI	Token Ring	125 MHZ	500	100km	4B/5B

TABLE IV

Similar considerations apply to 802.3, 802.4, 802.5, Arcnet and PC NET systems. All of these networks optimally make use of LED emitter technology. Flux budget and therefore fiber attenuation is the limiting factor on distance.

Different considerations apply to using POF in FDDI networks. This proposed standard [3] uses 1300 nm LEDs and PIN detectors in the Physical layer specification. Distances — limited by dispersion — of up to 2 km between nodes can be supported within the draft PMD document.

However, many FDDI nodes will be within a radius of 100 meters from a concentrater. Such work station hookups could be implemented with POF (although this is not part of the present draft PMD standard).

System design might be the following:

Optical Source: AlGaInP Laser Diode
Output Power: 0 dBm (1mw)
Fiber Attenuation: 170 dB/km
Receiver Sensitivity: -20 dBm
Dynamic Range: 20 dB
Distance: 0-50 meters

Distance limitations in this case result from optical bandwidth of the POF fiber. A numerical aperture of 0.47 gives a measured 3dB optical bandwidth of 96 MHZ (Figure 2) over 50 m. Using the formula $t_{10-90} = .48/f_{3dB}$, a risetime of 5 ns results. By using a fiber with smaller numerical aperture, the distance could be extended to 100 meters.

80 m POLYMER OPTICAL FIBER DATA LINK SYSTEM

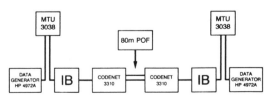

FIG. 3

TYPE	DESCRIPTION	COST PER METER
Twisted Pair	IBM Type 1 dual shielded twisted pair plenum grade	$0.977/meter 4.69 /meter
Coaxial	RG 62A/U, 93 ohm (Arcnet) plenum grade	0.61/meter 1.46/meter
	Ethernet PVC plenum grade	2.20/meter 5.68/meter
	Thin Ethernet RG58	0.43/meter
Glass Fiber Cable	100 micron core, duplex plenum grade	1.15/meter 1.70/meter
Plastic Fiber Cable	Duplex 500 micron core 1000 micron core	0.44/meter 0.78/meter

COMPARATIVE LAN MEDIA COSTS (8/4/88)

TABLE V

VI. Conclusions

In this paper we have tried to illustrate the status of POF technology and illustrate some possible applications to office and factory LANs. The fiber and related component technologies are all developing rapidly at the the present time. It is likely that installation of POF LANs will become pratical in the near future. As we noted above this will lower hardware costs for optical LAN systems. In addition a new phase in installation technology will begin as technicians can eliminate expensive epoxy and polish methods needed for glass optical fiber. POF LAN segments will address most effectively the work station interconnections of under 100 meter radius. Within this distance limitation POF systems will be able to handle data rates from 10 MHZ for Ethernet and MAP up to 125 MHZ for FDDI.

References

1. W. Groh, J. Coutandin, P.Herbrechtsmeier, J. Theis,
   "Material and System Concepts for Polymer Optical Fibers in LANs", FOC/LAN '88, Atlanta, Georgia.

2. B.V. Dutt, J.H. Racette, S.J. Anderson, F.W. Scholl, J.R. Shealy, "AlGaInP/GaAs Red Edge-Emitting LEDs for Polymer Optical Fiber Applications", to be published.

3. W.E. Burr, L. Zuqiu, "An Overview of FDDI", EFOC/LAN 88, June 1988, Amsterdam, The Netherlands.

Networks with polymer optical fibres
Communicating in new ways

As a result of the rapid development of microelectronics, information technology faces radical changes. Data processing, control systems and industrial automation will penetrate deep into existing structures, e.g. in the car and telecommunications industries. Optical techniques will become increasingly important in the future in information technology. The following article describes the application of optical fibres made of transparent plastics in local area networks and multiplex systems for the control of vehicles.

A concept for formalizing the exchange of data between different computers has become established in the last ten years, referred to by the term Local Area Networks (LANs). A LAN is a local intelligent network of different computers forming a group. Instead of one large central computer, LANs can combine and coordinate the intelligence of individual small computers, so that although each individual computer can deal with specific expert problems, a powerful whole is created by their communication with each other. The technique of computer communication in order to achieve the optimum exchange of data is referred to as the LAN technique.

The development of cheaper and smaller but more powerful microcomputers has stimulated the rapid spread of LAN technology, since the total intelligence of the system is immediately available to each subscriber through the exchange of data between the microcomputers or with the central computer. This is an advantage which is not only of great importance in linking work stations but will also lead to completely new potential solutions being found for control problems. The increasing use of microelectronic components in cars and the associated need for communication between the systems has led to LANs being designed for application in cars by leading vehicle manufacturers (such as General Motors) and established network manufacturers (such as AEG, Codenoll and Siemens).

Combining individual modules to form a complete system

Regardless of the specific field of application, there are basically three fundamental LAN configurations for running a system of multiple different computers, i.e. a LAN. The names of these configurations describe the topography of the corresponding systems, namely ring, star and BUS (see box headed "LAN configurations").

A LAN links together several user nodes, which are connected to each other via network modules in each node. This definition indicates how a LAN differs from the conventional stand-alone computer and that each network node on the network module is an integral part of the total system, which consists of the individual modules.

The accounting system of a company may be constructed in such a way that one node is responsible for the storage of all invoicing. A second node is used for processing incoming and outgoing invoices, which are stored on the first node, whilst a third node prints out the invoices, credit transfers etc. which are specified by the first node after being processed by the second node. The great advantage of a network is that it is possible to expand the system at will by adding new network nodes or replacing existing devices. In our example direct electronic transfer processing with the account-holding bank can be connected by a fourth node. This saves the awkward work of printing transfers and delivering them to the bank. If a greater processing capacity became necessary, an additional node could be used and invoice processing divided in such a way that one node handles incoming and one outgoing invoices via the data memory in node one.

All the computers participating in the network send their information into the network in formalized memoranda known as packets. Each packet, as in the postal service, has a destination address and address of origin and gives its contents, which in this case consist of data. The way in which the data are packed decides how a LAN functions. This principle of communication allows invoicing to be updated on the storage node, whilst another user in the same LAN system can have a cheque printed out with another node.

Since the data packets are transmitted one after the other, serially but at extremely high speed, the impression is given that the processes take place simultaneously. Transferred to a car, one computer in this way can control engine management, a second the anti-skid-braking system, a third the air conditioning or the instrument panel etc. Since the data packets are transmitted one after the other and at high speed, the control module in the instrument panel can instruct a front module to switch on the car lights, whilst the engine management module at the same time speeds up the injection pump to accelerate the vehicle.

A control system of this kind based on the linking of individual computers, allowing various data flows to be transmitted to many network subscribers almost simultaneously, is referred to as a multiplex system. The packaging of the necessary information in packets of the form of "data receiver-data origin-data" defines the type of LAN.

Advantages of fibre optic passive star LAN

The LAN configurations of ring, star and BUS (see box) show various weaknesses. In the case of the ring structure, in which the messages are passed on in steps by the network subscribers in between, not only is "network management" necessary, but the main drawback is that the entire system collapses if a cable breaks or a network node fails. Something similar can happen with an active star configuration: if the active star fails, the entire system is paralysed. Although this is ruled out in the case of the very common BUS architecture, in which the LAN network remains functional if one subscriber fails, the copper transmission cable, which makes all data transmission impossible if it fails or suffers interference, remains a weak point. A further drawback is that there is still a need for network management, which regulates who can talk to whom and when.

The problems and risks referred to are avoided by a LAN BUS system in passive star configuration. Optical waveguides make short circuits in the system impossible. Very great safety and reliability for the system are achieved with the star configuration developed by Codenoll Tech. Corp.

Optoelectronic converters such as transmitters (LEDs) and sensitive detectors and passive components such as star couplers are necessary in order to construct a star LAN. Great demands are made on the individual components. The production of light-emitting diodes of high efficiency for light emission is essential for optical communication, because the achievable signal level determines the quality of the transmission. The wavelength of the light used for transmission is chosen such that the optical fibre used exhibits maximum transmissibility in this range. Infrared light with wavelengths of 850–1500 nm is used for glass fibres, and visible red light in the 580–660 nm range for polymer optical fibres.

Codenoll Technology Corp. makes LEDs based on gallium arsenide and indium phosphide, developed in-house and manufactured at its own production plants, which are among the highest light-emitting transmitters in the infrared range on the market.

Another condition which has to be met for a LAN to be created in passive star configuration is the availability of extremely fast and sensitive photodetectors which convert optical into electronic signals (at signal levels of less than 10^{-6} watt). Passive optical star couplers consisting of quartz optical waveguides also have to be made. This is one of the specialities of Codenoll Technology Corp., USA, which has already installed more than 2000

LAN configurations
Ring, star and BUS

There are various configurations for constructing local networks. The fibre-optic passive star LAN is the system with the greatest inherent safety and reliability.

Fig. 1 shows the structure of a **LAN ring** in schematic form: 6 nodes each with a receiver and transmitter unit are linked together in a ring. If subscriber 1 wishes to communicate with subscriber 5 in this LAN, all the network subscribers in between, in other words nodes 2 to 5, must receive, process and pass on the memorandum from subscriber 1, which takes the form of a packet, to the next subscriber until the message reaches its recipient. The same process takes place in the opposite direction. It can be easily seen that this procedure, caused by the topography of the network, is clumsy and above all necessitates "network management", in other words not all subscribers can speak or transmit at the same time. The organization of this kind of network therefore requires communication via a special key, referred to as the token. Only the subscriber with the token can send messages into the network via the network node. When the transmission has finished, the token is passed on in the ring, so that the next node can transmit, if it wishes, or pass the token on. This procedure can be compared with a conference, at which an original without copies is read by the participants one after the other and only the one with the original can speak. This network protocol is known as the **token ring LAN** and has been internationally standardized. Apart from the evident complexity of the token ring LAN,

there are also limitations with regard to reliability. If a cable is broken or an individual node fails (possibly due to a power failure), the system collapses. This drawback cannot be tolerated for computer networks and is totally unacceptable for networks in mobile platforms such as aircraft or vehicles, since the failure of a non-vital system, such as the rear light in a car, leads to total collapse of the system.

The **active star configuration** for a LAN system is comparable with a telephone system, in which the subscribers are connected via a switchboard in a star shape (Fig. 2). Subscriber 1 in the network thus sends a data packet for subscriber 5 to the central switchboard. Here there is an active star coupler, which accepts the message through a receiving part, processes it and passes it on to subscriber 5 via its transmitting part. Depending on the cost and effort

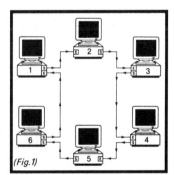
(Fig.1)

involved, the active star coupler can make several transmissions simultaneously or only one switching operation. In any case the complexity of the system is due to the fact that electrical and optoelectronic components such as the receiver and transmitter make the active coupler highly prone to interference. As in the case of the ring LAN, if the active star fails the whole system collapses.

Fig. 3 shows a **linear BUS configuration** for a LAN, as used in many copper cable BUS or LAN systems. The characteristic feature of the BUS system is that the data packets are sent at the same time to all subscribers (Fig. 3). Module 1 can thus communicate with module 5 on the BUS link by sending and receiving packets of data. In other words node 1 conveys its message to all subscribers via its transmitter without a node acting as mediator. This in principle has the advantage that if one subscriber fails, the LAN network continues to function. However, there is still the problem of network management, i.e. who can speak and when. There are three options for organization.

1. The token-key system, similar to the LAN ring. In this case the LAN is referred to as a **"token passing BUS"**. MAP and Arcnet LANs are examples of this token passing BUS system. Although they exhibit similar complexity to the token ring, the individual subscribers are not involved in transmitting all transactions in the network, so that the failure of one subscriber does not amount to system failure.

2. **CSMA/CD systems** stands for "carrier sense" (i.e. listening to discover whether the BUS system is occupied or free, or if another subscriber is talking, waiting until the end of the transmission), "multiple access" (when the network is quiet any subscriber can send) and "with collision detection" (if two subscribers talk at the same time, this is noted by the collision system and the transmission is consequently stopped. Each subscriber then waits for a statistically determined interval of time and starts sending again). This technique is similar to what happens at a meeting where there is no chairman. Everyone wants to speak, but only one can talk. If the next one wishes to speak, he waits until the first one has finished (carrier sense). At this moment several people wish to speak (multiple access). By chance all of them suddenly start talking (collision), with the result that an internal sequence is observed for one to speak after the other. This is similar to what happens in the LAN.

3. A serial, asynchronous **master-slave LAN,** comparable to a meeting with a chairman. The master module instructs the network modules (slaves) when an item of information is required, all the subscribers participating in the exchange of information.

BUS-LANs of this type or broadcast LANs are easy and cheap to install; in addition they are very reliable, since only individual systems can fail, except if the copper transmission cable fails. If the connection of a subscriber node to the cable is faulty, the characteristics of the cable are disturbed to such an extent that data transmission is no longer possible.

This severe failure problem does not apply to the fibre-optic passive star configurations, the structure of which is shown diagrammatically in Figure 4. A **LAN-BUS system in passive star configuration** (the trade name of Codenoll Technology Corp. is "Codestar") combines the advantages of a linear BUS system with the topography of a star network. The data information of each subscriber is converted in this system by optoelectronic converters into an optical signal and transmitted via an optical waveguide. Each optical signal of every subscriber runs across the passive star and is divided up equally between all the channels, i.e. subscribers, whose receivers convert the signal back from an optical to an electrical data flow. Short circuits are not possible in such a system because of the electrically insulating optical waveguide. There are no impedance, adaptation or reflection problems as there are in copper networks, because of the optical transmission. Consequently in the most extreme case, if a cable breaks or a connecting node fails, no more than one subscriber in the network will fail. The fibre-optic passive star LAN is therefore the system with the greatest inherent safety and reliability. For this reason it is suitable for particularly critical fields of application, such as mobile systems, and bank and factory automation.

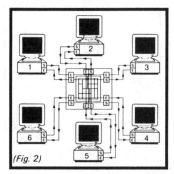
(Fig. 2)

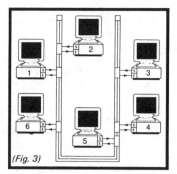
(Fig. 3)

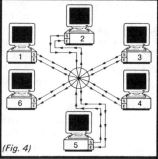
(Fig. 4)

Of the various LAN systems, the BUS system in passive star configuration (right) exhibits a number of advantages.

such Ethernet LANs for office and factory automation and for banks and engineering applications.

A major advantage is offered by the use of polymer optical fibres. This technique allows inexpensive LANs to be produced for short ranges, achieving high transmission speeds with passive optical stars. It will guarantee the high reliability necessary for application in cars, aircraft, spacecraft and marine systems. In the future it will also be possible to create "private BUS systems" based on polymer optical fibres.

Fibre-optic multiplex control in vehicles

The wiring of a car essentially fulfils the functions of power, signal and data transmission. The greater part of the wiring, around 80 %, is for power distribution, e.g. starter, light, fan etc., around 15-18 % is for status and sensor signal transmission, such as the level of fuel or windscreen washer fluid. 2-5 % of the wiring is for data transmission systems, such as anti-skid-braking system, fuel injection etc.

Since the proportion of electronic components in motor vehicles such as check control, anti-skidbraking systems and mobile telephones is continually rising and mechanical systems are increasingly being replaced by electronic ones (e.g. electronic accelerator pedal and electronic fuel injection) it can be anticipated that multiplex control systems will be used in the vehicles of the future.

Another development is leading to the substitution of complete car body parts of steel by plastic and thus to the loss of the electromagnetic screening of the electronics (the Faraday cage effect of the car body), making the system highly prone to interference.

The increasing electronic complexity and susceptibility to interference of electronic vehicle systems will therefore result in new technical requirements being established, which suggests that fibre-optic multiplex control will be a technologically advantageous alternative to the existing control systems.

Fibre-optic lighting and sensors have been used in car-making since the sixties. One example is the illumination of switches and displays with fibre ribbons, particularly at inaccessible points. An advantage is that several points can be fibre-optically illuminated simultaneously with one lamp; this means a considerable saving with the constantly increasing number of lamps in the vehicle. Fibre-optic sensors are used e.g. in washer fluid and fuel level indicators or warning lamps. Future applications for direct optical combustion monitoring or as an independent navigation aid are in the offing.

Fibre-optic data transmission in cars has so far only been used for point-to-point transmission in individual expensively priced models from Japanese manufacturers, such as for transmitting the air conditioning control data from the rear seats to the air conditioning system built into the engine compartment or for controlling a telecommunications system. The fibre-optic multiplex system for application in cars is considered below with reference to the possible network architecture and particularly with regard to the increasing use of microprocessor-controlled systems and the associated increasing demands made on the data rate which can be processed.

Increase in data rate

The data rate of modern microprocessor systems used in vehicles is now 4-10 kbps (thousand bits per second) and is rising constantly with the increasing use of complex systems, such as the anti-skid-braking system, collision avoidance system, electronic engine management, check control, navigation system etc. For this reason a data rate of more than 10 Mbps (10 million bits per second) is expected in the nineties, with the rise taking place in steps through the introduction of new microprocessor-controlled systems. The rising data rate is accordingly making ever greater demands on the data transmission system.

At present vehicle multiplex systems for experimental purposes still use conventional copper wiring. Future demands of increasing data rates and electrical screening for the multiplex control independent of the vehicle will inevitably lead to new technologies. The fibre-optic transmission system holds great promise for putting new developments into effect in cars, since only this system can provide the high transmission rates, immunity to electromagnetic interference and weight and size required, as well as stability to influences in the environment of the vehicle.

Car manufacturers such as General Motors have therefore already been working intensively for several years on the development of fibre-optic multiplex systems of this kind. They have not yet been actually introduced, however, owing to the costs and complexity of the system.

To take an example, the introduction of glass fibres in telecommunications has both increased the transmission capacity through the wide bandwidth and also simplified the technical application, since intermediate repeaters are needed at intervals of more than 50 km in glass fibre networks compared with approximately 1.5 km in copper networks, with the result that the number required is reduced. An optical transmission system, apart from the repeaters and converters, consists of the glass fibre cable, many kilometres in length, and the corresponding optical plugs, with the precision and quality of the optically polished end-surfaces of the fibre having a substantial effect on the quality of the transmission. In the car application, on the other hand, the requirement is to connect a large quantity of active electronic components, for instance in the engine compartment, doors, seats and boot, over relatively short (max. 4 m) but complicated paths with the console on the instrument panel and with each other via a large number of connectors and splitters, but without repeaters.

Conventional glass fibre technology is far too costly for such short-range applications with the large number of optical plugs, splitters, couplers etc. needed by the system. Moreover, the quartz fibres can easily break under mechanical strain during driving and their transmission quality can deteriorate in the aggressive environment. Furthermore, it is not technically possible at present to produce simple and inexpensive plugs and couplers.

Using transparent polymer fibres

The breakthrough by fibre-optic systems for use in vehicles is therefore closely linked to the development of optically transparent polymers. These can be advantageously processed by the extrusion spinning method to form optical fibres; the associated plugs and components can also be produced at low cost. This technique, development of which has only just started, allows transparent polymer fibres (POFs) to be cut to length and spliced in a similar way to copper cables today. This ensures that POF control systems can be industrially prefabricated using automatic assembling and manufacturing techniques and automatically fitted to vehicles by robots.

Advantages of this fibre system are its high mechanical strength, low weight and small cable diameter. The physical advantages of a fibre-optic system, such as high transmission rate and bandwidth, can be utilized in that instead of each electrical system being controlled individually and independently, all the systems are jointly controlled via a network, a central processing unit converting the control commands given by the driver into mechanical, optical or electrical functions according to a specified protocol.

The topography of such a network can be a star, ring or BUS system. These three alternatives have been examined for vehicles according to the proposals of General Motors (see R.E. Steele and H.J. Schmitt in SPIE Vol. 840, 2, 1987). This General Motors analysis shows that the ring and BUS systems require the shortest fibre lengths, around 30-40 m; the numbers of fibre ends at 36 and 68 respectively are also lower than in the star configuration, which has 73 fibre ends and line lengths of around 60 m. However, only a star system can offer the reliability required for the system, since the failure of a line does not result in collapse of the system. For this reason General Motors proposed a star configuration.

The vehicle configuration devised jointly by Hoechst AG and Codenoll Technology Corporation (an affiliated company of Hoechst Celanese Corp,) is shown in Fig. 5 with a double star architecture: one star for the front part of the vehicle with the instrument panel and engine management and one for the

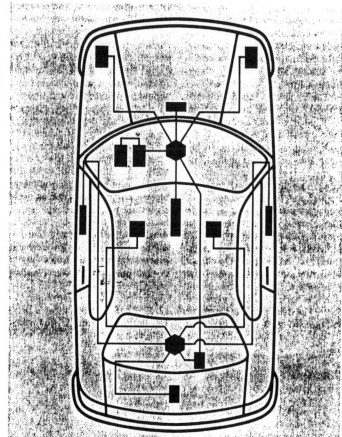

Fig. 5: Topography of vehicle multiplex system with central control, modules for the instrument readings, lights, doors, seats, centre console and central rear light unit.

interior and back. The multiplex system controlled via the two stars and a central processor drives a total of 12 modules which are located at the strategically electrically important positions in the vehicle.

The choice of this double star network combines various advantages, such as costs and reliability, since a passive or transmissive star configuration was selected. In this topography the signals on the transmitter and receiver channels are mixed onto the various channels by the star coupler without any intermediate transducers and repeaters. The electronic protocols for the control of the multiplex system can be freely selected. A serial asynchronous network was chosen which is simple and inexpensive to develop and easy to expand, and which is controlled by a minicomputer. The data rate in this system is 56 kbps and can be increased up to 10 Mbps, depending on the requirement.

Polymer fibre-optic multiplex system for the car

The significance of fibre-optic multiplex systems for car applications in the future is indicated by the savings which can be anticipated, not just on the system side but also in connection with the production process, e.g. by reducing assembly work. Subsequent servicing costs are another indirect source of savings.

Reliability and safety, which must be better than in conventional systems, are the major matters of concern to the user.

New opportunities presented by such a system, such as the use of complex electronic systems for self-diagnosis and maintenance and the ability to add on electronic components such as navigational systems, telephones, CD players etc. must be examined. Additional operating aids such as electronically controllable seat positioning and complex door units with automatic window winders, electronic locks and electronic mirror adjustment, which because of the considerable control requirement can only be accomplished with a multiplex system, should also be considered.

Accomplishing these tasks at reasonable cost depends essentially on whether success is achieved in producing polymer optical fibres in large quantities and at low cost according to the specifications for vehicle application. In addition, passive components such as optoelectronic plugs and couplers can be produced cheaply by the injection moulding method with thermoplastics, which allow high-precision processing, such as LCPs (liquid crystal polymers – the Hoechst trade name is Vectra®). The active components, i.e. the transmitter/receiver modules, are costly at present, but in the future it will be possible to produce them much more cheaply by using ASIC components (application-specific integrated circuits). However, this can only be achieved in mass production.

The ease with which a polymer-optic multiplex system can be assembled and its potential for automation are incomparably better than for conventional copper cable harnesses, which have up to two or more kilometres of copper cable and a weight of up to 50 kg, depending on the vehicle. The large openings in the car body through which the cable harnesses are laid are also disadvantageous as regards body stiffness. In this respect a polymer fibre-optic multiplex system is advantageous, since the optical fibre is extremely light (a few grammes per metre) and thin, i.e. less than 1 mm in diameter, and has high mechanical strength. The optical multiplex system must be simultaneously laid with the conventional copper cable power supply to operate the transmitters/receivers and the active vehicle components, since the multiplex system can only replace the control system.

Expectations on the reliability of the system hold out great promise, but they can only be assessed through extensive practical experience. The inherent advantages of this fibre-optic multiplex system with a passive star are that the system remains available if individual components fail and the fact that optical data transmission is totally immune to electromagnetic interference and reflections or impedance fluctuations. This gives the optical multiplex system a major advantage over conventional copper wiring. An additional plus point in relation to safety is the use of insulating, i.e. dielectric optical fibres in an optical multiplex system, in that short-circuits cannot occur.

The constantly increasing demands made of the data transmission rate due to the increasing electronic complexity of vehicles are at present still being met by conventional copper multiplex systems. In the future, however, it will only be possible to deal with data networks of several Mbps (million bits per second) reliably by using fibre optics. The bandwidth of polymer optical fibres allows data rates of more than 10 Mbps to be achieved without difficulty and can thus meet future needs.

The ability to add onto a multiplex system with a star coupler is ensured either by reserving additional channels in the star coupler or by providing free control contracts at the nodal points. Additional functions must, however, be implemented afterwards by re-programming the processor unit. The prospects are thus very good. The increasing demands made of the control systems of cars mean that new solutions have to be found. Polymer optical waveguides could make a major contribution towards technological progress in this area.

Dr. Ulrich von Alpen
Michael Coden

890203
Implementation of a Passive Star Based Fiber Optic Network for Full Vehicle Control

Richard H. Lefkowitz, Michael H. Coden
and Frederick W. Scholl
Codenoll Technology Corp.

Ulrich von Alpen
Hoechst AG

IMPLEMENTATION OF A PASSIVE STAR BASED FIBER OPTIC NETWORK FOR FULL VEHICLE CONTROL

Richard H. Lefkowitz, Michael H. Coden, Frederick W. Scholl

CODENOLL TECHNOLOGY CORPORATION
1086 NORTH BROADWAY
YONKERS, NEW YORK 10701

Ulrich von Alpen

HOECHST AG
P.O. BOX 60 03 20
6230 FRANKFURT/M. 80
WEST GERMANY

ABSTRACT

In late 1987 - early 1988 a fiber optic automobile Local Area Network (LAN) was designed and constructed. The project was a partnership of three companies:

- Hoechst provided a unique new low attenuation polymer optical fiber (POF) and a custom designed automobile chassis (see figure 1).

- VDO provided a new three screen color LCD dashboard display.

- Codenoll built the LAN (see figure 2) using its unique optoelectronics and Codenet® fiber optic LAN products and technology.

The local area network, designed to demonstrate the viability of plastic optical fiber LAN technology in a consumer item (i.e. an automobile) provided advantages of higher reliability and robustness, easier maintainability, EMC, reduction of EMI, expandability, size and weight reduction and, potentially, cost savings [1] [2]. This paper discusses the POF and LAN performance, some of the problems encountered in the LAN construction and their solutions. The demonstration vehicle - "Opto 1" - clearly shows the viability of a plastic optical fiber based network for use in controlling an automobile.

Figure 1a

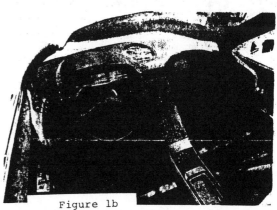

Figure 1b

0148-7191/89/0227-0203$02.50
Copyright 1989 Society of Automotive Engineers, Inc.

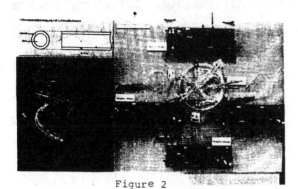

Figure 2

INTRODUCTION

As automobiles become more sophisticated and functionality increases, the automobile as a random interconnection of function modules becomes more complex. Treating the automobile as a system, however, provides opportunities for simplification, modularity and order. This paper describes the architecture of a working prototype - "Opto 1" - that demonstrates the advantages which can be obtained if the automobile system is treated as an analogue of an office or factory type local area network.

1. SYSTEM ARCHITECTURE

A. FIBER SYSTEM

A passive star architecture was chosen as the basic structure for the automobile local area network. This structure has several advantages [1] [2]:

1. This architecture uses a minimum number of transmitter and receiver pairs when compared with active star or double ring.

2. Passive star architecture is the most robust system architecture of all architectures considered [2].

3. Expandability of the system is available if it is planned for in advance.

4. Serviceability - Fault isolation to a spoke on the star is moderately easy.

5. Minimum latency - All receivers see data when it is transmitted.

Normally, the passive star architecture is considered to have several disadvantages. In a plastic fiber based automobile network, however, many of these disadvantages disappear:

1. Number of connectors - Each node requires eight physical connectors (two on each end of the duplex fiber for a total of four connectors on the fiber, one at the receive port of the star and one at the transmit port of the star, one at the receive port of the node, and one at the transmit port of the node). In a glass system, each connector is moderately expensive and is labor intensive. Connectorization requires some skill. Polishing each connection to reduce attenuation to an acceptable level is labor intensive.

   In a plastic optical fiber system, connectorization is simple and inexpensive. Plastic connectors can be easily crimped onto the fiber. The fiber is then simply cut with a sharp blade. Polishing is unnecessary.

2. A glass star is expensive to build. The light mixing element, whether fused biconical or a mixing rod, is expensive. In the case of the fused biconical star, the glass fusing is difficult with a fairly low yield of product. The mixing rod approach on the other hand, has fairly high yield (since a bad fiber or rod can be trimmed and/or repolished) but is labor intensive. Both approaches require high labor content for polishing and connectorizing the fibers.

   A plastic optical fiber star can be molded. Although the molding tool is expensive, the cost can be amortized over many parts. Yield from a mold is high. Connectorizing, using inexpensive plastic connectors, consists of crimping the connector onto the fiber and then cleaving the fiber with a sharp edge. No polishing is necessary. The cost of a plastic optical fiber star is low.

3. Attenuation in the fiber [3] - Plastic optical fiber has high loss (0.1 to 0.2 dB/m), even with a 660 nm visible light red LED (the wavelength that provides minimum attenuation). In an office type LAN, this presents a problem since flux budget is used up fairly quickly on long fiber runs.

   In an automobile LAN, maximum fiber length will be under 5 meters. A loss in the fiber of 0.5 to 1 dB is not considered a problem.

4. Attenuation in the star – A typical 32 port biconical glass star has a maximum attenuation of 21 dB. It is expected, based on measurements of smaller plastic stars, that a 32 port plastic star will have 21 dB or less attenuation. In order to reliably pass data through a 32 port star-based system, a flux budget (i.e. the difference between the output power of the LED and the input sensitivity of the receiver) must be high enough to overcome the attenuation due to the star, connectors and fiber.

In the worst case situation, the connectors at the transmitter would each attenuate the signal 1 dB (for a total of 2 dB). Total fiber loss along the longest path would be 0.5 to 1 dB (from the TX to the star) and 0.5 to 1 dB from the star to the receiver for a total of 1 to 2 dB fiber loss. Maximum loss in the star could be 21 dB. Allowing 3 dB of safety margin gives us a flux budget requirement of (2 dB + 1 dB + 21 dB + 3 dB) = 27 dB.

Currently, Codenoll ships a Codelink®-20BT transmitter base on our own patented Aluminum Gallium Arsenide LED, with an output power of -3 dBm into 100 um core glass fibers. This has been manufactured by us for 5 years and has proven to be reliable in over 20,000 network installations. We have developed 660nm LEDs based on the same device design which has equivalent power and reliability characteristics [4] [5]. A companion receiver product called the Codelink®-20CR has a sensitivity in excess of -37 dBm.

The above components provide our 10 Megabaud Codenet® Fiber Optic Ethernet LAN products for office automation [6][7] and factory automation [8][9] with an available flux budget of 34 dB. This is five times more than is required per the calculation above.

B. ELECTRONIC SYSTEM

The electronic system for the Opto-1 automobile network is based on a master/slave architecture. The Master Module, an 8051 based board, provides two fiber optic transceiver (transmitter/receiver) ports. One port of the Master Module is connected to a seven port optical star at the front of the automobile. The Master Module's second port was connected to a seven port star at the rear of the car. A block diagram showing the architecture of the Master Module is shown in figure 3.

Each seven port star is connected to five Slave Modules in addition to one of the ports of the Master Module. All Slave Modules

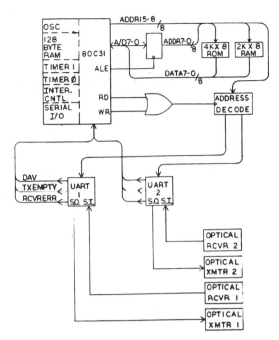

FIGURE 3 - BLOCK DIAGRAM OF THE MASTER MODULE

are identical. The Slave Modules provide the ability to read (and optically transmit) the status of up to 21 switch closure inputs. In addition to the input ports, each Slave Module provides 16 outputs. Each output is capable of driving a 0.5A/12V relay coil. The Slave Modules contain two sets of DIP switches. One DIP switch is set to the unique address of the module. 64 distinct addresses are allowed, providing system expansion up to 64 slave modules (i.e. 1344 inputs and 1024 outputs on each of the two ports of the master module). The second DIP switch allows selection of the number of input and output bits to be received and transmitted by the module. A block diagram showing the architecture of the Slave Module is shown in figure 4.

It should be noted that, in order to increase system efficiency, unused bits are not transmitted onto the network. A Slave Module can be enabled to receive and transmit 7, 14 or 16 output bits and to transmit data from 7, 14 or 21 input bits. In addition, an RS-232 compatible serial port is provided on each Slave Module to allow expansion at each node. This expansion capability is only utilized in one location in the car. In this case, the serial port is used to drive the dashboard display which was a three screen, full color

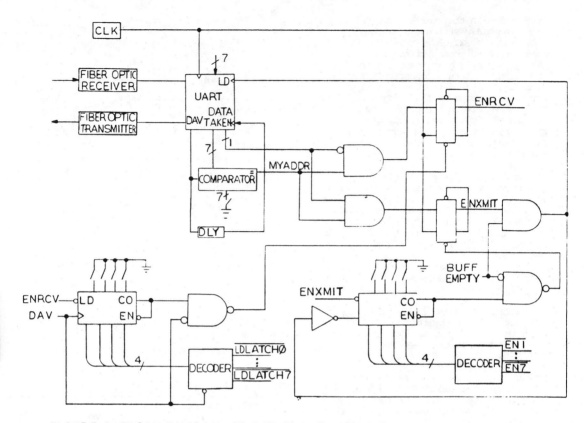

FIGURE 4 BLOCK DIAGRAM OF THE SLAVE MODULE

LCD display.

The network is interconnected as shown in figure 5. Boards are placed at locations which, in a standard car, would have a high concentration of discrete wires. These locations are:

1. The left and right front fenders - where wires are concentrated to control lights and the horn.

2. Under the hood - where ventilation fans and windshield wiper functions are controlled.

3. Under the dashboard and in the drivers console - where many switch contacts and indicators require a high concentration of wiring and interconnections.

4. In the doors - which contain switches motors and solenoids for control of seats, mirrors and door locks.

5. Under the seats - to control seat motors.

6. In the trunk - to control lamps at the rear of the car.

A complete list of functions which are sensed and controlled by the automobile network is contained in Table 1.

TABLE 1. CONTROL FUNCTIONS OF THE FIBER AUTOMOBILE NETWORK

This document lists the input/output functions connected to each of the ten Peripheral Modules in the Fiber Optic Automobile Network:

MODULE 1L (lft Front Fender):

Outputs:

Lft Low Beam Lamp
lft High Beam Lamp
lft Front Fog Lamp
lft Front Parking Lamp
lft Front Turn Signal
lft Front Hazard Lamp
lft Horn
lft Front Spare Lamp

Inputs:

Detect Lft Low Beam Lamp Failure
Detect lft High Beam Lamp Failure
Detect lft Front Fog Lamp Failure
Detect lft Front Parking Lamp Fail
Detect lft Front Turn Signal/
Hazard Lamp Failure

MODULE 1R (rt Front Fender):

Outputs:

rt Low Beam Lamp
rt High Beam Lamp
rt Front Fog Lamp
rt Front Parking Lamp
rt Front Turn Signal
rt Front Hazard Lamp
rt Horn
rt Front Spare Lamp

Inputs:

Detect rt Low Beam Lamp Failure
Detect rt High Beam Lamp Failure
Detect rt Front Fog Lamp Failure
Detect rt Front Parking Lamp Fail
Detect rt Front Turn Signal/
Hazard Lamp Failure

MODULE 2 (Engine Compartment):

Outputs:

Windshield Wiper (Intermittent)
Windshield Wiper (Low Speed)
Windshield Wiper (High Speed)
lft Ventilation Fan (Low Speed)
lft Ventilation Fan (Med.Speed)
lft Ventilation Fan (High Speed)
rt Ventilation Fan (Low Speed)
rt Ventilation Fan (Med.Speed)
rt Ventilation Fan (High Speed)
Engine Compartment Spare Lamp

Inputs:

TABLE 1 (Continued) CONTROL FUNCTIONS OF THE FIBER AUTOMOBILE NETWORK

MODULE 3A (Dashboard Controls):

Outputs:

Inputs:

Head Lamp (Low Beam) Switch
High Beam Switch
High Beam Flasher
Front Fog Lamp Switch
Parking Lamp Switch
lft Turn Signal Switch
rt Turn Signal Switch
Horn Switch
Windshield Wiper (3 posit.) Switch
Hazard Lamp Switch
Ignition Switch
Engine Start Switch
Rear Fog Lamp Switch
Passenger Compartment Light Switch

MODULE 3B (Dashboard Display):

NOTE: The following signals are provided to Module 3B as a serial bit stream via module 3A. Much of the data is provided as part of the demo package resident in the IBM PC which is connected to the Master Module.

Outputs:

Inputs:

Speedometer (7 bits)
Seat belt open
Engine RPM (5 bits)
Windshield Wash Level Low
Radiator Temperature (4 bits)
Anti-skip brake system failure
Lamp Failure (encoded as 5 bits by the Master Module)
Oil Temperature Warning
Oil Level Warning
Oil Pressure (3 bits)
Transmission Gear Position (3 bits)
Fuel Guage (4 bits)

TABLE 1 (Continued) CONTROL FUNCTIONS OF THE FIBER AUTOMOBILE NETWORK

MODULE 3B (Dashboard Display):

Outputs:

Rear Fog Lamps On indicator
Front Fog Lamps On indicator
High Beams On indicator
Parking Lamps On indicator
Hazard Lamps On indicator
rt Turn Signal On indicator
lft Turn Signal On indicator
rt Door Open
lft Door Open
Low Tire Pressure
Oil Change due
Inspection due
Hand Brake On
Battery Discharge
Time (12 bits)

Inputs:

MODULE 4 (Driver's Console):

Outputs:

Inputs:

Automatic Transmission Gear Position (3 bits)
lft Rearview Mirror Up
lft Rearview Mirror Down
lft Rearview Mirror Back
lft Rearview Mirror Forward
rt Rearview Mirror Up
rt Rearview Mirror Down
rt Rearview Mirror Back
rt Rearview Mirror Forward
Mirror Heater
lft Ventilation Fan High Speed
lft Ventilation Fan Medium Speed
lft Ventilation Fan Low Speed
rt Ventilation Fan High Speed
rt Ventilation Fan Medium Speed
rt Ventilation Fan Low Speed
Brake Pedal
Hand brake

TABLE 1 (Continued) CONTROL FUNCTIONS OF THE FIBER AUTOMOBILE NETWORK

MODULE 5L (lft Door):

Outputs:

lft Window Up
lft Window Down
Open lft Door Lock
Close lft Door Lock
Illuminate lft Hand Door Lamp
lft Hand Mirror Up
lft Hand Mirror Down
lft Hand Mirror Back
lft Hand Mirror Forward
Spare Lamp

Inputs:

lft Window Up
lft Window Down
Open Locks
Close Locks
lft Hand Door Open
rt Window Up
rt Window Down
Driver's Seat Heater
Driver's Seat Cushion Front Up
Driver's Seat Cushion Front Down
Driver's Seat Forward
Driver's Seat Backward
Driver's Seat Back Up
Driver's Seat Back Down
Driver's Seat Cushion Rear Up
Driver's Seat Rear Cushion Down
Driver's Seat Headrest Up
Driver's Seat Headrest Down
Driver's Seat Position 1
Driver's Seat Position 2
Driver's Seat Memory

MODULE 5R (rt Door):

Outputs:

rt Window Up
rt Window Down
Open rt Door Lock
Close rt Door Lock
Illuminate rt Hand Door Lamp
rt Hand Mirror Up
rt Hand Mirror Down
rt Hand Mirror Back
rt Hand Mirror Forward
Spare Lamp

Inputs:

rt Window Up
rt Window Down
Open Locks
Close Locks
rt Hand Door Open
rt Window Up
rt Window Down
Passenger's Seat Heater
Passenger's Seat Cushion Front Up
Passenger's Seat Cushion Front Down
Passenger's Seat Forward
Passenger's Seat Backward
Passenger's Seat Back Up
Passenger's Seat Back Down
Passenger's Seat Cushion Rear Up
Passenger's Seat Rear Cushion Down
Passenger's Seat Headrest Up
Passenger's Seat Headrest Down
Passenger's Seat Position 1
Passenger's Seat Position 2
Passenger's Seat Memory

TABLE 1 (Continued) CONTROL FUNCTIONS OF THE FIBER AUTOMOBILE NETWORK

MODULE 6 (Trunk):

Outputs:

Illum. lft Tail Lamp
Illum. rt Tail Lamp
Illum. lft Backup Lamp
Illum. rt Backup Lamp
Illum. lft Rear Turn/ Hazard Lamp
Illum. rt Rear Turn/ Hazard Lamp
Illum. lft Brake Lamp
Illum. rt Brake Lamp
Illum. lft Rear Fog Lamp
Illum. rt Rear Fog Lamp
Illum. Main Rear Lamp 1
Illum. Main Rear Lamp 2
Illum. Trunk Lamp
Illum. Passenger Compartment Lamp
Illum. Spare Lamp

Inputs:

Detect lft Tail Lamp Defective
Detect rt Tail Lamp Defective
Detect lft Backup Lamp Defective
Detect rt Backup Lamp Defective
Detect lft Rear Turn/Haz.Lamp Defect.

Detect rt Rear Turn/Haz.Lamp Defect.

Detect lft Brake Lamp Defective
Detect rt Brake Lamp Defective
Detect lft Rear Fog Lamp Defective
Detect rt Rear Fog Lamp Defective
Detect Main Lamp 1 Defective
Detect Main Lamp 2 Defective
Trunk Open

MODULE 7L (Driver's Seat):

Outputs:

Driver's Seat Cushion Front Up
Driver's Seat Cushion Front Down
Driver's Seat Declination Forward
Driver's Seat Declination Back
Driver's Back Rest Forward
Driver's Back Rest Back
Driver's Seat Heater On
Driver's Seat Rear Down
Driver's Seat Rear Up
Driver's Seat Position 1
Driver's Seat Position 2
Driver's Seat Memory
Spare Lamp

Inputs:

TABLE 1 (Continued) CONTROL FUNCTIONS OF THE FIBER AUTOMOBILE NETWORK

MODULE 7R (Passenger's Seat):

Outputs: Inputs:

Passenger's Seat Cushion Front Up
Passenger's Seat Cushion Front Down
Passenger's Seat Declination Forward
Passenger's Seat Declination Back
Passenger's Back Rest Forward
Passenger's Back Rest Back
Passenger's Seat Heater On
Passenger's Seat Rear Down
Passenger's Seat Rear Up
Passenger's Seat Position 1
Passenger's Seat Position 2
Passenger's Seat Memory
Spare Lamp

C. SOFTWARE ARCHITECTURE

The software routines which control the 8051 Microcontroller start by polling each module connected to a port and building a table of responses. Two control loops, one for each fiber optic port on the Master Module, run concurrently. This reduces the total time to poll each port since the five front Slave Modules are polled at the same time as the five rear Slave Modules are polled. Watchdog timers are provided for each of the two ports on the Master Module to ensure that a failed Slave Module will not disable the entire system. As data is received from the slave boards in response to the polls, the data is processed and a table of output bytes is constructed.

After all boards are polled, the output bytes which are constructed from the responses of the polling operations, are output to the modules.

The original system operated in a continuous mode at 38.4 kilobaud. The data packet required by the Master port which controls the front of the car is longer than the data packet for the Master port which controls the back of the car. Including polls, responses from the slave boards, and command outputs, the Master port which controls the front of the car passes 24 bytes. At a rate of approximately 250 microseconds per byte, total communication requires approximately 9 milliseconds. Allowing for software latency, the original design allowed each board in the system to be polled and updated 65 times per second.

The dashboard display module, supplied with its own internal microcomputer was, in its prototype form, only able to receive data at a rate of 9600 baud. This requirement slowed down the system to be updated at approximately 27 times per second.

Since the network was designed as a demo system, the speed of polling, followed by responses presented a unique demonstration problem. Each visible light LED was illuminated 27 times per second (54 times per second for the Master Module LEDs) as the module transmitted its output. This speed caused the LEDs to appear to be continuously illuminated. In order to be able to demonstrate the system and actually convince people that the LEDs were communicating data and not just staying continuously illuminated, the system was slowed to 10 poll/command cycles per second. At this speed the communicating LEDs obviously flash on when they are transmitting and have a clearly delineated off period. Moreover, this was sufficient for normal operation of the vehicle. Clearly, in a production vehicle, a fast polling capability of would be used.

2. CONSTRUCTION PROBLEMS AND SOLUTIONS

Since the boards were designed quickly using standard, off the shelf SSI and MSI parts, the boards were relatively large. This presented the major system construction problem. The problem, however was not as troublesome as it might have been. All modules were easily placed in clear areas in the fenders (modules 1L and 1R), under the hood (module 2), in the dashboard and console (modules 3 and 4), in the trunk (module 6) and under the seats (modules 7L and 7R). Placing modules in the doors (modules 5L and 5R) presented a problem.

Because of the electric windows and the heavy construction of the doors, there was very little room for the modules. In fact, initially there was mechanical interference when the windows were at their lowest position. Although an mechanically interference-free position was found, there have at least two occurrences where the board shifted in position, causing interference with the window and damage to the board.

In a real system these modules would be constructed using custom parts (ASICs). The reduced size would eliminate the interference problem.

Since the fibers were prototype material, they were constructed without jacketing. Therefore, they were much more fragile than a production fiber cable would have been. In installing the fibers, we jacketed them with tubing and tape. This allowed us to bundle the fibers where parallel runs occurred and to protect the fibers from being sharply bent. Once this was done, the fiber installation was easy. Production fiber would, of course, be jacketed with an extruded polymer of sufficient strength.

Figure 5 shows a schematic drawing of the placement of the stars. Star #1 was placed in the dashboard. This star services modules 1 through 4. Star #2 was place in the trunk immediately behind the rear seat. This star services modules 5 through 7. In addition, each star was connected to one port of the Master Module. Locating the stars in these positions limited the maximum fiber length to under 5 meters. The longest distance a signal had to travel along a fiber was between the Master Module and modules 5L and 5R. This 4-1/2 meter length created a signal attenuation of only 0.7 dB. Since there was a large flux budget available, this attenuation did not create a problem.

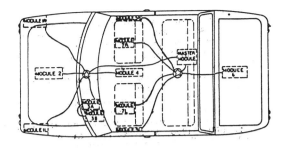

FIGURE 5 - SCHEMATIC DIAGRAM OF THE PLASTIC OPTICAL FIBER LAN SYSTEM.

3. ARCHITECTURAL PROBLEMS AND POTENTIAL SOLUTIONS

The system was built as a demonstration of the feasibility of using a plastic optical fiber passive star based Local Area Network (LAN) in an automobile. As such, it was extremely successful. The network provided sufficient bandwidth for the application. There was sufficient flux budget. Cost estimates show that, within a few years, a plastic optical fiber system will be cost effective as an alternative to bulky copper wiring harnesses and multiplexed twisted pair LAN systems.

As a prototype production system, the Master/Slave system built into this prototype car has one major shortcoming. It is not fault tolerant. Specifically, a failure of the Master Module will cause a catastrophic failure of the system. A failure of the star will also cause a catastrophic failure of the system. However, since the star is a totally passive component, only a catastrophic accident (e.g. a crash which actually cuts the center of the star) will cause star failure. Star failure is, therefore, not considered a likely event.

There are several star based architectures which will provide sufficient fault tolerance:

A redundant Master will provide some fault tolerance. A watchdog timer concept whereby the backup Master overrides the "live" Master if the watchdog times out is a commonly used concept in industrial control and transaction processing systems. A triple redundant system with voting logic will provide even more fault tolerance. This, however, is expensive to implement.

A distributed system is a better solution than a system where the Master is forced to carry the burden of fault tolerance. In a distributed system, each module contains intelligence. A module sends data to a specific receiving module. As an example, the door module could send window data to the other door and seat data to the seat modules. The modules would be programmed to follow a fixed communi- cations discipline such as CSMA/CD (of which CAN, J1850, Ethernet and Starlan are implementations) or a token passing discipline such as MAP. Since there are several modules that will generate little traffic (such as the door and seat modules) and other modules which will generate a great deal of traffic (such as the dashboard and engine modules), a token passing discipline is probably less desirable. A CSMA/CD structure, where the backoff algorithm varies depending on the degree of criticalness of the data would probably be most desirable.

4. CONCLUSION

A system has been built which replaces the copper wire bundles used for vehicle control with a plastic optical fiber LAN. The power supply connections and all electrical system connections have been wired with copper. Cost projections show that, with the added complexity of present and future automobiles, the weight, fuel, reliability and EMI constraints and the need to simplify troubleshooting and repair of the automobiles, a plastic optical fiber LAN system will be cost effective and technically feasible in the near future.

5. REFERENCES

1. U. von Alpen, M. Coden, Communicating in New Ways, pp. 18-24, Hoechst High Chem Magazine 5 (1988)

2. R. E. Steele, H. J. Schmitt, Development of Fiber Optics for Passenger Car Applications, SPIE Proceedings, Volume 840, pp 1-9 (1987)

3. F. W. Scholl, Plastic Optical Fibers Take Aim at LANs, Electronic Engineering Times, Issue 474, February 22, 1988.

4. F. W. Scholl, M. H. Coden, S. Anderson, B. Dutt, J. Racette, W. B. Hatfield, Reliability of Components for Use in Fiber Optic LANs, pp. 108 - 117, SPIE Vol. 717, Reliability Considerations in Fiber Optic Applications (1986).

5. B.V. Dutt, J. H. Racette, S.J. Anderson, F.W. Scholl, J.R. Shealy, AlGaInP/GaAs Red Edge-Emitting Diodes for Polymer Optical Fiber Applications; pp 2091-2092, Applied Physics Letters (21 November 1988)

6. D. E. Stein, High-Rise Optical Communications Networks, pp. 92 - 96, Ninth International Fiber Optic Communications and Local Area Networks Exposition (1985)

7. Advertising Agency Installs Fiber LANs, Lightwave Magazine, October 1988

8. W. B. Hatfield, F. W. Scholl, M. H. Coden, R. W. Kilgore, J. H. Helbers, Fiber Optic Ethernet in the Factory Automation Environment, pp. 297 - 302, SPIE (Aug 1987)

9. P. Musich, GTE's Net Saves $10M, Network World, January 11, 1988

HIGH SPEED POLYMER OPTICAL FIBER NETWORKS

D. Bulusu, T. Zack, F. Scholl and M. Coden
Codenoll Technology Corporation
1086 N. Broadway, Yonkers, N. Y. 10701

and

R. Steele, G. Miller and M. Lynn
Packard Electric
Division of General Motors
P.O. Box 431, Warren, Ohio 44486

ABSTRACT

The purpose of this paper is to compare the cost and performance trade-offs of polymer optical fiber networks with glass fiber networks and networks on other media. The discussion also includes distance versus attenuation and bandwidth limits for higher data rate transmission over polymer fibers. Finally, we describe the performance of a 10 Mb/s Ethernet system built and tested using a modular approach incorporating a prototype snap-lock connection system designed for low cost high volume applications. Low cost is achieved through the use of larger diameter polymer optical fiber, which facilitates the use of molded plastic components and simple crimp-on fiber terminations, integrated receivers and visible red LEDs.

1. INTRODUCTION

Over the past three years, interest has been rapidly growing in polymer optical fiber (POF) as a medium of choice for data communications[1,2] at data rates upwards of 10Mb/s for desk top applications at distances of the order of 10 meters and for office automation applications as well at 50 to 100 meter lengths. In addition, automotive industry is looking at POF based networks in passenger cars for signal, control and monitoring functions[3]. Typically the distances involved in automobiles vary from 2 to 20 meters. Needless to say the driving force for the growing interest is low cost. Besides the cost of the medium, the cost of the connectors and other components, both passive as well as active that make up the system is significantly lower. Moreover, the relative ease of installation of POF is also a favorable factor in driving the costs down.

In the following, we briefly review the properties of the currently available polymer fibers, indicate limitations from the systems point of view, compare POF network costs with networks on other media and point out directions for future research and development both in POF and in related optoelectronic devices.

Finally, we describe the performance of a 10 Mb/s Ethernet system built and tested using a modular approach incorporating a snap-lock connection system designed for low cost high volume applications.

2. POLYMER FIBER PROPERTIES

Currently available POF is a step-index fiber and is predominantly based on polymethyl methacrylate (PMMA) core. The refractive index of PMMA is 1.492. The material used for the cladding depends on the manufacturer. It is generally fabricated from copolymers made up of methyl methacrylate (MMA) and different fluoropropyl methacrylates with refractive indices varying from 1.408 to 1.438[4]. Accordingly, the resultant numerical aperture (NA) varies from .49 to .40. The PMMA based fibers are commercially available with nominal diameters of 250, 500, 750, 1000, 1500 and 2000 micrometers. The cladding in POF is typically of the order of 10 to 15 micrometers thick.

Figure 1 shows typical attenuation properties of one of the commercially available fibers. The best transmission window is centered at the greenish yellow wavelength of 580 +/- 20 nm with an attenuation range of 110 to 150 dB/km. The other two windows are centered at the green wavelength of 530 +/- 30 nm and the red wavelengh of 650 +/- 10 nm. The attenuation ranges corresponding to the green and red regions are 160 to 180 dB/km and 150 to 200 dB/km respectively. Theoretical models based on molecular vibration limited attenuation seem to predict a value of 5 - 10 dB/km[4,5]. As far as we are aware, the best reports to date from research laboratory samples appear to indicate a value of about 25 to 50 dB/km at 650 nm[4].

Figure 2 draws a comparison of the typical bandwidth characteristics measured on samples of PMMA core fibers with a simple theoretical model. The model based on pulse spreading at the end of a step index fiber at the end of a fiber of length, L and core index n_c gives a 3dB bandwidth of

$$f_{3dB} = 1.2 \ c \ n_c/L \ (NA)^2$$

where c = velocity of light in vacuum. Note that the measured bandwidths are higher than the theoretical estimates. This is perhaps due to the selective excitation of the lower order axial modes and attenuation of the higher order cladding modes. The chromatic dispersion limit for POF is approximately 60 MHz-km. Thus the experimental measurements seem to indicate the limit due to modal dispersion. At 100 meters, the bandwidth is 60 MHz in fibers with a numerical aperture of 0.47 and is somewhat limited for high speed networks.

There is currently active research in progress in several research laboratories to improve the attenuation, bandwidth and thermal properties of POF. On an optimistic note, fibers with an attenuation of 50 dB/km and a bandwidth of 25 MHz-km may be expected to be available as the datacom applications abound and the market demand grows within the next few years[5]. Select laboratory research results offer the promise of graded index polymer fibers too with higher bandwidth and lower attenuation in not too distant a future[5]. Table I shows a comparison of the POF with other media costs.

TABLE I: COMPARISON OF COST OF POF WITH OTHER MEDIA

Medium	Cost/meter, $
POF 500 micron core	0.22
POF 1000 micron core	0.40
62.5/125 micron silica fiber	1.30
Twisted pair: unshielded	0.52
shielded	2.00
Thin Coaxial cable, RG 58	0.43

3. OPTOELECTRONIC DEVICES AND CONNECTORS

3.1. Sources

The available III-V semiconductor based visible emitters, matching the best transmission window region of the PMMA fibers, namely, the wavelength region of 530 to 580 nm, are not very efficient. Because of this, all the present emphasis on sources for the PMMA based fiber is concentrated at the red 650 nm transmission window[6]. The popular AlGaAs/GaAs structures have the advantage of a very mature technology. Although such device structures are not totally ruled out, the active layer AlGaAs alloy composition corresponding to the wavelength region of 650 nm appears to have a direct to indirect band transition, which competes with the band to band recombination. This may pose limitations in developing very efficient and high speed emitters from the AlGaAs/GaAs alloy structures. Recent focus for fast and efficient red emitters has moved towards the $(Al_xGa_{1-x})_{0.5}In_{0.5}P$ alloy structures lattice matched to GaAs substrates[7-12], because of favorable band structure in this materials system. Even un-optimized devices without proper facet coatings showed 5 ns speeds and −10 dBm coupled powers into 500 micron core PMMA fibers with a numerical aperture of 0.48[6].

3.2. Detectors

Silicon detectors predominantly used in the 850 nm wavelength region are also good for the 650 nm red region although the sensitivity is lower by about 1 dB. These are relatively inexpensive and available in a variety of packages commercially.

3.3. Connectors

There are two potential advantages of POF as a medium for connector intensive datacom networks. One is the low cost of connectors and the other is the much simpler fiber termination in the connector. The larger diameter polymer fiber is easily terminated in a plastic connector, by merely stripping the jacket, introducing the fiber in a ferrule, trimming the excess fiber with a knife edge and "polishing" the fiber by finishing the trimmed end face on a hot plate. With experience the whole termination procedure takes several seconds to about a minute and requires rather simple tools and accessories. Table II shows a comparison of the approximate connector costs, insertion loss and termination costs associated with currently available glass and polymer fibers.

The prototype connection system developed and used for the Modular Ethernet described in Section 6 later has estimated costs and performance comparable to or better than the POF connectors listed in Table II.

TABLE II: COMPARISON OF CONNECTORS FOR GLASS AND POLYMER FIBERS

CONNECTOR	INSERTION LOSS	COST/TERMINATION	AVG.PRICE
SMA	<1 dB	$5-10	$4-6
ST*	<1 dB	$5-10	$6-8
BULKHEAD	<1 dB	N/A	$3-4
PLASTIC CONNECTOR	<2 dB	$0.5-1.0	$0.50
PLASTCI BULKHEAD	<2 dB	N/A	$0.50
PLASTIC DEVICE RECEPTACLE	<2 dB	N/A	$0.50

*ST is a registered trade mark of AT&T.

4. SYSTEM CONSIDERATIONS

For polymer optical fiber to gain entry into the high speed communication networks, one of the first things the system designer would like to know is how far and how fast can the data be transmitted without repeaters over the POF. Figure 3 shows, for Non Return To Zero (NRZ) data, the transmission distance as a function of data rate for the presently available POF. Also included in the figure is distance vs. data rate for a fiber with an attenuation of 50 dB/km and a bandwidth of 16 MHz-km. Such a POF fiber with reduced attenuation and improved bandwidth may become available as the datacom markets for POF grow[5]. At data rates less than 60 Mb/s, the fiber attenuation is a limiting factor in how far the data can be sent, while for data rates equal to or greater than 60 Mb/s the bandwidth sets the limit. Therefore for data rates up to 60 Mb/s, the POF attenuation dictates how far the data can be transmitted and is independent of the coding used for transmitting the data. For example, for 10 Mb/s Ethernet protocol using Manchester coding, which results in a clock rate of 20 Mb/s, the distance is determined by the attenuation of the POF rather than the bandwidth. However, as one enters the bandwidth limited regime, namely, for data rates faster than 60 Mb/s, the coding used for the data transmission, such as for example, Return to Zero (RZ) actually imposes a clock rate that is double the date rate and therefore halves the distance transmitted. Similarly, the FOUR-TO-FIVE bit conversion 4b/5b used for 100 Mb/s Fiber Distributed Data Interface (FDDI) networks effectively converts the data rate to 125 Mb/s clock rate and calls for shortening the NRZ distance from Fig. 3 by a factor of 20 percent.

5. DIRECTIONS FOR FUTURE RESEARCH

5.1. Polymer fibers

It is clear from the system considerations discussed above that for a wider acceptance and application in high speed data communications networks, there are a number of areas where the polymer fibers need improvement. For applications with data rates less than 60 Mb/s, the critical parameter is attenuation. Currently, with 660 nm AlGaAs/GaAs LED based transmitters, taking into account a spectral width of about 40 nm for the LED emission pattern, the effective attenuation of the PMMA fibers is about 0.25 dB/m. With an available flux budget of 19 dB (Transmitter power of -12 dBm and Receiver sensitivity of -31 dBm) from the commercially available optoelectronic sources and detectors, the maximum transmission distance is limited to about 65 meters. While the larger diameter POF is good for light collection at the transmitter port, it poses problems at the receiver interface, as the sensitive area of the p-i-n detector and its junction capacitance need to be kept smaller for high speed operation. At present, this seems to further limit the distance. In order to meet the goal of greater than 100 meter transmission, it is imperative that the effective fiber loss be less than 0.16 dB/m in the range of 640 to 680 nm. This means the loss at 650 nm wavelength should not exceed about 0.12 dB/km. This figure is good for systems with active star configuration or for point to point applications only.

In order to reduce system costs and improve reliability, passive star based networking is very attractive. Polymers can be injection molded and this offers a viable approach to manufacture inexpensive passive components such as couplers for network applications[13]. However, passive components, wherever inserted in a network for branching purposes, introduce further loss. For example, typical insertion loss of a 7-port coupler is about 12 dB. With a flux budget of 19 dB, a margin of 3 dB would limit POF link lengths to 16 meters only with an effective attenuation of 0.25 dB/m. Thus to transmit 10 Mb/s Ethernet data over 100 meters in a 7-port passive star based network configuration, the maximum loss in the 640 to 680 nm spectral range should not exceed 40 dB/km.

For applications with data rates faster than 60 Mb/s, such as the 100 Mb/s FDDI the bandwidth limits the transmission distance to 48 meters because of limited 6 MHz-km bandwidth of the available PMMA fibers. This is adequate for short distance desktop applications. For inter or intra office links over 50 or 100 meters, the bandwidth needs improvement.

Other areas of improvement for POF are higher operating temperatures upto 150 to 200^0C and the fire retardance properties. Information on the fire retardance properties of the commercial fibers is unavailable in most cases, although we notice some momemtum on the part of the manufacturers to gather such data[14].

5.2. Optoelectronic devices

As mentioned earlier, the available sources matching the lower attenuation spectral windows at 530 and 580 nm of the PMMA fiber are rather poor emitters. The broad area display LEDs are unsuitable for fiber applications. Besides broad emission patterns, which are hard to capture and couple into fibers, these LEDs are too slow for high speed datacom applications. Simultaneous with the research effort on polymer fibers, a search for efficient and faster sources in this wavelegth region would be very fruitful towards realizing low cost POF networks. Devices based on III-V alloy structures consisting of $(Al_xGa_{1-x})_{0.5}In_{0.5}P$ for confining as well as for active layers in double heterostructures or quantum well structures may be potential sources suitable for the spectral range of 530 and 580 nm. Experience gained by optoelectronic community so far on the design of fiber optic sources using several approaches such as restricted junction areas, quantum wells, or strained layer superlattices may result in the desired sources and detectors as well.

For receivers, one can use the silicon p-i-n detectors, although the sensitivity will be lower at the 530 and 580 nm compared to that at 660 nm. Perhaps III-V alloy based detectors may give the improved sensitivity. However, for both the sources and the detectors, while these alternatives may be attractive, one has to pay attention to the overall costs. Unless high volumes are involved cost of these special devices may be too high.

6. MODULAR 10 Mb/s ETHERNET SYSTEM USING THE SNAP-LOCK CONNECTORS

Figure 4 shows a photograph of the polymer fiber optic module of the modular 10 Mb/s Ethernet system developed. The polymer fiber optic module fits onto different product specific modules containing the rest of the electronic circuitry to perform the Ethernet protocol. Depending on the choice of the product specific module used, the combination results in either (1) a stand-alone transceiver or (2) a card that plugs into a Personal Computer or into a Multi-Media Multi-port repeater. In the latter case, a full compatibility with all other media choices, such as for example glass fiber, or coaxial cable or twisted pair based plug-in cards, is realized. The network schematics for the two test cases are illustrated in Figures 5 and 6.

System performance tests were carried out using the test set-up shown in Figure 5 containing two stand-alone tranceivers and LAN (Local Area Network) analyzers. The results of tests conducted over a 137 hr period are summarized in Table III below:

TABLE III: SUMMARY OF RESULTS OF NETWORK PERFORMANCE TESTS ON POLYMER FIBER OPTIC 10 Mb/s ETHERNET TRANSCEIVERS

Fiber Size: 1000 um, Maximum Link Distance Tested: 50 meters

Peak Network Utilization

50.11 percent
4971 kbits/s
789 frames/sec

Frame Parameters

Average size: 776 Bytes
Maximum Size: 1472 Bytes
Minimum Size: 76 BYTES
Total frames: 3.864×10^8
Total bytes: 3.05×10^{11}

NO ERRORS WERE RECORDED.

7. SUMMARY

In summary, it appears that for data communications at data rates less than 60 Mb/s the currently available polymer fiber attenuation is the limiting factor. Ethernet Link lengths are limited to approximately 60 meters with a flux budget of 19 dB available from low cost 660 nm LED based transmitters and p-i-n detector based receivers. Higher data rate FDDI applications are bandwidth limited to about 45 meters. In order to realize longer link lengths lower fiber attenuation and higher bandwidth are required. Also, another avenue for transporting data over longer distances is to increase the flux budget through the use of lasers and perhaps some effort to optimize the silicon p-i-n detectors at 660 nm for better sensitivity and speed. Certainly, the lower cost of the components both passive and active required for the POF LANs and the lower installation costs will be major driving forces in the very near future for low cost networking. Moreover, efforts to integrate more of the electronic functions onto fewer chips will drive down the costs further.

8. ACKNOWLEDGEMENTS

We gratefully acknowledge stimulating discussions with R. Lefkowitz, S. Anderson, H. Jayawickrama, B. Wislocki, A. Koszyn, J. Racette, J. Olin, J. Orsine, J. Yurtin and D. Messuri. We are indebted to J. P. Chalmin and B. Ramsey for excellent help and discussions on the design of connectors and issues related to marketing of the POF network products. We thank W. Hatfield for critical comments on the manuscript.

9. REFERENCES

1. F. W. Scholl, "Plastic Optical Fibers Take Aim at LANs', EE Times, issue 474, Monday, Feb 22, 1988.

2. F. W. Scholl, M. H. Coden, S. J. Anderson, B. V. Dutt, "Applications of plastic optical fiber to local area networks", pp 190-195, Proceedings SPIE Vol 991 Fiber Optic Datacom and Computer Networks (1988).

3. L. K. Di Liello, G. D. Miller and R. E. Steele, "Passive star based optical network for automotive applications", Paper No. 900627 presented at the SAE International Congress, Detroit, Michigan, USA, Feb 26 - March 2, (1990)

4. Private communication from P. Herbrechtsmeier (Nov, 1988).

5. Private communication from B. Chiron (Aug, 1990).

6. B. V. Dutt, J. H. Racette, S. J. Anderson, and F. W. Scholl, "AlGaInP/GaAs red edge-emitting diodes for polymer optical fiber applications", pp 2091-92, Appl. Phys. Lett. 53 (21) (1988).

7. M. Ikeda, Y. Mori, M. Sato, K. Kaneka and N. Watanabe, p 1027, Appl. Phys. Lett. 47, (1985).

8. M. Ishikawa, Y. Ohba, H. Suguwara, M. Yamamoto, and T. Nakanisi, p 931, Appl. Phys. Lett. (1986).

9. K. Kobayashi, S. Kawata, A. Gomyo, Y. Watanabe, H. Nagasaka, pp 153-156, Extended Abstracts of the 18th International Conference on Solid State Devices and Materials, Tokyo, 1986 (Komiyama, Japan).

10. M. Ikeda, A. Toda, K. Nakano, Y. Mori, and N. Watanabe, p 1033, Appl. Phys. Lett. 50 (1987).

11. H. Tanaka, Y. Kawamura, S. Nojima, K. Wakita, and H. Asahi, p 1713, J. Appl. Phys. 61 (1987).

12. D. P. Bour and J. R. Shealy, p 1658, Appl. Phys. Lett. 51 (1987).

13. M. H. Coden and B. V. Dutt, patent application pending (1989).
14. Private communication from several manufacturers of POF from Japan (1990).

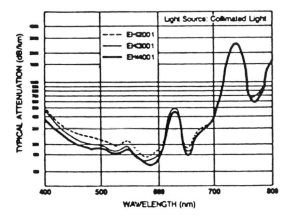

Figure 1
Typical attenuation of PMMA fibers
(courtesy of Mitsubishi Rayon Corp)

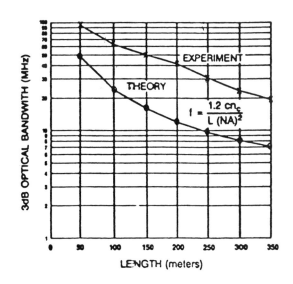

Figure 2
Optical bandwidth of PMMA fibers
(Courtesy of Mitsubishi Rayon Corp)

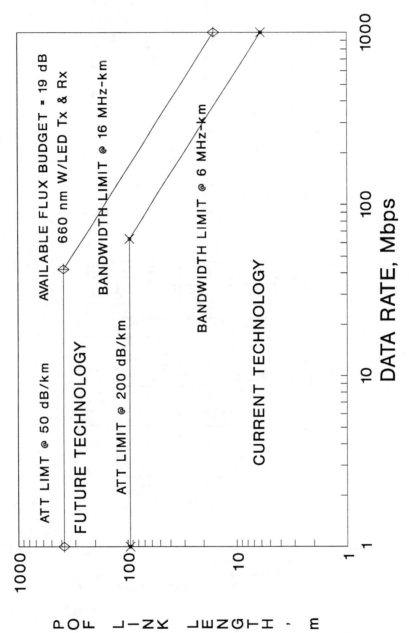

FIGURE 3. POF LINK LENGTHS vs DATA RATE

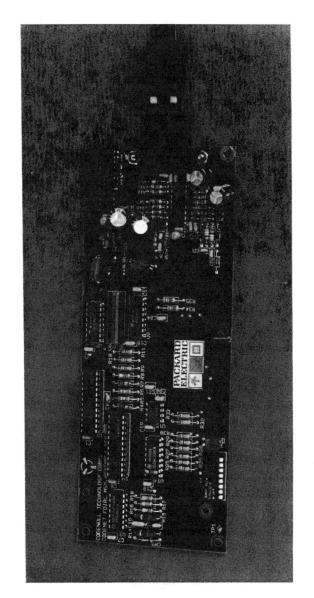

FIGURE 4

Photograph of the polymer fiber optic module with the snap-lock connector

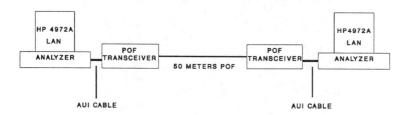

Figure 5

Schematic of test set-up used
to ascertain performance of the
POF Ethernet Transceiver

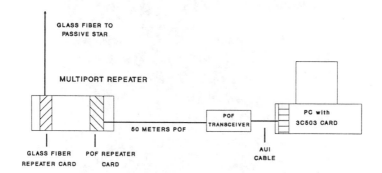

FIGURE 6

SCHEMATIC OF NETWORK CONFIGURATION
USED TO ASCERTAIN PERFORMANCE AND VERIFY
COMPATIBILITY OF POF MODULAR ETHERNET SYSTEM
IN A MULTI-MEDIA ENVIRONMENT

V
341

Glossary of Terms

Glossary of Terms

-A-

Absorption - In an optical fiber, loss of optical power resulting from conversion of that power into heat. Intrinsic causes of absorption in a fiber involve tails of the ultraviolet and infrared absorption bands. Extrinsic components causing loss include (a) impurities, e.g., the OH-ion and transition metal ions and, (b) defects, e.g., results of thermal history and exposure to nuclear radiation. See also: Attenuation

Acceptance Angle - The half-angle of the cone within which all incident light is totally internally reflected by the fiber core. Acceptance angle is related to fiber NA: O = sin -1 NA. **Note:** For graded-index fibers, acceptance angle is a function of position on the entrance face of the core. In that case, the local acceptance angle is: $\sin^{-1}[n^2(r) - n_2^2]^{1/2}$, where n(r) is the local refractive index and n2 is the minimum refractive index of the cladding.

Access - The ability to use a computer or program to store or retrieve information.

Access List - A security system that will only permit users on an internally maintained list to access specific resources or to have specific powers.

Access Method - Way to determine which workstation or personal computer will be next to use the LAN. A set of rules used by network software and hardware that direct the traffic over the network. Examples of access methods are token passing and Carrier Sense Multiple Access with Collision Detection (CSMA/CD).

Active Circuit - A circuit or device that requires electrical power to operate.

Active Fiber Optic Segment (AFOS) - A link segment providing a ppont-to-point connection between Active Stars and Active MAUs, between two Acrive Stars, or between two Active MAUs. See Segment.

Active Port Diameter - On a light source or detector the diameter of the area in which light can be coupled to or from an optical fiber.

Active Star - An active device that connects two or more fiber optic link segments. Optical signals received on the input fiber of any input/output port pair are converted to electrical signals. These are relayed to the outputs of all other input/output port pairs, and converted to optical signals which are transmitted on the output fibers. Collisions are detected and enforced by the Active Star. An Active Star may have two types of ports: (1) Asynchronous Ports capable of receiving FOIRL compatible signals, and (2) Synchronous Ports that transmit and receive only those optical signals synchronized to the 10BASE-F synchronous Active Idle signal. Synchronous ports must be used to connect Active Stars.

Address - A number specifying a particular user device attachment point... The location of a terminal, a peripheral device, a node, or any other unit or component in a network.... A set of numbers that uniquely identifies something - a workstation in a LAN, a location in computer memory, a packet of data traveling through a network. Similar to the address of a house.

AdvanceNet - An Ethernet-based local area network from Hewlett Packard, Palo Alto, Ca.

Algorithm - A "recipe" for making a computer do something. A sequence of steps followed by a computer to accomplish a task.

Alias - Another name that you can be known by on the network.

Allocations - The assignments of frequencies by the FCC for various communications uses (for example: television, radio, land-mobile, defense, microwave, etc.). The assigned frequencies are to achieve a fair division of the available spectrum and to minimize interference among users.

Aloha - The name comes from a method of telecommunications whereby signals are beamed at satellites whenever transmission is ready to go. If it got through, fine. If it didn't, the sender tries again. Now, ALOHA describes a system using a "transmit at will" access method of transmission was first used by satellite dishes in Hawaii beaming at communications satellites over the equator and used for communicating with dishes in other Pacific Ocean countries.

Alphanumeric - Describing a collection of characters that contains letters and numerals. ALPHA, as in alphabet, and NUMERIC as in number.

Alternate Buffer - In a data communications device, the section of memory set aside for the transmission or receipt of data after the primary buffer is full. This helps the device control the flow of data so transmission is not interrupted because there's no place to put the incoming or outgoing data.

Amplifier - A device used to boost the strength (dB level) of an electronic signal. Amplifiers are spaced at intervals throughout a cable system to rebuild the strength of TV or data signals that weaken as they pass through the cable network. Midsplit configurations use a forward and a reverse amplifier in the same enclosure to boost signals in both directions. A device which boosts the strength of a signal. Used in broadband networks to prevent the attenuation [deterioration] of transmitted signals.

Amplitude - The distance between high or low points of a waveform or signal. Also referred to as the wave "height". See: Amplitude Modulation.

Amplitude Modulation - A method of adding information to an electronic signal where the height (amplitude) of the wave is changed to convey the added information. In the case of LANs, the change in the signal is registered by the receiving device as a 1 or a 0. A combination of these conveys difference information, such as words, numbers or punctuation marks.

Analog - A Data format using continuous physical variables such as voltage amplitude of frequency variations... In communications, the description of the continuous wave or signal (such as the human voice) for which conventional telephone lines are designed.

Analog Signal - An electrical signal that varies continuously over an infinite range of voltage or current values, as opposed to a digital signal, which varies discretely between two values, usually one and zero. It is easiest to think of analog signals as sine waves or various sizes. Compare with Digital Signal.

Angle of Incidence - The angle between an incident ray and the normal to a reflecting or refracting surface. See also: Critical Angle; Total Internal Reflection.

Angstrom (A) - A unit of optical length (obsolete). $1A = 10^{-10}$ meter. **Note:** The angstrom has been used historically in the field of optics, but it is not an SI (International System) unit.

Angular Alignment - The alignment of two optical fibers with respect to the angle formed by their axes. Angular Misalignment Loss - The optical power loss caused by angular deviation from the optimum alignment of source to optical fiber, fiber-to-fiber, or fiber-to-detector. See also: Extrinsic Joint Loss; Lateral Offset Loss.

Angular Misalignment Loss - The optical power loss caused by angular deviation from the optimum alignment of source to optical fiber, fiber-to-fiber, or fiber-to-detector. See also: Extrinsic joint loss; Intrinsic joint loss; Lateral offset loss.

ANSI - Abbreviation for American National Standards Institute. A voluntary organization that helps set standards and also the U.S. in the International Standards Organization (ISO).

ANSI X3T9 - The American National Standards Institute's sub-committee that formulated the FDDI product that ANSI has since hailed as the standard for fiber optic 100Mbps token-passing rings. Formulated in October of 1982.

APPC - Advanced Program-to-Program Communications. An IBM-designed strategy for facilitating co-processing under SNA (Systems Network Architecture). Allow applications on a PC to share data with applications on an IBM mainframe. Also referred to as Logical Unit (LU) 6.2.

AppleTalk - A network architecture developed by Apple Computer and implemented on the Macintosh and other computers and peripherals. The protocols defined by AppleTalk may run on a number of physical media, including shielded or unshielded twisted pair (LocalTalk, PhoneNet) and standard and thin Ethernet cable.

AppleTalk Network System - An overall networking system (and any software that supports it) linking computers and peripheral devices such as the Laser Writer II.

Application Layer - Layer seven of the ISO reference model; provides the actual user interface.

Application Program - A tool to manipulate information and help you perform your work. Sometimes called an Application or a Program.

Architecture - The manner in which hardware or software is structured. Architecture typically describes how the system or program is constructed, how its components fit together; also refers to the protocols and interfaces modules or components of the system. Network architecture defines the functions and description of data formats and procedures used for communication between nodes or workstations.

ARCnet - Abbreviation for Attached Resource Computer Network. Datapoint's 2.5 Mbps local area network. One of the earliest and most popular LANs ever invented.

ARP - Address Resolution Protocol specified in RFC-904, used to resolve the mapping of 32 bit Internet addresses into physical addresses on a particular LAN.

ARPANET - Department of Defense Advanced Research Project Network which ties together many university, government, and business users and computers.

ASCII - Abbreviation for the American Standard Code for Information Interchange. A way of encoding characters into binary - on/off - bits. ASCII is the predominant, seven-bit character set used for data communications and data processing in smaller machines. You can represent 128 different characters in ASCII - (2 X 2 X 2 X 2 X 2 X 2 X 2). In contrast, IBM uses an 8-bit coding scheme for coding characters on its mainframes (but not on its PCs). It is called EBCDIC (pronounced Eb-si-dick). You can represent 256 characters with EBCDIC eight bit coding - (2 X 2 X 2 X 2 X 2 X 2 X 2 X 2).

Assembler - A computer language close to machine language.

Asynchronous - A method of transmitting data. A low-cost alternative to synchronous communications. One or more bits are added to the beginning and the end of each data character in asynchronous communications. This allows the receiver of the signal to recognize the characters being sent. Asynchronous is the simplest form of communication since it does not require the sender and receiver to each have a clock to time each other to "stay in synch." However, the addition of these extra bits (called "framing" bits) means that more bits have to be transmitted in asynchronous communications to get the same message across than in other methods, like synchronous communications. Asynchronous techniques are popular with mini- and micro-computers.

Asynchronous Transmission - Transmission in which each information character is individually synchronized, usually by means of start and stop elements. Also called start-stop transmission.

Attachment Unit Interface (AU Interface, AUI) - The physical interface between the DTEs' PLS sublayer and the MAUs' Physical Medium Attachment (PMA) sublayer. The AUI carries encoded control and data signals.

Attenuation - The decrease in power from one point to another expressed in dB. The difference [loss] between transmitted and received power due to transmission loss through equipment, lines, or other communications devices... A reduction in the strength of an electrical signal during transmission, measured in decibels. Opposite of gain. Decibels are measured logarithmically. The general term used used to denote the decrease of power from one point to another. In fiber optics, the optical power loss per unit length is expressed in decibels per kilometer (dB/km) at a specific wavelength. Note: In optical fibers, attenuation results from absorption, scattering, and other radiation losses. Attenuation is generally expressed in dB without a negative sign. Calculations and equations involving loss show and use the negative sign. Attenuation is often used as a synonym for attenuation coefficient, expressed in dB/km. This assumes the attenuation coefficient is invariant with length.

Attenuation Allowance - See Flux Budget

Attenuation Coefficient - A factor expressing optical power loss per unit of length, expressed in dB/km.

Attenuation-Limited Operation - The condition prevailing when the received signal amplitude (rather than distortion) limits performance.

Attributes - Information associated with text you type. Text attributes include the size and style as well as the font you use.

Authentication - A network security mechanism providing a guaranteed identification of a particular user, host or service on a network in a manner that cannot be counterfeited.

Avalanche Photodiode (APD) - A photodiode that shows gain in its output power that it receives through avalanche multiplication of photo current. Note: As the reverse-bias voltage approaches the breakdown voltage, hole-electron pairs created by absorbed photons acquire sufficient energy to create additional hole-electron pairs when they collide with ions; thus, a multiplication (signal gain) is achieved. See also: Photon; PIN photodiode.

Axial Ray - A light ray that travels along the optical fiber's axis. See also: Meridional Ray; Skew Ray.

-B-

Babbling Tributary - A workstation that continuously transmits meaningless messages.

Backbone Node - One of the six major concentration points of fibers in our planned HSDN topology. All major active electronic components will be located at the backbone node locations.

Background Task - A secondary job performed usually while the user is performing a primary task. For example, many network servers will carry out the duties of the network (like controlling who is talking to whom) in the background, while at the same time the user is running their own application (like word processing).

Back Panel - See Connector Panel.

Back up - To make a spare copy of a disk or of a file on a disk. Backing up disks and files ensures information won't be lost if the original is lost or damaged.

Backup Server - A second machine on a network that has copies of all files so at least two up-to-date copies always exist in case the primary server fails.

Backscattering - That portion of scattered light which returns in a direction generally reverse to the direction of propagation. See also: Rayleigh Scattering; Reflectance; Reflection.

Bandwidth - The difference, expressed in Hertz (Hz), i.e., cycle per second, between the highest and lowest frequencies of a transmission channel. Bandwidth varies with the type and method of transmission. The range of frequencies that can pass over a given circuit. Generally, the greater the bandwidth, the more information that can be sent through the circuit in a given amount of time... The bandwidth of an optical fiber is defined as the lowest frequency where the magnitude of the baseband frequency response in optical power has decreased by 3 dB compared to zero frequency. See also: Fiber Bandwidth.

Bandwidth-Limited Operation - The condition prevailing when the system bandwidth, rather than the amplitude (or power) of the signal, limits performance. The condition is reached when material and modal dispersion distort the shape of the waveform beyond specified limits. See also: Attenuation-Limited; Distortion-Limited Operation; Material Dispersion; Modal Dispersion.

Baseband - A signaling technique in which the signal is transmitted in its original form and not changed by modulation. Local Area Networks as a whole, fall into two categories - broadband and baseband. The simpler, cheaper, and thus less sophisticated of the two is baseband. In baseband LANs, the entire bandwidth of the LAN cable is used to transmit a single digital signal. In broadband networks, the capacity of the cable is divided into many channels, which can transmit many simultaneous signals. While a baseband channel can only transmit one signal and that signal is usually digital, a broadband LAN can transmit video, voice and data simultaneously. The electronics of a baseband LAN are simpler. The digital signals from the sending devices are put directly onto the cable without modulation of any kind. Only one signal is transmitted at a time. Multiple "simultaneous" transmissions can be achieved by a technique called time division multiplexing (see multiplexing). In contrast, broadband networks (which typically run on coaxial cable) need more complex electronics to decipher and pick off the various signals transmitted. Ethernet and IEEE 802.3 are generally implemented as baseband transmissions.

Baseband Coaxial System - A system whereby information is directly encoded and impressed on the coaxial transmission medium. At any point on the medium, only one information signal at a time can be present without disruption. See Collision.

Baseband LAN - A local area network employing baseband signaling. An example of a baseband LAN is "Ethernet" - bus topology with a CSMA/CD access control technique.

Baseband System - Provides transmission of digital data over various topologies.

Basic - Beginner's All-Purpose Symbolic Instruction Code, a popular computer language used by many personal computer systems.

Batch Processing - A type of data processing where related transactions are grouped, transmitted, and processed together by the same computer at the same time. A type of processing where time is not critical and no user input is needed while the processing (i.e., the "crunching") takes place. An example of batch processing is the corporate payroll. All the employees' weekly times are processed at one time. The other type of data processing is called real time. An example might be the entering of sales orders by clerks, who answer 800 In-WATTS lines and take orders for product. A LAN can handle both types of data processing.

Baud - A measurement of the signaling speed of a data transmission device. The speed in baud is equal to the number of times the line condition (frequency, amplitude, voltage or phase) changes per second. At low speeds (under 300 bits per second), bits per second and baud are the same. But as speed increases (baud is different to bits per second) - because several bits are typically encoded per baud.

Baud Rate - Measurement of data transmission speed, expressed in bits per second or bps.

Beacon Token Process (FDDI) - A process defined by SMT ulsed to locate ring failures. When a station notes such a failure, it enters the beacon process and begins to continuously transmit "beacon frames". The station will continue to transmit unless it receives a beacon from a station "upstream", iln which case it will stop sending ilts own beacon and repeat the beacon from the upstream station. Soon, only one station, the station immediately "downstream" from the failure, will be beaconing. When a beaconing station receives its own beacon, it indicates that the ring has been restored and that station will stop beaconing and issue a claim.

Beam Diameter - The distance between two diametrically opposed points at which the irradiance is a specified fraction of the beam's peak irradiance; most commonly applied to beams that are circular or nearly circular in cross section. Synonym: Beamwidth. See also: Beam Divergence.

Beam Divergence - The increase in beam diameter with increase of distance from the source.

Beamsplitter - A device for dividing an optical beam into two or more separate beams; often a partially reflecting mirror. See also: Coupler; Splitter.

Beamwidth - See Beam Diameter.

Bell 103 - AT&T specification for a modem doing asynchronous originate and answer at speeds up to 300 bits per second on a dial-up phone line.

Bell 201C - AT&T specification for a modem doing synchronous transmission at 2,400 bits per second on both leased line and dial-up phone lines.

Bell 208 A - AT&T specification for a modem doing synchronous at speeds of 4,800 baud, leased line only.

Bell 208B - AT&T specification for a modem doing synchronous at speeds of 4,800 baud on a dial-up phone line.

Bell 209 - AT&T specification for a modem doing synchronous at 9,600 bits per second on both leased line and dial-up phone lines.

Bell 212 - AT&T specification for a modem doing asynchronous transmission at speeds of 1,200 baud on both leased line and dial-up phone lines.

Bell 212A - AT&T specification for a modem doing asynchronous originate and answer at speeds up 1,200 bits per second on a dial-up phone line.

Bend Loss - Increased attenuation occurring when the fiber is curved around a restrictive radius.

Bend Radius - Radius of curvature that a fiber can bend without breaking.

Binary - A numbering system that allows only two values, zero and one - 0 and 1. Binary is the way most computers store information, in combination of ones and zeros. Voltage on. Voltage off. See also: Bit.

Binary Synchronous Communication (BSC/BISYNC) - A method of transmitting data - a half-duplex, character-oriented, synchronous data communications transmission method originated by IBM in 1964. See also: Synchronous.

BIOS - Basic Input Output System.

Bit - A binary unit of information that can have either of two values, 0 or 1. The basic way of storing and transmitting information. Contraction of BInary digiT. The smallest unit of information in a binary system of notation. Data bits are used in combination to form characters; framing bits are used for parity, transmission synchronization, and so on.

Bit Buffer - A section of memory capable of temporarily storing a single binary unit (bit) of information. Used to make data transmission accurate or consistent.

Bit Error Rate (BER) - The percentage of received bits that are in error. Usually expressed as some number referenced to a power of 10. For example, if you refer to a bit error of ten to the minus seven, it means that (theoretically) only one bit will be in error every ten million bits transmitted. Phone lines typically have ten to the minus four errors, whereas satellites and LANs are around ten to the minus nine. BER's of 10^{-9} (1 error in 1 billion) are typical... In digital applications, it is the ratio of bits received in error to bits send. BERs of 10^{-9} (one error bit to a billion sent) are typical.

Bitmap - A dot-by-dot representation of a text character or graphic image.

BITNET - Acronym for Because Its There Network.

Bit Rate - The rate of data throughput on the medium in bits per second.

Bit Time - The duration of one bit symbol (1/BR).

Block - A collection of transmitted information which is seen as a discrete entity. Usually has its own address, control, routing and error checking information. See also packet and packet switching.

Blocking - A connection can't be made. A call can't be completed. There are many reasons for blocking. Sometimes not enough equipment. Sometimes not sufficient lines, or sufficient room on the lines.

BNC Connector - The connector on the back of the computer for a T-connector or thin Ethernet cable.

Booting - The process of loading a computer's memory with the essential information so it can function. The word comes from "pulling oneself up by one's bootstraps". Booting comes in two versions - cold boot and warm boot. A "cold boot" happens when you turn your computer's power on. There is nothing in memory. You now load the operating system into the computer's memory. A "warm boot" occurs when you reload the operating system into memory. In CP/M, applications programs sometime overwrite parts of memory where the CP/M operating system was hiding. You "warm boot" CP/M by hitting control C or by exiting an application. CP/M automatically reloads its missing parts into memory. Since no applications programs over-write any of the MS-DOS operating system, you actually never do a "warm boot" in MS-DOS. When you hit the "reset" button, or simultaneously hit the Control, Alt., and Delete buttons on an IBM or IBM compatible, your throw the MS-DOS operating system out of memory and (hopefully) reload a "new" version.

Bps - Bits per second, a measure of speed in serial transmission. (See also baud.) There are many ways to measure bits per second. So don't assume that just because one LAN has a faster bits per seconds, it will transmit your information faster. You have to factor in speed of writing and reading from the disk, and the accuracy of transmission. Some datacom schemes have better error-checking systems. Typically such systems force a retransmission of data if a mistake is detected. You might have a fast, but "dirty" (i.e. lots of errors) transmission media, which may need lots of retransmissions. Thus the effective bps of that LAN may actually be quite low. Also used to describe hardware capabilities, as in a 9600-bps modem.

BR - The rate of data throughput (bit-rate) on the trunk coaxial medium expressed in hertz.

BR/2 - One half of the BR in hertz.

Branch - An intermediate cable distribution line in a broadband coaxial network that either feeds or is fed from a main trunk. Also referred to as a feeder.

Branch Cable - The AUI cable interconnecting the DTE and MAU system components.

Break-Out Boxes - Plug your computer cable in one side. Plug your printer cable in the other side. With a break-out box you can connect different cable together and see what works with what. Some break-out boxes have LEDs on them, which show which pins are "live" (with electricity). Break-out boxes are mainly used with RS-232-C cables.

Bridge - Equipment which connects different LANs, allowing communication between devices on separate LANs. As in "to bridge" several LANs. Bridges connect LANs that use the same protocol. Sometimes you need bridges because some LANs can only have a maximum number of workstations. Compare with gateway, which connects LANs using different protocols. Simply stated: Equipment which allows the interconnection of LANs, allowing communication between devices on separate networks using similar protocols. Bridges are viewed in the OSI model as operating at the MAC sublayer of level 2.

Broadband - There are typically two methods by which LANs use to convey information. The first is baseband and the second is broadband. The simpler, cheaper and thus, least sophisticated of the two is baseband. In baseband LANs, the digital signals from the sending devices are put directly onto the cable without modulation of any kind. Only one signal is transmitted at a time. Broadband is more complex, more expensive and more sophisticated. You can usually carry several signals simultaneously on a broadband LAN - perhaps several data channels, maybe a video channel. The channels on a broadband network are typically kept separate with a technique called frequency division multiplexing (see multiplexing). Each channel is made to occupy (modulated to) a different frequency slot on the cable. At the receiving end it is demodulated down to its original frequency. This technique is how 50 channels of TV occupy one coaxial cable on a cable TV (CATV).

Broadband LAN - LAN which uses FDM (frequency division multiplexing) to divide a single physical channel into a number of smaller independent frequency channels. The different channels created by FDM can be used to transfer different forms of information - voice, data, and video.

Broadband System - Provides multiple channel transmission over one medium using frequency division multiplexing or separate frequency allocations for each channel.

Broadcast - The act of sending a signal from one station on a LAN to all other stations, all of which are capable of receiving that signal.

Broadcast Message - A message from one user sent to all users. Just like a TV station signal. On LANs, all workstations and devices receive the message. The broadcast address for Ethernet or IEEE 802.3 networks consists of 32 bits of all ones.

Buffer - A software program, storage space in RAM or a separate device used to compensate for differences in the speed of data transmission. A printer buffer is the simplest example. The computer can send data to be printed many times faster than the printer can printer can print it. There are two alternatives. Slow the computer down (the typical alternative). Or, put a buffer between the computer and the printer. Have the computer dump to the buffer. And the printer take from the buffer as it's ready. Exactly like a water dam. Also, protective coating over the fiber. See also: Fiber Buffer.

Bug - An error that occurs in a computer program or in the computer's electrical system.

Bulletin Board System - A fancy name for an electronic message system typically running on a microcomputer. Call up, leave message, read messages. The system is like a physical bulletin board. (That's where the name come from.) Some people call bulletin board systems "electronic mail systems". You can use any computer, any communications software and virtually any 300 or 1,200 bps modem and typically have 15 minutes before the system cuts you off.

Bus - A network topology which functions like a single line which is shared by a number of nodes.

Bus Network - A one cable LAN, in which all workstations are connected to a single cable (also called "a transmission medium"). On a bus network all workstations hear all transmissions on the cable. Each workstation then selects those transmissions addressed to it based on address information contained in the transmission. The simplest and currently the most common LAN topology.

Bypass Cabling or Relays - Wired connections in a ring network that permit traffic to travel between two nodes that are not normally wired next to each other. Such bypass cabling might be used in an emergency. Or it might used while other parts of the system are being serviced. Usually such bypass relays are so arranged that any node can be removed from the ring and the two nodes on either side of the removed node can then talk.

Bundle - A group of individual glass fibers contained within a single jacket acting as one transmission channel.

Bundling - The practice of selling both hardware and software as a single package.

Bypass Switch - An FDDI station option that ensures ring integrity. The bypass switch has a default state when power is lost. When a power failure occurs, optical connections are made within the DAS that reconnect the incoming primary ring fiber with the primary ring fiber. Similar connections are made on the secondary ring. This allows network operation to continue without the need for reconfiguration, which can still be employed for emergency situations.

Byte - A generic term that indicates an eight-bit sequence of binary digits often used to represent a character; used to measure computer and disk-storage capacity.

-C-

Cable Fiber Optic - A jacketed fiber in a form that can transmit optical signals.

Cable Loss - The amount of RF (radio frequency) signal attenuated (loss) by coaxial cable transmission. The cable attenuation is a function of frequency, media type, and cable distance. For coaxial cable, higher frequencies have greater loss than lower frequencies and follow a logarithmic function. Cable losses are usually calculated for the highest frequency carried on the cable. There are many reasons for cable loss, including the cable's shape, size, length and what it's made of.

Cable Terminator - See SCSI Cable Terminator.

Cabling - Wire or groups of wire capable of carrying voice or data transmission. Popular types in networking include coaxial, twisted pair, fiber optic, and RS-232 cable.

Cache - Typically part of the computer's RAM set aside for a "permanent temporary" memory of the last few items read from disk. The theory is the user is likely to want to use those last few things. And therefore it's worth the software and hardware to load them into this temporary RAM. By them being there and instantly available, it gives the computer user the impressions the computer and its programs are actually running a lot faster than they are theoretically capable of. The value of a cache is directly related to how much you will use the last few items of information you have just accessed from your hard or soft disk, or from a distant file server.

CAD - Abbreviation for Computer-Aided Design. A computer system and its related software which is used to design things. A CAD program might be as simple as having computerized drafting tools. It might be as complex as having layouts of integrated circuits, speciality designs which can be pulled up and included into your own design. CAD may also allow for the simulation and testing of various products and inventions before they are actually manufactured. CAD is one of the applications typically run on sophisticated LANs.

Call - A request for connection or the connection resulting from such a request. In data communication, typically it's a conversation between devices. (The same as a telephone call.)

CAM - Abbreviation for Computer-Aided Manufacture. The actual production of goods implemented and controlled by computers and robots. Often used in conjunction with CAD. Currently only a few factories are completely automated. Usually, there is some human intervention in the actual construction of the product, often to make sure a part is placed in the robot correctly.

Camera-Ready - Ready for offset reproductions with no modifications.

Carrier Signal - A continuous waveform (usually electrical) whose properties are capable of being modulated or impressed with a second information-carrying signal. The carrier itself conveys no information until altered in some fashion, such as having its amplitude changed (amplitude modulation), its frequency changed (frequency modulation), or its phase changed (phased modulation). These changes convey the information.

Carrier Detect Circuitry - Electronic components which detect the presence of a carrier signal and thus determine if a transmission is about to happen.

Carrier Sense - The signal provided by the physical layer to the access sublayer to indicate that one or more stations are currently transmitting on the trunk coaxial cable.

Cassette Tape - Units used to store information for mini and micro- computers.

CATV - Abbreviation for Community Antenna Television, a method of delivering quality television reception by taking signals from a well-sited high, central antenna in the community and delivering them to people's homes by means of a coaxial cable network. The electrical components for such systems are also used to create broadband LANs.

CCITT - Abbreviation for Comite Consulatif International de Telegraphie et Telephonie. An international communications standards. Based in Geneva, Switzerland. Concerned with devising and proposing recommendations for international telecommunications.

CCS - Hundred (Roman numeral C) call seconds, a measure of communications traffic. Often used to compare the capacity of a network, especially a telephone system. For example, one conversation of 100 seconds is one CCS.

Center Wavelength - The average of the two optical wavelengths at which the spectral radiant intensity is 50% of its maximum value.

Central Processing Unit - See CPU.

Centronics Printer Standard - The Centronics standard was developed by a company which makes printers called, the Centronics Company. The Centronics plug is a 36-pin single interface with eight of the 36-pin carrying their respective bits in parallel. There are several types of Centronics male and female plugs and receptacles. The pinning - the location of and function of the 36-individual wires is standard from one Centronics cable to another. The Centronics Printer Standard has been adopted by many printer companies. It has now become an effective standard. It is much less susceptible to being modified. In contrast, serial interfaces and cables are always being modified.

Channel - A physical or logical path allowing the transmission of information. The data communications equivalent of a road. 1. A path for electrical transmission. Also called a circuit, facility, line, link, or a path. 2. A specific and discrete bandwidth allocation in the radio frequency spectrum (for example, in a broadband LAN) utilized to transmit one information signal at a time. In general, a path for transmission (usually one way) between two or more points. Through multiplexing, several channels may share common equipment.

Channel Bank - A device used at each end of time-division-multiplex transmission systems to divide the bandwidth into separate channels and to provide control of those channels. Similar to frequency division multiplexing.

Channel Set - All the letters, numbers and characters which a computer can use. The symbols used to represent data.

Channel Translator - Device used in broadband LANs to increase carrier frequency, converting upstream (towards the head-end) signals into downstream signals (away from the head-end).

Character - Any coded representation of an alphabet letter, numerical digit, or special symbol.

Cheapernet - See Thin-net

Chip - The integrated circuit and its package of coded signals.

CICS - Customer Information Control System. An IBM product that provides an environment for on-line database transactions such as the entering of insurance claims.

CIM - Acronym for Computer Integrated Manufacturing, a catch phase implying unity and integration of all the information processing technologies required to make a product, from concept and design to detailed drafting, prototyping, machine and numerical control as well as a wedding of computer aided design, computer aided manufacturing and networking.

Circuit - A communications path between two points...The physical medium on which signals are carried across the AUI. The data and control circuits consist of an A circuit and a B circuit forming a balanced transmission system so that the signal carried on the B circuit is the inverse of the signal carried on the A circuit.

Circuit Switching - There are basically three types of switching - circuit, time division and packet (or message). In circuit switching, the two people or machines communicating are joined together by a temporary physical circuit ("temporary" being for the length of the conversation). Circuit switching is the method of switching phone calls. It is also used in star-configuration LANs, such as data PBXs. See also multiplexing... A switching technique in which an information path (i.e., circuit) between calling and called stations is established on demand, for exclusive use by the connected parties until the connection is released.

Cladding - A low refractive index material that surrounds the core and provides optical insulation and protection to the core... Optical cladding promotes total internal reflection for the propagation of light in a fiber.

Cladding Mode - A mode that is confined by virtue of a lower index medium surrounding the cladding. See also: Mode

Cladding Mode Stripper - A device that encourages the conversion of cladding modes to radiations modes; as a result of its use, cladding rays are stripped from the fiber. A cladding mode stripper often uses a material having a refractive index equal to or greater than that of the waveguide cladding to induce this conversion. See also: Cladding; Cladding Mode.

Client - In general, an entity on a network which requests some service of another entity on the network. Entities may be computers, printers, workstations, or other devices.

Clock - Exactly as it sounds. An oscilliator-generated signal that provides a timing reference for a transmission link. Used in computers to synchronize certain procedures, such as communication with other devices. (See synchronous) It simply keeps track of time, which allows computers to do the same things at the same time so they don't "dump into each other". Simply stated, a device that generates precisely-spaced timing pulses (or the pulses themselves) used for synchronizing transmissions and recording elapsed times.

Clocked Data One (CD1) - A Manchester encoded data "1". A CD1 is encoded as a LO for the first half of the bit-cell and a HI for lthe second half of the bit-cell.

Clocked Data Zero (CD0) - A Manchester encoded data "0". A CD0 is encoded as a HI for the first half of the bit-cell and a LO for lthe second half of the bit-cell.

CMIP - (ISO DIS 9596-2) Common Management Information Protocol.

CMIS - (ISO DIS 9595-2) Common Management Information Services.

CMX Series - Ungermann-Bass' 3270 multiplexer product line.

Coaxial Cable - A type of electrical cable in which a copper wire is surrounded by insulation and then surrounded by a tube of metal whose axis of curvature coincides with the center of the piece of wire, hence the term coaxial. (A cable which consists of an outer conductor concentric with an inner conductor; the two are separated from each other by insulating material.) Coaxial cables have wide bandwidth and can carry many data, voice and video conversations simultaneously. CATV runs on coaxial cable. So do some LANs. One problem with coaxial-based LANs is the difficulty of connecting to and removing workstations from the LAN. Coaxial cable is a cumbersome, thick medium, suited to more permanent installations such as CATV installations in households... A physical network medium which offers large bandwidth and the ability to support high data rates with high immunity to electrical interference and a low incidence of errors.

Coaxial Cable Interface - The electrical and mechanical interface to the shared coaxial cable medium either contained within or connected to the MAU. Also known as MDI (Medium Dependent Interface).

Coaxial Cable Segment - A length of coaxial cable made up from one or more coaxial cable sections and coaxial connectors, and terminated at each ind in its characteristic impedance.

Code - A specific way of using symbols and rules to represent information.

Codebeam - *Tradename for Codenoll's point-to-point line of sight lightbeam transmission products capable of sending information through the air (line-of-sight) using optical sources and detectors (light beam) and interfaces.*

Codec - An assembly comprising an encoder and a decoder in the same equipment.

CodeNet - *Tradename for Codenoll's fiber optic Ethernet Local Area Network, made possible by Codenoll's high power LED technology. Introduced in 1982, CodeNet LANs have been installed in more than 2,000 major fiber optic networks in 22 countries throughout the world representing over 30,000 fiber optic LAN connections. Also the tradename for Codenoll's line of fiber optic Local Area Network products for use with personal computers, minicomputers, mainframes and communications gateways.*

Codelink - *Tradename for Codenoll's point-to-point fiber optic transmission products requiring connection of only two points in Local Area Network. Used for both commercial and military data communications plus general telecommunications.*

Code Rule Violation (CRV) - A sequence of successive analog waveform amplitudes that are not the result of the valid Manchester encoded output of a single optical transmitter. CRVs are caused by the collision of two or more optical transmissions or by the intentional encoding of the transmitted preamble.

Coherent - Light source (Laser) in which amplitude of all waves is exactly equivalent, and rise and fall together.

Collimation - The process by which a divergent or convergent beam of radiation is converted into a beam with the minimum divergence possible for the system (ideally a parallel bundle or rays). See also: Beam Divergence.

Collision - The result of two workstations trying to use a shared transmission medium (cable) simultaneously (overlapping). The electrical signals, which carry the information they are sending, bump into each other (a collision). This ruins both signals (interference) meaning they both have to retransmit their information. In most systems, a built in delay will make sure the collision does not occur again. The whole process takes fractions of a second. Collisions in LANs make no sound.

Collision Detection - The process of detecting that simultaneous (and therefore damaging) transmission has taken place on a shared medium. Typically, each transmitting workstation that detects the collision will wait some period of time and try again. Collision detection is an essential part of the CSMA/CD access method. Workstations can tell that a collision has taken place if, having sent data, they do not receive an acknowledgment from the receiving station.

Collision Detector - An optoelectronic circuit that monitors the signals received via the fiber optic cable from the passive **Codestar** and sends a "collision presence" signal to the host when more than one data signal is detected coming from the **Codestar**.

Collision Presence - A signal provided by the physical layer to the media access sublayer (within the data link layer) to indicate that multiple stations are contending for access to the transmission medium.

Combiner - A passive device in which optical power from several input fibers is collected at a common point. See also: Coupler.

Common Carrier - An organization licensed by some public regulatory authority, e.g. the FCC, to provide a specific set of services for a specific set of rates. Common means the carrier is obligated to carry for everyone. Carrier means they convey something - freight, data, etc. - for their customers. Examples are GTE Sprint, Yellow Freight, AT & T Communications, MCI, American Airlines.

Common File System - A combination of hardware and software that provides all users of a network with access to the same information. See file server.

Compatibility - The ability of a computer system to accept and process data prepared by another similar system without having to adapt it.

Compatibility Interfaces - The MDI coaxial cable interface and the AUI branch cable interface, the two points at which hardware compatibility is defined to allow connection of independently designed and manufactured components to the baseband transmission system.

Compilation - The translation of programs written in a language understandable to programmers into instructions understandable to the computer. Think of programmers writing in every language but Greek and computer understanding only Greek. In this case, Greek is called machine language. The other languages (the programmer languages) are called things like COBOL, FORTRAN, Pascal, dBASE. A compiler is a special program that translates from all these other languages into machine language.

Compiler - A program that translates a high-level language (such as C, Pascal or Fortran) to assembly code.

Compression - The use of any of several techniques to reduce the number of bits needed to represent information in data transmission or storage. This saves storage space on magnetic storage devices such as hard disk, tape drives and floppy disks. It also saves transmission time.

Concentrator (FDDI) - See Wiring Concentrator.

Conditioning - The "tuning" or addition of equipment to improve the transmission characteristics of a leased voice-grade line so that it meets the specifications for higher-speed data transmission. Voice grade lines often have too much "noise" on them. By altering the equipment at both ends of the line, this noise on the line can be overcome. This allows transmission of data, which is much more sensitive to noise than voice.

Connector - A junction which allows an optical fiber or cable to be repeatedly connected or disconnected to a device such as a source or detector... Hardware installed on cable ends to provide physical and optical cable attachment to a transmitter, receiver or another cable.

Connector Panel - The rear surface of a computer or peripheral device, which includes the connectors for peripheral devices or for the computer. Also called Back Panel.

Connect Time - The time a circuit is in use. Connect time typically refers to circuit-switched systems, such as telephone lines.

Connection Management (CMT) [FDDI] - Manages a station's PHY layer components and their interconnections in order to achieve locical connection to the ring. CMT also manages the configurations of MAC and PHY entities within a station. Additionallyl, CMT includes the detection and isolation of faults at the PHY layer plus monitoring and link quality.

Contention - A way to determine how separate workstations can access a cable (shared transmission medium). In this case, each workstation tries to access the network at will. If the network is busy, they must wait and try again. Think of it as "first come, first served"... A "dispute" between two or more devices over the use of a common channel at the same time.

Control Character - A non-printing character used to start, stop or modify a function. Control characters are often added to transmissions to direct certain procedures performed by the information transmitted. These may simply "tell" the transmission where to go, or they may instruct a device to perform some function on the transmission, say, packetize it for transmission over a packet-switching network. Control characters act as engineers on trains, driving and directing the message as it travels.

Controller - A device which acts as the electrical and logical interface between a host system and a local area network. Often it is a plug-in addition to the equipment and involves software as well as hardware. In standard Ethernet, the controller is attached to the network bus by way of a transceiver. For Thin-net, the controller and transceiver are usually combined. A device between the host and terminals that relays information between them. It administers their communication. Controllers may be housed in the host or on a file server. Typically one controller will be connected to several terminals. The most common controller is the IBM Cluster Controller for their family of mainframes.

Control Signal One (CS1) - An encoded control signal used on the Control In and Control Out circuits. A CS1 is encoded as a signal at half the bit rate (BR/2).

Control Signal Zero (CS0) - An encoded control signal used on the Control In and Control Out circuits. A CS0 is encoded as a signal at the bit rate (BR).

Coprocessor - An additional processor that takes care of specific tasks to reduce the load of the CPU. These tasks may vary.

Copy Protect - To make a disk uncopyable. Software publishers frequently try to copy protect their disks to prevent them from being illegally duplicated by software pirates.

Core - The light conducting portion of a fiber defined by its high refractive index. The core is normally the center of a fiber bounded by concentric cladding... The central light transmission part of the fiber with a refractive index higher than that of the cladding.

COS - Abbreviation for Corporation for Open Systems...an industry-sponsored group of computer manufacturers and users that promotes the adoption of international standards in data communications.

Coupler - A device that distributes optical power among two or more ports or concentrates optical power from two or more fibers into a single port. Couplers may be active or passive. See also: Combiner; Splitter; Star Coupler.

Coupling Efficiency - The fraction of available output from a radiant source which is coupled and transmitted by an optical fiber. The coupling efficiency for a Lambertian radiator is usually equal to the sin2O max. for the optical fiber being used. See also: Lambertian Radiator.

Coupling Loss - The power loss suffered when coupling light from one optical device to another. See also: Angular Misalignment Loss; Extrinsic Joint Loss; Insertion Loss; Intrinsic Joint Loss; Lateral Offset Loss.

COW - Character-Oriented Windows Interface. An SAA-compatible user interface for OS/2 applications.

CP/M - Abbreviation for Control Program for Microcomputers, a popular but increasingly outdated microcomputer operating system. **CPU** - Abbreviation for Central Processing Unit, the part of a computer that includes the circuitry for interpreting and executing instructions. (Synonymous with mainframe.)

CRC - See Cyclical Redundancy Check... An error-checking algorithm which is included in a packet before transmission. The receiver checks the CRC on each packet it receives and strips it off before giving the packet to the station. If the CRC is incorrect, there are two options: either discard the packet, or deliver the damaged packet, with an appropriate status indicating a CRC error.

Critical Angle - The smallest angle at which a meridional ray may be totally reflected within at the core-cladding interface. When light propagates in a homogeneous medium or relatively high refractive index (n_1) onto an interface with a homogeneous material of lower index (n_2), the critical angle is defined by $\sin^{-1}(n_1/n_2)$. See also: Acceptance Angle; Angle of Incidence; Meridional Ray; Reflection; Refractive Index (of a medium); Total Internal Reflection.

Cross Modulation - A form of signal distortion in which modulation from one or more RF carrier(s) is imposed on another carrier.

Crossover Cable - Same as null modem cable.

Crosstalk - The introduction of signals from one communication channel into another. Often happens when cables cross each other. Also the name of a popular asynchronous communications software program made by Microstuf, Roswell, Georgia... The phenomenon of light leakage or information transfer from a waveguide to one adjacent. Also called optical coupling... Interference or an unwanted signal from one transmission circuit, detected on another (usually parallel) circuit.

CRT - Abbreviation for Cathode Ray Tube, an electronic vacuum tube, such as a television picture tube, that can be used to display images.

CSMA (Carrier Sense Multiple Access) - A contention technique which allows multiple stations to gain access to a single channel.

CSMA/CD - Abbreviation for Carrier Sense Multiple Access with Collision Detection, a method of having multiple workstations access a transmission medium (multiple access) by listening until no signals are detected (carrier sense), then transmitting and checking to see if more than one signal is present (collision detection). The means of access control in Ethernet. Each workstation attempts to transmit when they "believe" the network to be free. If there is a collision, each workstation attempts to retransmit after a preset delay, which is different for each workstation. It is one of the most popular access methods for PC-based LANs. Think of it as entering a highway from an access road, except that you can crash and still try again. Or think of it as two polite people who start to talk at the same time. Each politely backs off and waits a random amount of time before starting to speak again. Most Ethernet-based LANs use CSMA/CD... A contention technique which allows multiple stations to successfully share a broadcast channel by avoiding contention via carrier sense and deference, and managing collisions via collision detection and packet retransmission. See also: CSMA; Collision Detection (CD).

Current Startup Disk - The disk that contains the system files the computer is currently using.

Customizable - A program characteristic that allows users to modify resources (such as menus) and integrate their own programming tools into a specific computer environment.

Cyclical Redundancy Check - A method of detecting errors in a message by performing a mathematical calculation on the bits in the message and then sending the results of the calculation along with the message. The receiving workstation performs the same calculation on the message data as it receives it and then checks the results against those transmitted at the end of the message. If the results don't match, the receiving end asks the sending end to send again.

-D-

Daisy Chain - A group of devices linked together sequentially.

Daisy Wheel - A letter quality printing mechanism whose printing element is made up of a flat metal or plastic wheel with letters molded at the ends of the spokes.

Dark Current - The external current that, under specified biasing conditions, flows in a photodetector when there is no incident radiation. The average or DC value of this current is identified by the symbol Id.

Data - The basic elements of information that can be processed or produced by a computer... A representation of facts, concepts, or instructions in a formalized manner suitable for communication, interpretation, or processing; any representations, such as characters, to which meaning may be assigned.

Database - Data stored in computer-readable form, and usually indexed or sorted in a logical order. Users can use the index or logical arrangement to find the item of data they need. Used to store names, addresses, order entry data and so on. A typical data base is inventory. Data bases in a central file server are one of the most common LAN applications.

Data Collection - The act of bringing data from one or more points to a central point.

Data Communications - The movement of encoded information by an electrical transmission system. The transmission of data from one point to another.

Data Frame - Consists of the Destination Address, Source Address, Length Field, LLC Data, Pad and Check Frame Sequence.

Datagram - A particular type of information encapsulation at the network layer of the adapter protocol. No explicit acknowledgment of the information is sent by the receiver. Instead, transmission relies on the "best effort" of the link layer... A packet that includes a complete destination address specification (provided by the user, not the network) along with whatever data it carries. It is a one-way construct much like a telegram... A transmission method in which sections of a message are transmitted in scattered order and the correct order is reestablished by the receiving workstation. Used on packet-switching networks.

Data Link - The second level of the OSI model of data communications, one of the emerging LAN standards. It is the level that puts messages together and coordinates their flow. See OSI model. Also used to refer to a connection between two computers over a phone line.

Data Link Service - A service which guarantees transmission between two stations sharing the same physical medium.

Dataphone - A trademark of the AT & T Company to identify the data sets manufactured and supplied by the Bell System.

Data Processing - (Same as "information processing.") The execution of a systematic sequence of operations performed upon data.

Data Processing System - A network of machine components capable of accepting information, processing it according to a plan, and producing the desired results.

Data Rate - The maximum number of bits or information which can be transmitted per second; usually expressed in megabits per second - Mbs... A measure of the signaling rate of a data link.

Data Set - A device containing the electrical circuitry necessary to connect data processing equipment to a communications channel, usually through modulation and demodulation of the signal.

Data Sink - The equipment which accepts the transmitted data.

Data Source - The equipment which supplies the data signals to be transmitted.

Data Stream - Generally, the flow of information being transmitted in a communications system, or path along which it flows.

dB - An abbreviation for decibel, used as a relative unit of measure between two signals on a logarithmic basis. dB is an expression of a ratio between an input level and an output level... The standard unit used to express the relative strength of two signals. When referring to a single signal measured at two places in a transmission system, it expresses either a gain or loss in power between the input and output devices. The reference level must always be indicated, such as 1 milliwatt for power ration... A measure of power in communications: the decibel referenced to one milliwatt. Zero dBm = 1 milliwatt, with a logarithmic relationship as the values increase.

dBmV - An abbreviation for decibel millivolt. The level at any point in a system expressed in dB's above or below a 1 millivolts (dBmV). Zero dBmV is equal to 1 millivolt across 75 ohms. The level at any point in a system expressed in dBs above or below a 1 millivolt/75 ohm standard is said to be the level in decibel-millivolts or dBmV. Zero dBmV is equal to 1 millivolt across 75 ohms.

DCE - Abbreviation for Data Communications Equipment. In common usage it is synonymous with modem. Often used in conjunction with DTE, Data Terminal Equipment. DCE is the equipment that sits between end devices (DTE) and the network. It does the work of helping the terminal equipment communicate over the network. It establishes, maintains and terminates the connection in a data conversation. It also provides any encoding or conversion necessary.

DDCMP (Digital Data Communications Message Protocol) - A byte oriented synchronous protocol developed by Digital Equipment Corporation that supports half or full duplex modes, and either point-to-point or multipoint lines in a DNA (Digital Network Architecture) network.

DDS - Abbreviation for DataPhone Digital Service, a trademark of the AT & T Company to identify a private-line interstate service for digital data communications.

Decibel (dB) - The standard unit used to express gain or loss of optical power. $dB = 10 \log_{10}(P_2/P_1)$

DECnet - Proprietary network architecture and protocols used by Digital Equipment Corporation. DECnet is a suite of protocols and software supporting communications over a variety of local and wide area network media.

DECnet Phase IV - The most recently released version of DECnet, the officially supported standard in wide use today. It is characterized by two level addressing hierarchy where a network can consist of from 1 to 63 "areas", each of which can contain from 1 to 1023 individual computers.

DECnet Phase V - The next version of DECnet, still under development, which is claimed to be fully compatible with OSI networks.

Dedicated Line - A leased telephone line, reserved for the exclusive use of one customer.

Default - The default value of a setting is the original one, which is in effect until other instructions are entered...A preset response to a question or prompt. The default is automatically used by the computer if you don't supply a different response. Default values prevent a program from stalling or crashing if no value is supplied by the user.

Default Route - In TCP/IP networks, a pointer to a gateway that will be able to make connections to the correct destination.

Demodulation - The process of separating a data (digital) signal from an analog carrier signal. Most of the time the signal is coming in over a phone line. Demodulation is the opposite of modulation. Here, the changes made upon the carrier signal during modulation are detected by the receiving device, often a modem. Analog signals from the telephone or satellite transmission medium are turned into digital signals understood by computers...The process of retrieving an original signal from a modulated carrier wave. The technique used in data sets to make communication signals compatible with business machine signals.

Density - The amount of information that can be stored on one sector of one track of a disk.

DES - Data Encryption Standard. A National Bureau of Standards-approved scheme that encrypts data for security purposes. DES is the data-communications encryption standard specified by Federal Information Processing Systems (FIPS) Publication 46.

Desktop Publishing - A system that provides the user with the ability to produce publication-quality documents with a computer and a laser printer.

Destination - Receiver of data; data sink.

Destination Address - That part of a message which indicates for whom the message is intended. Usually a collection of characters or bits. Just like putting a destination address on an envelope.

Detector - A transducer that provides an electrical output signal in response to an optical signal. The current is dependent on the amount of light received and the type of device. See also: Receiver.

Device - Frequently used as a short form of Peripheral Device.

Diagnostic - A test program that runs an electronic device over a range of operating conditions. The program compares the results with previous calculations and corrects any discrepancies. The procedure tests how well the system will operate in the field. Many personal computers have diagnostic programs that check to make sure all is in working order. For example, some diagnostic programs check each section of memory in the computer and devices, the floppy and hard disks.

Dialogue Box - A box that appears on the computer's screen containing a message requesting more information. Sometimes the message warns you that you're asking the computer to do something it can't do or that you're about to destroy some of your information.

Digital - In data communications, the description of the binary (off/on) output of computer or terminal. Modems convert the pulsating digital signals into analog waves for transmission over conventional telephone lines.

Digital Signal - A signal that is either zero (off) or one (on) (or one "something"), rather than as a continuum of voltages. Think of digital signals as a collection of ones and zeros flowing down the cable. Think of analog signals as various size waves with no exact shape.

DIOSS - Distributed Office Support System. An IBM host-based system that translates, exchanges and routes information via store-and-forward techniques.

DIP Switch - Dual In Line Package. Little "on-off, yes-no", switches typically mounted on a printed circuit board and used to configure various options for the device - for example, if the printer automatically puts a line feed at the end of every line, or waits for one from the computer.

Directional Coupler - A passive device used in a cable system to divide or combine unidirectional RF power sources.

Disk - A storage medium for digital data. Can be hard or floppy. Disks store information as magnetic pulses, which are created by short bursts of voltage. Different sections of the disk are electronically magnetized of charged in specific patterns that represent the information being stored, be it letters, number or symbols.

Disk Drive - A peripheral device that stores information on disks.

Disk Server - A device equipped with disks and a program that permits users to create and store files on the disks. Each user has access to sections of the disk. It gives users disk space which they would not normally have at their own personal computers. Some sort of connection between the disk server and PC is needed, like a LAN. Compare with file server, which allows users to share files.

Disk Space - The amount of space available on a disk for storing or processing a document or application.

Dispersion - The cause of bandwidth limitations in a fiber. Dispersion causes a broadening of input pulses along the length of the fiber. Two major types are: a) mode dispersion caused by differential optical path lengths in a multimode fiber, and b) material dispersion caused by a differential delay of various wavelengths of light in a waveguide material... Distortion of an electromagnetic signal caused by different propagation characteristic of different wavelengths, and the differing path lengths of modes in a fiber. See also: Material Dispersion; Modal Dispersion.

Display - What you see on the screen of your computer.

Distortion - The unwanted change in waveform that occurs between two points in a transmission system. AMPLITUDE vs FREQUENCY DISTORTION is caused by the non-uniform gain or attenuation of the system with respect to frequency. DELAY VS FREQUENCY DISTORTION is caused by differences in the transit time of frequencies within a given bandwidth under specified conditions. NON-LINEAR DISTORTION is a deviation from the normal linear relationship between the input and output of a system or component.

Distortion-Limited Operation - The condition prevailing when distortion of a received signal, rather than its amplitude (or power), limits performance. The condition reached when a system distorts the shape of the waveform beyond specified limits. In a fiber optic system, it usually results from material and modal dispersion. See also: Attenuation-Limited Operation; Bandwidth Limited Operations; Material Dispersion; Modal Dispersion; Multimode Distortion.

Distributed Data Processing - The processing of information in separate locations equipped with independent computers. The computers are connected by a network, even though the processing is geographically dispersed. Often a more efficient use of computer processing power since each CPU can be devoted to a certain task. A LAN is the perfect example of distributed processing.

Distributed File Server - A system by which files residing on disks located throughout a network may be made available to workstations also located throughout a network.

Distributed Processing - A general term usually referring to the use of intelligent or programmable terminals for processing at sites remote from a company's main computer facility.

DLC - Data-Link Control. The set of rules used by two nodes, or stations, on a network to perform an orderly exchange of information over the network. A data link includes the physical transmission medium, the protocol, and associated devices and programs, so it is both physical and logical.

DMA - Abbreviation for Direct Memory Access. A fast method of moving data from a storage device directly to RAM which speeds processing.

DNA (Digital Network Architecture) - Digital Equipment Corporation's overall specification for networking with DEC computers. (Digital Equipment Corporation's layered data communications protocol). Abbreviation also used by Network Development Corporation for their network.

Document - Whatever you create with application programs - a file you can open, modify, view or save. Compare File.

Domain - In networks, the technical term for a subdivision of the hosts on a network, the division can be physical, as in separate building LANs, or logical, as in giving the hosts in a particular administrative area their own name group even though they are on the same physical network.

Dopant - A material, usually germanium or boron oxide, added to silica to change its index of refraction.

DOS - Abbreviation for Disk Operating System. A program or set of programs that instruct a disk-based computing system to schedule, supervise work, manage computer resources and operate/control peripheral devices, including disk drives, keyboards, screens and printers. Comes in different types from different vendors. The most popular operating system for PCs is MS-DOS from Microsoft, Bellevue, WA. PC-DOS is MS-DOS from IBM.

Dot-Matrix - A printing method utilizing a matrix pattern of ink dots.

Double-Density - Storing twice as much information on a floppy disk as other standard disk systems.

Double-Sided - A floppy disk system that can store on both sides of a disk, doubling its storage capacity.

Downline Loading - A system in which programs are loaded into the memory of a computer system via the same communication line(s) the system normally uses to communicate with the rest of a network. This system is in contrast to systems in which all programs are loaded into the computer from a disk or tape associated with the computer. A PC connected to a LAN may use this type of loading when it is first turned on in the morning to get the information it needs from a file server.

Download - The process of loading software into the nodes of a network from one node or device over the network media.

Draft Quality - Type from a dot-matrix printer with clearly discernible individual dots.

Driver - See Resource.

Drop Cable - The cable which allows connection and access to and from the trunk cables of a network. The cables that connect individual PCs to the bus on a bus LAN.

DTE - Abbreviation for Data Terminal Equipment. Workstations and PCs are data terminal equipment. Often used on conjunction with DEC, Data Communications Equipment. DTE must connect with DCE for a data conversation. DCE takes the signals from DTE, converts or encodes them, and puts them on the line for transmission. DTE and DCE may be contained in the same device. An example is a modem inside a PC. It is easiest to think of DTE as the end points of a communication link, while DCE are devices on the link. For example, a PC, a printer and a PBX are DTE, while a modem, a cluster controller, a multiplexer and a line driver are DCE. (User equipment) The end-user machine [terminal, computer, controller, etc.] which plugs into a unit which is the termination point of the communications circuit (DCE).

Dual Attach Station (DAS) [FDDI] - A full FDDI station that attaches to both the primary and secondary rings. The DAS contains circuitry to reconfigure the dual ring in the event of a fault in either both of the rings.

Dual Cable - A two-cable system in broadband LANs in which the coaxial cable provides two physical paths for transmission, one for transmit and one for receive, instead of dividing the capacity of a single cable.

Duct System - Pipes or tubing of plastic and metal that hold wires and cables. This enables them to pass under or through the floors of a building or between buildings without being damaged by heat, water, dirt, etc.

Dumb Terminal - A workstation that doesn't have a computer in it and therefore can't do much with the data you feed it. Typically just a keyboard and a screen. The DEC VT 100 is the world's most popular dumb data terminal. The world's most popular terminal is the single line telephone.

Duplex - Characteristic of data transmission. Either full or half duplex. Full permits simultaneous, two-way communication. Half means only one side can talk at once.

Duplex (Two Position) Connectors [FDDI] - Used as the physical connections between stations on the FDDI ring to connect fiber optic cables. The connectors are polarized to prevent the transmitting and receiving fibers from becoming inadvertently interchanged.

Dynamic Routing Strategy - A way to route messages through a network. If one route is disabled or too busy, another route is chosen automatically. A packet switching network has dynamic routing strategy.

-E-

EBCDIC - Extended Binary Coded Decimal Interchange Code. An 8-bit data-exchange code used in computer systems and associated communications equipment. See ASCII.

Echo - The return of transmitted data. Echo means what you send and what is received at the other end is bounced back to you and is what you see on your screen. The idea of "echo" is that the sender can be

sure the data has reached its destination because it is being sent back by the receiving device which had to receive it to send it back. The above is an explanation of full duplex echo. There's also half-duplex echo where you see what you type, not necessarily what is received at the other end. This is called local echo. Echo is often confused with full and half duplex. This is because full duplex is also called echoplex, a misleading term.

ECMA (European Computer Manufacturers Association) - Standards organization dedicated to the development of data processing standards; not a trade organization. ECMA was the first group to define the OSI transport Layer Protocol.

EIA - Abbreviation for Electronics Industries Association. U.S. trade organization that issues its own standards and contributes to ANSI... The U.S. national organization of electronic manufacturers. It is responsible for the development and maintenance of industry standards for the interface between data processing machines and data communications equipment.

EIA Interface - A standardized set of signal characteristics (time duration, voltage, and current) specified by the Electronic Industries Association.

Electrical Connector - Provides fiber optic transceiver access to AC power (+12 to +15 VDC), then transmits this power source to the unit's power converter.

Electronic Mail - A service for the transmission of messages among users of computers and workstations.

Emulation - The imitation of one device by another device. For example, it is possible for a PC or microcomputer to act like one of several dumb terminals. Terminal emulation is the most common form of emulation. There's also printer emulation, where XY-maker printers act like JH-maker printers.

Encapsulation - The provision of end-to-end support of communication using a network protocol X across a network which only supports protocol Y by packaging protocol X packets in headers in the data portion of protocol Y packets.

End Finish - Quality of the surface at an optic-fiber's end, commonly described as mirror, mist, hackle, chipped, cracked, or specified by final grit size used in polishing ($1\mu m$, $.3\mu m$, etc.).

End Separation Loss - The optical power loss caused by distance between the end of a fiber and a source, detector, or another fiber. See also: Extrinsic Joint Loss.

Environment Variables - Variables that are set by users of system shells to streamline the process of using the shell.

EPROM - Abbreviation for Erasable Programmable Read-Only Memory. An EPROM is a little memory chip with a hole in the top. Through the hole you shoot ultraviolet light and clear its memory. Then you shoot electricity through its little pins and the EPROM has new and different material in memory. EPROMs are good because they don't lose their memory when you turn off the electricity. They are bad because you usually have to ship them back to the factory to get new memory. There is also EEPROM, which stands for Electrically Erasable Programmable Read-only Memory. It is just like an EPROM except you use electricity to erase it.

Equalization - A means of modifying the frequency response of an amplifier or network, thereby resulting in a flat overall response. It is slope compensation done by a module within an amplifier enclosure.

Equilibrium Length - For a specific excitation condition, the length of multimode optical waveguide necessary to attain stable distribution of power among propagating modes.

Equilibrium Mode Distribution (EMD) - The condition in a multimode optical fiber in which the relative power distribution among the propagating modes is independent of length. Synonym: Steady-State Condition. See also: Equilibrium Length; Mode; Mode Coupling.

Error Detection - Checking for errors in data transmission. A calculation is made on the data being sent and the results are sent along with it. The receiving workstation then performs the same calculation and compares its results with those sent. The calculation may be as simple as the number of 1s in one part of the message or it may be as complicated as a cyclical redundancy check... Code in which each data signal conforms to specific rules of construction so that departures from this construction in the received signals can be automatically detected. Any data detected as being in error is either deleted from the data delivered to the destination, with or without an indication that such deletion has taken place, or delivered to the destination together with an indication that it has been detected as being in error.

Error Detecting and Feedback System - A system wherein any signal detected as an error automatically initiates a request for retransmission of the data in error. (Also called "decision feedback system", "requests repeat systems", and "ARQ system").

Error Detecting Code - A code in which each data signal conforms to specific rules of constructions so that departures from the norm - errors - are automatically detected. (Synonymous with "self-checking code.") Such codes require more signal elements than are necessary for conveying the fundamental information.

Error Detecting System - A system which employs an error detecting code and is so designed that the errors detected are either automatically deleted from the delivered data (with or without an indication that such deletion has taken place), or delivered together with an indication that an error has been detected.

Error Message - A statement flashed on the computer screen indicating that the user has erred.

Error Rate - The ratio of incorrectly received data (bits, elements, characters, or blocks) to the total amount of data transmitted.

Ethernet - A 10Mbit/sec CSMA/CD standard, utilizing coaxial cable, developed at Xerox, Rochester, NY. One of the most popular baseband LANs in use. (A baseband local area network specification developed jointly by Xerox Corporation, Intel Corporation, and Digital Equipment Corporation to interconnect computer equipment using coaxial cable and "transceivers".) The term is often used interchangeably with IEEE 802.3, though there are minor differences between the two standards.

Ethernet Address - A unique, hexadecimal Ethernet number that identifies a device (such as Macintosh II) with installed EtherPort II cards, on an Ethernet network.

Ethernet Standards - Not all Ethernet and IEEE 802.3 standards are identical. In fact, in some instances, there are enough differences in Ethernet standards to cause major problems. In a network environment (such as Ethernet Version 1.0 and IEEE 802.3) nodes can coexist and communicate properly on a network but the important link is overall transceiver-to-node integrity.

Ethernet Transceiver Cable - An eight-wire (four twisted pair) copper wire interface between an Ethernet controller card and a cable transceiver, using two 15-pin connectors, over which data is transmitted and received.

EtherPort - Refers to the Kinetics EtherPort SE and EtherPortII, which are internal Ethernet controllers for the Macintosh E and the Macintosh II, respectively.

EtherSE - A Kinetics product that connects a device with a SCSI port to Ethernet using the SCSI connector.

EtherTalk - Software for the Macintosh II that routes information between the Macintosh and Ethernet cable.

Expansion Slot - A slot in the PC (or other computer) into which an option card is installed.

External Modem - A modem in its own little box connected to a computer through the computer's serial port. Compare with an internal modem, which typically comes on one printed circuit card and is placed into one of the computer's expansion slots and thus connects to the computer through the computer's "backplane". Internal modems cost less. But because they're mounted in the computer, it's harder to see what they're doing.

Extinction Ratio - The ratio of the low optical power level to the high optical power level within a data stream.

Extrinsic Joint Loss - Loss caused by imperfect alignment of fibers in a connector or splice. Contributors include, angular misalignment, lateral offset, end separation, and end finish. Generally synonymous with insertion loss. See also: Angular Misalignment Loss; End Separation Loss; Intrinsic Joint Loss; Lateral Offset Loss.

-F-

F Connector - A Type of connector used by the CATV industry to connect a coaxial cable to equipment... A low cost connector used by the TV industry to connect coaxial cable to equipment.

FastPath Gateway/Bridge - A Kinetics product that connects an AppleTalk network to Ethernet. FastPath is a programmable device, and may be used as a bridge between two logical AppleTalk networks or as a gateway between two dissimilar network architectures.

Facsimile - (Also called FAX) - The transmission of photographs, maps, diagrams, and other graphic data by communications channels. The image is scanned at the transmitting site, transmitted as a series of impulses, and reconstructed at the receiving station to be duplicated on paper.

FDDI - (ANSI X3T9) An emerging standard for a 100 Mb it/sec local area network, based upon fiber optic media configured as dual counter rotating token rings. Proposed as ANSI standard X3T9.5.

FDDI LAN - A network based on a backbone of dual, counter-rotating 100Mbps fiber optic rings, one of which is normally designated as the primary ring. The dual ring consists of a primary ring and a secondary ring. Because the various stations are connected only to their adjacent neighbors on the ring and to no other stations, the network can continue to operate in the event that one of the stations on the ring fails, or if even one of the point-to-point fiber optic segments becomes disabled. The counter-rotating ring is connected to single fiber "slave" rings through concentrators. Bypassing of inactive stations is accomplished with fiber optic switches. The concentrator can also pass stations on the "slave" ring electronically.

FDM - Frequency Division Multiplex. See Frequency Division Multiplexing. Method by which the available transmission frequency range is divided into narrower bands, each used for a separate channel. As utilized by broadband technology, the frequency spectrum is divided up among discrete channels, to allow one user or a set of users access to single channels.

Feedback - The return of part of the output of a machine, process, or system to the input, especially for self-correcting or control purposes.

Ferrule - A component of a fiber optic connection that holds a fiber in place and aids in its alignment.

Fiber - Any filament or fiber, made of dielectric materials that guides light characterized by a core and cladding synonym: optical waveguide. (A single, separate optical transmission element, characterized by a core and a cladding.) See also: Fiber Bundle.

Fiber Bandwidth - The range of frequencies over which light intensity exiting a waveguide can be varied before attenuation varies 3dB from the mean expressed in megahertz (MHZ). The frequency at which the magnitude of the fiber transfer function decreases to a specified fraction of the zero frequency value. Often, the specified value is one-half the optical power at zero frequency.

Fiber Buffer - Material used to protect an optical fiber or cable from physical damage, providing mechanical isolation or protection. Fabrication techniques include both tight jacket, or loose tube, buffering, as well as multiple buffer layers. See also: Buffer; Fiber Bundle.

Fiber Bundle - An assembly of unbuffered optical fibers. Usually used as a single transmission channel, as opposed to multifiber cables, which contain optically and mechanically isolated fibers, each of which provides a separate channel.

Fiber Optics - A data transmission medium consisting of glass fibers. Light-emitting diodes send light through the fiber to a detector which then converts the light back into electrical signals. Fiber optics will be the predominant media for LANs in the future. Fiber optic LANs offer immense bandwidth, absolute protection from eavesdropping, electromagnetic interference and radioactivity. When the bomb drops, fiber optic LANs will still work, if they're running unattended... A technology that uses light as an information carrier... Fiber optic cables (light guides) are a direct replacement for conventional coaxial cable and wire pairs. The glass-based transmission facility occupies far less physical volume for an equivalent transmission capacity; the fibers are immune to electrical interference... Light transmission through optical fibers for communication of signaling.

Fiber Optic Cable - A cable containing one or more optical fibers.

Fiber Optic Inter-Repeater Link (FOIRL) - A link segment providing a point-to-point connection between FOIRL MAUs. See Segment.

Fiber Optic Link - Any optical transmission channel designed to connect two end terminals or to be connected in series with other channels.

Fiber Optic Medium Attachment Unit (FOMAU) - The portion of the physical layer between the FOMDI and AUI (or repeater unit PLS where the AUI is not implemented) which contains the electronics which transmit, receive and manage the encoded signals impressed on, and recovered from the optical fiber cable link segment.

Fiber Optic Medium Dependent Interface (FOMDI) - The mechanical and optical interface between the optical fiber cable link segment and the FOMAU.

Fiber Optic Physical Medium Attachment (FOPMA) - The portion of the FOMAU that contains the functional circuitry.

Fiber Optic Transceiver - A device that converts electronic signals to optical signals, then drives them on to the fiber cable.

Fiber Pair - A number of terminated optical fibers, interconnected to provide two continuous light paths. Each end of the two light paths is terminated in an optical connector.

File - A collection of related information that is given a specific name and considered a single unit by the computer. It can contain both data and programs... An ordered collection of data, stored on a disk or tape. Files are of various sizes and contain any type of information. Files are made of characters. Some files are created by characters. There are two types of "things" on computer disks - files and applications programs.

File Conversion Software - Programs that translate text formats from one application into text formats for another application.

File Server - A thing containing files, which are shared by everyone connected to the LAN. In some LANs the device is a microcomputer. In other LANS, this device is a special computer "thing" with a huge disk driver and some specialized program. File servers can offer anything from simple data storage, to gateways and protocol conversion. A file server usually has software rules for allowing LAN users to get into and out of the files/data bases on the file server.

File Transfer - The copying of a file from one node to another on a local area network.

Filter - A circuit that selects one or more components of a signal depending on their frequency. Used in trunk and branch lines for special cable service such as two-way operation.

Firewall - A point of interconnection between two portions of a network, usually implemented by an intelligent router, where information propagating into one network from the other can be restricted for reasons of fault isolation and routing control.

Firmware - Programs kept in semipermanent storage, such as various types of read-only memory. These programs can be altered but with difficulty. Firmware is used in conjunction with hardware and software. It also shares characteristics of both. Compare with hardware and software. See EPROM, which is where firmware is usually stored.

Fixing Rollers - The rollers inside a laser printer that fuse toner to paper.

5ESS - An AT&T telecommunications switch.

Flag - A program-readable symbol that indicates data are available, that space is available to store data, or that some operation has been completed. Also used in packet transmission to indicate the beginning and end of the packet. It refers to a group of bits or bytes which provides some instruction to computer devices.

Floppy Disk - A small inexpensive disk used to store and record information together with a disk drive.

Flow Control - Hardware or software mechanisms employed in data communications to turn off transmission when the receiving workstation is unable to store the data it is receiving. Various methods of regulating the flow of data during a data conversation. Buffers are an example of flow control... The capability of network nodes to manage buffering schemes in order to allow devices of differing data transmission speeds to communicate with each other.

Flux Budget (a/k/a Attenuation Allowance, Loss Budget and Optical Power Budget) - Optical power attenuation permitted between any two transceivers. This attenuation allowance is for optical connector losses, optical cable attenuation and the optical cable attenuation and the optical power division in a Codestar passive fiber optic coupler. The sum of these attenuations, losses and divisions must not exceed the flux budget.

FOCUS - *Acronym for Codenet Fiber Optic Cabling Universal System whereby a single fiber optic cabling system can be installed and changed at a later date (from any type of network to another), at any time, without have to install additional cable.*

FOIRL BER - Mean beat error rate of the FOIRL.

FOIRL Collision - Simultaneous transmission and reception of data in a FOMAU.

FOIRL Compatability Interfaces - The FOMDI (Fiber Optic Medium Dependent Interface) and the AUI (optional), the two points at which hardware compatibility is defined to allow connection of independently designed and manufactured components to the baseband optical fiber cable link segment.

FOMAU (Fiber Optic Medium Attachment Unit) - A fiber optic device that provides for the connection to all Ethernet and IEEE 802.3 compatible DTEs via baseband transceivers with full compliance to Ethernet and IEEE 802.3 standards. An integrated FOMAU mounts in a PC; provides the same function as the modem-type unit.

FOMAU's Transmit Optical Fiber - The optical fiber into lwhich the local FOMAU transmits signals.

FOMAU's Receive Optical Fiber - The optical fiber from which the local FOMAU receives signals.

Format - The plan or arrangement by which information is stored.

Forward Direction - The direction of signal flow away from the head-end in a broadband LAN. High frequencies travel in this direction.

Forward Error Correction - Code incorporating sufficient additional elements to enable the nature of some or all of the errors to be indicated and corrected entirely at the receiving end.

Frame - A group of bits sent over a communications channel, usually containing its own control information, including address and error detection. The exact size and make-up of a frame depends on the protocol used.

Frame-Based Services and Functions [FDDI] - Provided by SMT to permit higher-level management entities to gather information about, and to control, the FDDI network.Specific capabilities include network statistics gathering; network fault detection, isolation and resolution; and FDDI configuration and operational parameter(s) fine tuning.

Freeware - Software that is available for anyone to use at no charge.

Frequency - The number of times an electromagnetic signal repeats an identical cycle in a unit of time, usually one second. One Hertz (Hz) is one cycle per second; a MHz (Kilohertz) is one million cycles per second; a GHz (Gigahertz) is one billion cycles per second.

Frequency Division Multiplexing - The splitting of a communication line into separate frequency bands each capable of carrying information signals. This allows several messages to be sent at the same time over the same transmission medium. See also broadband and multiplexing. A method of dividing a communication channel bandwidth among several sub-channels with difference carrier frequencies. Each sub-channel can carry separate data signals. A system of transmission in which the available frequency transmission range is divided into narrower bands, so that separate messages may be transmitted simultaneously on a single circuit.

Frequency Modulation - Modulation is the process of using a medium to carry information. We could "modulate" a flashlight beam by turning it on and off, thus sending digital information. An electrical sine wave traveling down a twisted wire pair can also be modulated to carry information. A sine wave is defined by its frequency, amplitude and phase. These are the only three parameters of a sine wave that can thus be changed to carry information. Frequency defines how many times a second a sine wave cycles. If you change the frequency, you can modulate the signal to carry information. This is called frequency modulation. The other common forms of modulations are amplitude and phase.

Frequency Plan - Specification of how the various frequencies of a broadband cable system are allocated for use.

Fresnel Reflection - Reflection of a portion of the light incident on a plantar interface between two homogeneous media having different refractive indices. Fresnel reflection occurs at the air-glass interfaces at entrance and exit ends of an optical fiber. Resultant transmission losses (on the order of 4% per interface) can be virtually eliminated by use of antireflection coatings or index matching material.

Frequency Response - The change of gain with frequency.

Frequency Shift Keying (FSK) - A method of putting data on an analog line. You modulate a carrier signal by shifting its frequency up or down from a mean value. Frequency shifts occur when there is a change from one binary value to another. In other words, the shifts come between two discrete values. In a sense this is changing analog signals to digital signals. See also: Modems.

Frequency Translator - In a mid-split configuration, an active electronic circuit in the headend picks up information signals on one 6 MHz channel coming in from the reverse direction and converts them to another 6 MHz channel above the mid-split frequency and sends them out in the forward direction. Also see Channel Translator.

Fresnel Reflection - Reflection of a portion of the light incident on a planar interface between two homogeneous media having different refractive indices. Fresnel reflection occurs at the air-glass interfaces at entrance and exit ends of an optical fiber. Resultant transmission losses (on the order of 4% per interface) can be virtually eliminated by use of antireflection coatings or index matching materials.

FTAM - File Transfer, Access and Management, OSI Version of FTP. Based on dissimilar systems.

FTP - The File Transfer Protocol of the DARPA Internet protocol suite, specified by RFC-1011...An upper-level TCP/IP service that allows copying files across a network.

Full Duplex - A connection on the network that allows transmission in both directions at the same time... Used to describe a communications system or component capable of transmitting data simultaneously in two direction.

-G-

Gain - Increased signal power, usually the result of amplification.

Gateway - A computer system and its software that permit two networks (or a network and a computer) using different protocol to communicate with each other... A special node that interfaces two or more dissimilar networks, providing protocol translation between the networks...An intelligent electronic device used to interconnect two or more networks that operate at a level above the Network Layer in the OSI model.

Gateway Configuration - Refers to setting up the gateway device with the data it needs to assign addresses and route packets on a network.

Gigahertz (GHZ) - A unit of frequency equal to one billion hertz.

GOSIP - Government OSI Profile. The federal government's version of the OSI network.

Graded Index - An optical fiber core whose refractive index is changed in a systematic way from center to edges to decrease modal dispersion... A type of fiber where the refractive index of the core varies smoothly with the radius. This type of fiber provides high bandwidth capabilities.

Graphics - (1) Information presented in the form of pictures or images. (2) The display of pictures or images on a computer's display screen. Compare Text.

-H-

Half Duplex - Transmission in two directions, one direction at a time... Used to described a communications system or component capable of transmitting data alternately, but not simultaneously, in two directions.

Halfstep Signaling (a/k/a/ Return To Zero Code) - This is the name given to the signaling which returns to zero volts after the end of the packet. It differs from simple DC coupling in that the first transition from the quiescent state at the beginning of the packet is only one half of the peak-to-peak amplitude of the signal. Thus it contains no DC component which could cause signal skew during the early part of the preamble of the packet. (Signal skew is due to AC coupling used in some interface circuits.)

Handshaking - An exchange of predetermined signals for purposes of control when a connection is established between two data sets... A preliminary procedure, usually part of a communications protocol, to establish a connection.

Hard Copy - A printed version of the machine's output in an easily readable form.

Hard Disk - A rigid disk of magnetically coated material that rotates in a sealed housing. It's used as a recording and playback system for data and computer programs.

Hardware - The physical apparatus that makes up a computer, including silicon chips, transformers, and boards and wires. The term is also used to describe pieces of equipment like the printer, modem and CRT.

Hayes Interbridge - A device that connects an AppleTalk network to a remote AppleTalk network through a modem.

HDLC (High-Level Data Link Control) - The International Standards Organization physical link protocol. Various manufacturers have their own derivative of HDLC, the most common of which is IBM's SDLC (Synchronous Data Link Control)... Abbreviation for High-Level Data Link Control. The international standard communications protocol.

Head-End - A central point in broadband networks that receives signals on one set of frequency bands and retransmits them on another set of frequencies. Viewed as a central hub.

Header - That part of a message, at the beginning, which contains destination, address, source address, message numbering, and possible other information. It helps direct the message along its journey.

HEPNET - High Energy Physics Network.

Hertz (Hz) - Synonymous with cycles per second: a unit of frequency, one Hertz is equal to 1 cps.

Hesoid - The name server mechanism used by MIT's Project Athena to provide naming for services and data objects in a distributed network environment.

Hierarchical File System - HFS - A system of grouping folders to facilitate the management of disk files. HFS can be compared to a company's corporate chart in which the president is at the highest level, the vice president a level below and so on.

High Frequencies - Frequencies from 160 MHz to 400 MHz allocated for the forward direction in a mid-split system.

High-Split - A broadband cable system in which the bandwidth utilized to send toward the head-end (reverse direction) is approximately 6 MHz to 180 MHz, and the bandwidth utilized to send away from the head-end (forward direction) is approximately 220 MHz to 400 MHz. The guard band between the forward and reverse directions (180 MHz to 220MHz) provides isolation from interference.

Host - A computer system that provides computer service for a number of users. Usually a mainframe.

Host Concept - Many protocols such as IBM's SNA, for example, employ some large data processing facility as part of the network.

Host System - Any device which acts as the course of, or a destination for, data; for example a printer, a computer terminal, or a computer.

Hub - The center of a star topology network or cabling system. File servers often act as the hub of a LAN. They house the network software and direct communications within the network. They may also act as the gateway to another LAN.

Hybrid Network - A LAN with a mixture of topologies and access methods. For example, a network that includes both a token ring and a CSMA/CD bus.

-I-

ICMP - Internet Control Message Protocol, one of the Internet protocols used for exchange maintenance and control information...The TCP/IP process that provides the set of functions used for network-layer management and control.

Icon - An image that graphically represents an object, a concept, or a message.

Idle (IDL) - A signal condition where no transition occurs on the transmission line is used to define the end of a frame and ceases to exist after the next LO to HI transition on the AUI circuits. An IDL always begins with a HI signal level. A driver is required to send the IDL signal for at least 2 bit times and a receiver is required to detect IDL within 1.6 bit times.

IEEE - Abbreviation for the Institute of Electrical and Electronic Engineers, a publishing and standards-making body responsible for many standards used in LANs, include the 802 series.

IEEE-802/802.3 - Standards for the interconnection of local area networking computer equipment. The IEEE-802 standards deal with the Physical and Link Layers of the ISO Reference Model for OSI ...Refers to 10 Mbit/sec CSMA/CD standard for baseband local area networks, very similar to Ethernet.

IEEE 802.5 - An IEEE standard published in 1985 for a ring architecture LAN using token passing as an access method (also specified by the IO as Draft International Standard 8802/5).

IEEE 802.6 - An emerging (still under development) IEEE standard for a Metropolitan Area Network spanning many kilometers using a distributed queuing access control method and a dual bus architecture.

Impudence - A measure of the electrical property of resistance, expressed in ohms. Different cable systems have different resistance levels: broadband utilizes CATV standard 75 ohm cable, and baseband Ethernet utilizes 50 ohm.

In-Band signaling - The transmission of signaling information at some frequency or frequencies that lie within a carrier channel normally used for information transmission.

Incremental Linking - A technique that keeps a partially linked application together. Incremental linking is generally used to manipulate an existing object rather than to create a new one.

Index Matching Material - A Material, often a liquid or cement, whose refractive index is nearly equal to the core index, used to reduce Fresnel reflections from an optical fiber's end face. See also: Fresnel Reflection; Refractive Index.

Index of Refraction - The ratio of the velocity of light in a vacuum to the velocity of light in a given medium.

Index Profile - In a graded-index optical fiber, the refractive index as a function of radius.

Information Bit - A bit used as part of a data character within a code group (as opposed to a framing bit).

Infrared (IR) - The bank of electromagnetic wavelengths between the visible part of the spectrum (about 750nm) and microwaves (about 30μm).

INIT - A stand-alone code resource executed at system start-up. An INIT specifically expands the system heap before a PC finishes initializing.

Injection Laser Diode (ILD) - A semiconductor device consisting of at least one P-N junction capable of emitting coherent stimulated radiation under specified conditions.

Inline Assembler - An assembler that allows users to integrate assembly code with the high-level language directly in a program, i.e. linking is not necessary. Programmer use this to increase the speed of a program.

Insertion Loss - Total optical power loss caused by the insertion of optical component such as a connector or coupler.

Install - To add information to the system file or to add new system files to the system folder of a startup disk.

Integrated Detector/Preamplifier - A single chip which contains a detector and an amplifier which converts optical signals to usable electrical output.

Integrated Software - One program that contains other programs and permits their simultaneous use or the transfer of data between them.

Intelligent Terminal - A "programmable" terminal which is capable of interacting with the central site computer and performing limited processing functions at the remote site... A terminal which can be programmed.

Intellipath - A NYNEX system for telephone and data service.

Interactive - A way of operating a computer where the computer lets you change things as you go along, asks you questions, lets you enter data directly, etc. Almost all applications software for personal computers word processors, spreadsheets etc. - are interactive. Compare with batch processing and real-time processing.

Interface - A shared connection or boundary between two devices or systems. The point at which two devices or systems are linked. Common interface standards include EIA Standard RS-232B/C, adopted by the Electronic Industries Association to ensure uniformity among most manufacturers: MIL STD 188B, a mandatory standard established by the Department of Defense; and CCITT, the world recommendation for interface, mandatory in Europe and closely resembling the American EIA standard... A demarcation between two devices, where the electrical signals, connectors, timing and handshaking meet. Often the procedures, codes and protocols that enable the two devices to interact are included or carried out by the interface. An example is an RS-232-C port. Some of its 25 pins are used to send different information and makes sure devices can talk to each other. The pins carry different messages, like "request to send," "acknowledgment" and others.

Internet - The network of networks that were originally connected together by the ARPANET, now expanded to include those networks connected to the NSFnet.

Internetwork - Between two distinct networks.

Inter Repeater Link (IRL) - A mechanism for interconnecting two and only two repeater units.

Interrupt - An access method used in some bus and ring networks where workstations can "interrupt" a server to get service. The server will temporarily suspend whatever it is doing, give the workstation a file or whatever it wants, then return to its previous task. Interrupt is preferred over polling, an older, less efficient, easier-to-implement method of getting the server (central computer's) attention. See Polling.

Intranetwork - Within one network

Intrinsic Joint Loss - Loss by fiber-parameter (e.g., core dimensions, profile parameter) mismatches when two nonidentical fibers are joined. See also: Extrinsic joint loss; Lateral offset loss.

I/O - Abbreviation for Input/Output.

IP Router - A device which joins Ethernet networks and routes IP data packets.

ISDN - Integrated Service Data Network. A proposed set of protocols that combines circuit-switched, predominantly voice service with packet-switched, predominantly digital service into a totally digital network. The subsequent digital network would then be capable of carrying voice, computer data, facsimile and video signals.

ISO - International Standards Organization. An independent international body formed to define standards for multivendor network communications. Its seven-layer OSI reference model specifies how different vendor's products communicate with each other across a network.

ISO OSI - The International Standards Organization's architecture for Open Systems Interconnection, a scheme for a universal standard architecture and protocol suite (see below).

ISO Reference Model for OSI - A standard approach to network design which introduces modularity by dividing the complex set of functions into more manageable, self-contained, functional slices. These layers, from the innermost layer, are as follows:
(1) **Physical Layer** - concerned with the mechanical and electrical means by which devices are physically connected and data is transmitted.
(2) **Link Layer** - concerned with how to move data reliably across the physical data link.
(3) **Network Layer** - provides the means to establish, maintain, and terminate connections between systems; concerned with switching and routing of information.
(4) **Transport Layer** - concerned with end-to-end data integrity and quality of service.
(5) **Session Layer** - standardizes the task of setting up a session and terminating it; coordination of interaction between end-application processes.
(6) **Presentation Layer** - relates to the character set and data code which is used, and to the way data is displayed on a screen or printer.
(7) **Application Layer** - concerned with the higher level functions which provide support to the application or system activities.

Isolation Management - The ability to limit the damage caused by malfunctioning network components by remotely partitioning the network to isolate failing or misbehaving sections, so that the operation of the remainder of the network is not impaired while the failing subset is corrected.

IVDT - An Integrated Voice Data Terminal. A cross between a telephone and an intelligent terminal or a personal computer. Most have telephone auto dialers, the ability to get electronic mail, some word processing ability for electronic messages and other features which the inventors hope will make the IVDT gadget sell. So far, IVDTs have not fulfilled their inventors' lofty aspirations.

-J-

Jabber - Continuously sending random data.

Jam - A short encoded sequence emitted by a node to ensure that all other nodes have detected a collision.

Justification - Making all full lines of text the same length in order to create an even right edge.

-K-

K - A standard quantity measurement of computer storage. A K is loosely defined as one thousand bytes. In fact, it is 1,024 bytes, which is the equivalent of two raised to the tenth.

Kbps - Abbreviation for kilobits per second. 1,000 bits per second.

Kerberos - The authentication system for open network computing environments developed and used at MIT's Project Athena.

Kermit - see Protocol.

-L-

Lambertian Radiator - An optical source which has radiance uniform in all directions, proportional to the cosine of the angle from the normal.

LAN - Local Area Network.

LANA - Local Area Network Adapter.

Language - Any set of characters used to form related commands or instructions that combine into meaningful communications acceptable to a computer.

Laser - A device that produces monochromatic, coherent light through stimulated emission. Most lasers used in fiber optic communications are solid-state semiconductor devices... An acronym for Light Amplification by Stimulated Emission of Radiation, a device which transmits an extremely narrow and coherent beam of electromagnetic energy in the visible light spectrum. See also: Injection Laser Diodes, Stimulated Emission.

Lasing Threshold - The lowest excitation level at which a laser's output is dominated by stimulated emission rather than spontaneous emission. See also: Laser, Spontaneous Emission, Stimulated Emission.

LAT - Local Area Terminal protocol, a proprietary protocol used by Digital Equipment Corporation for support of terminal-to-host communication over a local Ethernet.

Lateral Offset Loss - An optical power loss caused by transverse or lateral deviation from optimum alignment of source to optical fiber, fiber-to-fiber, or fiber-to-detector.

Launch Angle - The angle between the light ray and the optical axis of an optical fiber or fiber bundle.

Layer - In the OSI model, it refers to a collection of network processing-functions that together compose one layer of a hierarchy of computing functions. Each layer performs a number of functions, essential for successful data communication. See OSI model.

Leased Channel - A point-to-point channel reserved for the sole use of a single leasing customer.

Light - 1) In a strict sense the visible spectrum, nominally covering the wavelength range of 400 nm to 750 nm. 2) In the laser and optical communication fields, the much broader portion of the electromagnetic spectrum that can be handled by the basic optical techniques used for the visible spectrum extending from the near-ultraviolet region of approximately 0.3μm, through the visible region, and into the mid-infrared region to 30μm. See also: Infrared (IR), Ultraviolet (UV).

Light Emitting Diode (LED) - A semiconductor device which emits incoherent light from a p-n junction (when biased with an electrical current). Light may exit from the junction strip edge or from its surface (depending on the device's structure).

Lightwave Data Communication - Made possible by fiber optic technology. Based on the fact that pulses of light transmitted over fiber will enable computer networks to communicate over greater distances at higher rates of speed, with complete immunity to electrical interference of any type, offer greater reliability at lower cost than can electrical, copper-based networks, with either coaxial cable or twisted pair wiring. The conversion of electrical signals to lightwave signals is accomplished via transceivers containing special compound semiconductors made of Gallium Arsenide and Indium Phosphide. These two compounds have enabled the implementation of lightwave communication at extremely high data rates. This technology, when implemented in an Optical Bus Passive Star cabling system, gives the ultimate in reliability for any network. Furthermore, the absence of electrical connectivity between computers avoids grounding problems, ground loops, electromagnetic interference and a host of related problems that are associated with transmission over copper wire or cable. Fiber optic networks, configured in Passive Star topologies, are rapidly becoming known as the most cost-effective, thoroughly reliable, easily-maintained networks in the world.

Line Driver - A circuit designed to transmit data outside the enclosure of a computer system, but not more than a few hundred feet. Data is not changed in any way when sent.

Line Speed - The maximum rate at which signals may be transmitted over a given channel, usually measured in bauds or bps. Line speed varies with the capabilities of the equipment used.

Line Turnaround - The delay in a communications link between the time one block of data has been sent and received and the next block can be transmitted.

Link Layer - Layer two of the ISO reference model; also known as the data-link layer.

Linker - A program that builds a complete, executable file from several separately compiled files.

Load - The process of putting data into the computer or its memory.

Local Area Network (LAN) - A network that is located in a localized geographical area (e.g., an office, building, complex of buildings, or campus), and whose communications technology provides a high-bandwidth, low-cost medium to which many nodes can be connected... A data communications network spanning a limited geographical area, a few miles at most. It provides communication between computers and peripherals, some switching to direct messages.

LocalNet - The broadband architecture used in all of Sytek's networks. Also the product name of their network. Sytek is in Sunnyvale, Ca.

Local Session Number - The number assigned to each session established by an adapter. Each session receives a unique number that distinguishes it from any other active sessions. \

LocalTalk - Shielded twisted pair wiring which supports AppleTalk protocols for networking Macintoshes.

LocalTalk Cable - The insulated wire used to join LocalTalk connector boxes.

LocalTalk Connector Box - A piece of equipment consisting of a small white box with a built-in cable that links a device to a LocalTalk cable system.

Locking - Preventing several people getting to and changing the same data simultaneously. Locks may be permanent and prevent access completely, or they may be "advisory" - a user is warned the data is being used by another user. Locks prevent the destruction of data that can occur if two people access a file at the same time.

Login - The process of identifying and authenticating oneself to a computer system. Used to control access to computer systems.

Long-Haul Network - A network most frequently used to transfer data over distances of from several thousand feet to several thousand miles. These networks can use the international telephone network to transport messages over most or part of these distances.

Loopback - Type of diagnostic test in which the transmitted signal is returned to the sending device after passing through a data communications link or network. This allows a technical (or built-in diagnostic circuit) to compare the returned signal with the transmitted signal and get some sense of what's wrong. Loopbacks are often done, excluding one piece of equipment after another. This allows the user to figure out logically what's wrong.

Loopback Tests - A test procedure in which signals are looped from a test center through a modem or loopback switch and back to the test center for measurement.

Loop System - Generally, the hardware configuration in a closed series transmission circuit with a fixed number of terminal points.

Loss - See Absorption, Angular Misalignment Loss, Attenuation, Backscattering, End Separation Loss, Extrinsic Joint Loss, Insertion Loss, Intrinsic Joint Loss, Lateral Offset Loss, Material Dispersion, Microbending, Rayleigh Scattering, Reflection, Transmission Loss.

Loss Budget - See Flux Budget.

Low Frequencies - Frequencies from 5 MHz to 116 MHz allocated for the return direction in a mid-split system.

LSI - Large-scale integration, the art of putting many thousands of transistors onto a single small circuit chip.

LSN - See Local Session Number.

LU - Logical Unit. The port, or network-addressable device, that provides access to a device on the network. For example, LU 6.2 is the IBM protocol that provides peer-to-peer communication over an SNA network; LU 6.2 is also referred to as APPC.

-M-

Macintosh O/S - Apple Computer's single-user proprietary operating system.

Macrobending - In an optical fiber, all macroscopic deviations of the axis from a straight line, distinguished from microbending. See also: Microbending Loss.

Mainframe - A large computer normally supplied complete with peripherals and software by a single large vendor, often with a closed architecture. Mainframes almost always use dumb terminals connected in star configurations.

MAN - Metropolitan Area Network, a test network that Harvard has been using, connecting the Harvard Cambridge campus with the Observatory, the School of Public Health and Massachusetts General Hospital.

MAP - Abbreviation for Manufacturing Automation Protocol, a token-passing bus LAN designed for factory environments and perhaps one day a standard on the factory floor. Often used in CAD/CAM... An OSI-related network application used widely in manufacturing environments; promulgated primarily by General Motors.

Master Station - The unit which controls all the workstation on a LAN, usually through some type of polling. The master station on a token-passing ring allows recovery from error conditions, such as lost, busy or duplicate tokens, usually by generating a new token. Sometimes severs are referred to as master stations.

Material Dispersion - Light impulse broadening caused by various wavelengths of light traveling at differing velocities through a fiber. Material dispersion increases with the increasing spectral width of the source.

MATV (Master Antenna Television System) - A small, less expensive cable system usually restricted to one or two buildings such as hospital, apartments, libraries, hotels, office buildings, etc.

MAU - A Medium Attachment Unit - also called a transceiver - on Ethernet cable to which network devices are connected using a transceiver cable.

MBit/sec - Megabits per second, a measure of network bandwidth.

Mbps - Abbreviation for megabits per second. One million bits per second.

MDEF - A menu definition procedure that users write to define their own menu type.

Mean Time Between Failure - See MTBF.

Medium - A copper wire or microwave transmission signal.

Media Access - Refers to the ability of a station or device to access and transmit data on a LAN.

Media Access Control (MAC) [FDDI] - Defines token-passing protocol for FDDI networks, as are packet formation, addressing and recovery mechanisms. The maximum packet size is 4,500 bytes. MAC controls the flow of data on the ring. Acting like a switch, MAC normally sources IDLE (IDL) control symbols for transmission on the ring. When a start delimiter arrives on the ring, MAC will monitor each packet. If the packet is destined for another station, MAC will simply repeat the packet on the ring, only noting if a transmission error has occurred. If the packet is addressed to MAC's station, the packet will be copied into its buffers while simultaneously repeated on the ring. If the packet was sourced by MAC, the packet will be absorbed and not retransmitted. A fragment of the packet (about 6 bytes) is repeated by MAC. When MAC receives a token and has data to transmit, the token is absorbed. Switching into a sourcing mode, MAC encapsulates its data into proper control symbols for an FDDI packet, and inserts this packet into the ring. MAC continues to insert packets into the ring until it has completed its data transmission or token holding time has expired. At the completion of the transmission, MAC will issue a token that allows multiple packets from different stations to be on the ring concurrently, which increases the effective bandwidth utilization.

Media Interface Connector (MIC) [FDDI] - Mating interface at the bulkhead receptacle. The MIC is the interface to the cable plant.

Medium Attachment Unit (MAU) - The portion of the physical layer between the MDI and AUI that contains the electronics which send, receive, and manage the encoded signals impressed on, and recovered from the fiber optic medium.

Medium Dependent Interface (MDI) - The mechanical and optical interface between the fiber optic medium and the MAU.

Megabyte - 1,048,576 bytes. The basic unit of mass storage and data-transfer rates.

Megahertz (MHz) - Unit of frequency equal to one million hertz.

Memory - The temporary internal storage of information in a computer.

Menu - A list of commands that appears on your screen which will perform functions in your program when selected.

Meridional Ray - A ray that passes through the optical axis of an optical fiber (in contrast with a skew ray, which does not). See also: Axial Ray; Numerical Aperture; Skew Ray.

Message - A logical partition of the user device's data stream to and from the adapter.

Message Switching - The technique of receiving a message, storing it (if necessary) until the proper outgoing line is available, and retransmitting it toward its destination automatically.

MIB - (RFC-1066) Management Information Base.

MIS - (RFC-1065) Management Information Structure.

Microbend Loss - A form of increase attenuation caused by: a) having the fiber curved around a restrictive radius of curvature, or b) microbends caused by minute distortions in the fiber imposed by externally induced perturbations. Excessive bend loss may result from poor drawing or cable manufacturing techniques. In an optical fiber, loss caused by sharp curvatures involving local axial displacements of a few micrometers and spatial wavelengths of a few millimeters. Such bends may result from fiber coating.

Micron - (μm) Micrometer. Millionth of a meter = 10^{-6} meter.

Microprocessor - A CPU on a single chip. See CPU.

Microsecond - One-millionth of a second.

Microwave - Any electromagnetic wave in the radio frequency spectrum above 890 megaHertz.

Mid-Split - A broadband cable system in which the cable bandwidth is divided between transmit and receive frequencies. The bandwidth utilized to send toward the head-end (reverse direction) is approximately 5 MHz to 100 MHz, and the bandwidth utilized to send away from the head-end (forward direction) is approximately 160 MHz to 300 MHz. The guard band between the forward and reverse directions (60 MHz to 100 MHz) provides isolation from interference.

Minicomputer - An intermediate-sized computer system between a microcomputer (P.C.) and a large computer (mainframe) in size... A small or medium computer accessed by dumb terminals. A minicomputer is bigger and more powerful than personal computer, a PC. A mini is for many. A personal computer is for one person.

Modal - This mode gives users only a predetermined subset of choices. For example, a dialogue box gives users the option of continuing or canceling. Users must do one or the other and cannot deviate from these choices.

Modal Dispersion - See: Multimode Distortion.

Mode - A permitted electromagnetic field pattern within a waveguide fiber... A method of operation as in binary mode, alphameric mode, etc... An electromagnetic field pattern within a waveguide fiber. In any cavity or transmission line, one of those electromagnetic field distributions that satisfies Maxwell's equations and the boundary conditions.

Mode Coupling - In an optical fiber, the exchange of power among modes. The exchange of power may reach statistical equilibrium after propagation over a finite distance that is designated the equilibrium length. See also: Equilibrium Length; Mode; Mode Scrambler.

Mode Filter - A device to remove high order modes to simulate equilibrium mode distribution in a short length of optical fiber. See also: Equilibrium Mode Distribution.

Mode Scrambler - A device for inducing mode coupling in an optical fiber.

Modem - A device that converts serial digital data from a transmitting terminal to a signal suitable for transmission over a telephone (analog) channel. At the other end another modem reconverts the analog signal to digital data for use by the computer. The word modem come from MOdulator- DEModulator. Installed in pairs at each end of an analog communications line. The modem at the transmitting end modulates digital signals received locally from a computer or terminal; the modem at the receiving end demodulates the incoming analog signal, converting it back to its original (i.e., digital) format, and passes it to the destination device. See also: Modulation, DCE, and DTE.

Modem Eliminator - A small device that can replace a modem when the distance of the data link is short. It does not need any external electrical power. It takes some power out of the line.

Modulation - The process by which a characteristic of one wave is varied in accordance with another wave or signal as in modems, which transform computer signals into waves that are compatible with communications facilities and equipment. Types of modulation include: AMPLITUDE MODULATION (AM), in which the amplitude of the carrier is varied in accordance with the instantaneous value of the frequency by an amount proportional to the instantaneous value of the modulating signal: PHASE MODULATION, in which the angle relative to the unmodulated carrier angle is varied in accordance with the instantaneous value of the amplitude of the modulating signals; PULSE AMPLITUDE MODULATION, modulating signal; DIFFERENTIAL MODULATION, in which the choice of the significant condition for any given signal element is dependent upon the choice of the previous signal element; FREQUENCY MODULATION (FM), in which the instantaneous frequency of a sine wave carrier departs from the carrier in which the amplitude of the pulse carrier is varied in accordance with successive samples of the modulating signal; PULSE CODE MODULATION, in which the modulating signal is sampled, quantized, and coded so that each element of the information consists of different kinds and/or numbers of pulses and spaces... Any systematic change to a carrier signal to encode and convey information. The changes or differences in the values of the properties (frequency, phase or amplitude) convey the information. For example, digital signals from a computer device are altered to analog signals for transmission over telephone lines.

Monitor - The computer's screen that displays vivid characters and on which information stored in the computer can be read.

Motherboard - The main circuit board in a personal computer. It accepts other add-on boards, including network interface cards, which are plugged into the motherboard.

Mouse - A small device that rolls around on a flat surface next to a computer that, when moved, programs the computer to perform certain functions.

MS-DOS - The single-user operating system developed by Microsoft for the IBM Personal Computer and compatibles.

MS OS/2 LAN Manager - The multiuser network operating system co-developed by Microsoft and 3Com. LAN Manager offers a wide range of network management and control capabilities unavailable with existing PC-based network operating systems.

MTBF - Mean Time Between Failure. Used by manufacturers to measure reliability of equipment. Almost always measured in hours. Keep in mind that an MTBF figure for a LAN card or a computer is always longer than what you will find in your own experience, because testing methods by manufactures in laboratories do not reproduce office environments very well. Obviously, "reliability" does not measure "availability". The two must not be confused. If you want extremely high "availability", but duplicate, redundant systems and keep them both running simultaneously. See also: Reliability.

Multicasting - Directing a message or packet to some subset of all the stations on a network by the use of a special destination address.

Multi-Channel Cable - An optical cable having more than one fiber.

Multi-Drop - A communications circuit with multiple terminals and peripherals. Viewed as a branch off the bus of a LAN.

Multi Fiber Cable - An optical cable that contains 2 or more fibers, each of which provides a separate information channel. See also: Fiber Bundle; Optical Cable Assembly.

Multi Link - A low cost software LAN from the Software Link, Atlanta, Ga.

Multi Mode Distortion - In multimode fiber, the pulse distortion resulting from differential mode propagation rates.

Multi Mode Fiber - A fiber that supports propagation of more than one mode of a given wavelength...This cable causes the light signal to propagate incoherently, causing dispersion effects that limit the bandwidth and distance of communication. Nearly all current commercial fiber optic communications products use multi mode cable, the de facto standard having a core diameter of 62.5μm.

Multiplex - Putting two or more signals into a single channel... To interleave, or simultaneously transmit, two or more messages on a single channel...The use of a common physical channel in order to make two or more logical channels, either by splitting of the frequency band transmitted by the common channel into narrower bands, each of which is used to constitute a distinct channel (frequency-division multiplex), or by allotting this common channel in turn, to constitute different intermittent channels (TDM).

Multiplexing - Sending several signals over a single line and separating them at the other end. This is done by varying the physical characteristics (frequency, amplitude or phase) of the signals to prevent them from interfering with each other. It is also possible to separate them in time. That is called time-division-multiplexing... The process of dividing a transmission facility into two or more channels.

Multiplexor - Equipment that permits simultaneous transmission of multiple signals over one physical circuit.

Multipoint Circuit - A circuit that interconnects three or more stations.

Multitasking - The simultaneous performing of two or more tasks by a computer.

Multiuser - Refers to a system which can provide service to more than one user simultaneously.

Multi-Window - Refers to the capability of opening more than one terminal session on a workstation or terminal.

-N-

Name Server - Software that provides name to address mapping services for other systems on some portion of a network. The use of name servers eliminates the need for individual systems to maintain their own lists of names to address mappings.

Nanometer (nm) - One billionth of a meter = 10^{-9} meter.

Nanosecond (ns) - One-billionth of a second expressed as 10^{-9} sec.

NetBIOS - Network Basic Input/Output System. Software developed by IBM that provides the interface between a PC's operating system, the I/O bus and the network; a *de facto* network standard.

Net/One - The family of local area network products, bridges, gateways, network interfaces and software from Ungermann-Bass, Santa Clara, CA. Local area network for heterogeneous device interconnection, which is available in baseband, Ethernet, broadband, and fiber optic versions.

NetWare - A popular LAN operating system from Novell, Orem, UT.

Network - A series of points interconnected by communication channels. The switched telephone network consists of public telephone lines normally used for dialed telephone calls; a private network is a configuration of communication channels reserved for the use of a sole customer. A series of nodes connected by communications channels... See also LAN.

Network Access Control - Electronic circuitry that determines which workstation may transmit next or when a particular workstation may transmit.

Network Architecture - The structures and protocols of a computer network. See architecture.

Network File Systems (NFS) - Extends the file system of a UNIX or other computer so that the set of files on this remote host appears as a local file system on a client machine.

Network Interface Controllers - Electronic circuitry that connects a workstation to a network. Usually a card that fits into one of the expansion slots inside a personal computer. It works with the network software and computer operating system to transmit and receive messages on the network... A communications device that allows interconnection of information processing devices to a network.

Network Interface Unit (NIU) - The Ungermann-Bass, Inc. trademarked name for its network interface controller.

Network Layer - The third layer of the OSI model of data communications. It involves routing data messages through the network using alternative routes. See OSI Standards.

Network Management - Administrative services performed in managing a network, such as network topology and software configuration, downloading of software, monitoring network performance, maintaining network operations, and diagnosing and troubleshooting problems.

Network Management Console (NMC) - The Ungermann-Bass device which provides the execution environment for Net/One network management software and utility programs, and storage for software to be downloaded into Net/One components in the network.

Network Management System - A comprehensive system of equipment used in monitoring, controlling, and managing a data communications network. Usually consists of testing devices, CRT displays and printers, patch panels, and circuitry for diagnostics and configuration of channels, generally housed together in an operator console unit.

Network Service - An application available on a network, e.g., file transfer.

Network Terminator - A device (50 ohm resistor) which must be attached to each end of an Ethernet cable in order for a network to function properly.

Network Topology - The geography of a network. For example, whether it is mesh, star, bus or ring.

NFS - Network File System. An extension of TCP/IP that allows files on remote nodes on a network to appear locally connected.

NIC - Network Information Center.

NOC - Network Operations Center, a site that contains hardware and software that can be used to manage a network.

Nodes (a/k/a/ Stations) - In computer-based LANs, at least two intelligent systems needing to share data are required. These intelligent systems, when connected to the network, are referred to as Nodes, and are able to communicate with each other via a network topology such as Star, Bus, Ring or Tree. A node usually consists of a personal computer, an adapter with a cable and other adapters to the personal computer (such as: disk drives, printers, and plotters). In addition to the personal computer hardware, all necessary software must be available... Points in a network where service is provided, service is used or communications channels are interconnected. Nodes are sometimes used interchangeably with workstations... A station.

Noise - The word "noise" is a carry-over from audio practice. Refers to random spurts of electrical energy or interference... Random electrical signals, generated by circuit components or by natural disturbances, that make up transmitted data inaccurate by introducing errors. Noise can come from lightning, crossed cables, electrical motors... Generally, any disturbance that tends to interfere with the normal operation of a communication device or system. Random electrical signals, introduced by circuit components or natural disturbances, which degrade the performance of a communications channel.

Noise Equivalent Power (NEP) - The root-mean-square (rms) value of optical power which is required to produce an rms signal-to-noise ratio of 1; and indication of noise level which defines the minimum detectable signal level.

Noise Measurement Units - A series of terms used to express both weighted and unweighted circuit noise, as stated in dBrn (decibel rated noise). Noise measurement units vary with the procedures used for noise weighting.

Noise Weighting - A method of assigning a specific value, in numerical readings, to the transmission impairment due to the noise encountered to an average user operating a particular class of telephone subset. Noise weightings generally in use have been established by the agencies concerned with public telephone service, and they represent successive stages of technological development.

Non-Proprietary LAN - A LAN that can connect the equipment of many vendors. See Proprietary LAN.

Nonvolatile - A term used to describe a data storage device that retains its contents when power is lost. A hard disk is an example.

Novell PC net - A specific vendor's PC LAN.

Null Modem Cable - An RS-232 cable in which pins 2 and 3 are reversed, fooling the two computers connected at each end into thinking they are talking through modems. Not to be confused with modem eliminators.

Numerical Aperture (NA) - A characteristic parameter of any given fiber's light gathering capability. Defined by the maximum angle of light which is relative to the fiber's axis, and which is propagated through the fiber. The sine of the vertex angle of the largest cone of meridional rays that can enter or leave an optical system or element, multiplied by the refractive index of the medium in which the vertex of the cone is located. The acceptance angle of the fiber defined as: $NA = (n_1^2 - n_2^2)^{1/2} = \sine 0\ max$ where n_1 and n_2 are respectively, the refractive index of the core and the cladding.

Numeric Keys - Keys on the right side of computer keyboards that let the user perform numeric entry and calculation quickly. Sometimes called a 10-key pad.

-O-

Object Code - Executable machine code. Programs that have been compiled or assembled. These are the programs in a language understandable to the computer.

Off-Line - A general description of equipment of devices not under direct control of the CPU, or terminal equipment which is connected to a transmission line.

On-Line - A general description of equipment of devices which are under the direct control of a CPU, or terminal equipment which is connected to a transmission line.

On-Line System - A system in which the data to be input enters the computer directly from the point of origin (which may be remote from the central site) and/or the output data is transmitted directly to the location where it is to be used.

Open - To make available. You open files or documents in order to work with them.

Open System Interconnection (OSI) Model - See OSI Standards.

Operating System - The software of a computer that controls the execution of programs, typically handling the functions of input/output control, resource scheduling, and data management. See also: DOS.

Operating Time - In data communications, the total time required to dial a call, wait for the connection to be established, and complete the transaction with the personal or equipment at the receiving end.

Optical Cable Assemblies - A cable complete with connectors. Generally, a cable that has been terminated by a manufacturer and is ready for installation.

Optical Connectors - Are used to attach the transmit and receive optical fibers in the fiber optic cable to the fiber optic transceiver. The optical connectors are designed to make connection by simply hand tightening the nut on the external optical connector to the connectors on the fiber optic transceiver.

Optical Filter - A device that selectively transmits certain optical wavelengths and blocks a range of wavelengths. An element that selectively transmits certain optical wavelengths and blocks a range of wavelengths. A filament-shaped optical waveguide made of dielectric materials.

Optical Fiber Cable Interface - See FOMDI.

Optical Fiber Cable Link Segment - A length of optical fiber cable containing two optical fibers and comprising one or more optical fiber cable sections and their means of interconnection, with each optical fiber terminated at each end in an optical connector plug.

Optical Idle Signal - The signal transmitted by the FOMAU into its transmit optical fiber during the idle state of the DO circuit.

Optical Power (LED) - Radiant power, expressed in watts.

Optical Power Budget - See Flux Budget

Optical Receiver - A device that receives optical signals from an optical transmitter via the receiver fiber of the fiber optic cable. It converts optical signals to electrical signals which are then conditioned and transmitted through the fiber optic transceiver interface cable to the controller and the host.

Optical Star Coupler - See Star Coupler.

Optical Time Domain Reflectometry (OTDR) - A method for characterizing a fiber wherein an optical pulse is transmitted through the fiber and the resulting backscatter and reflections are measured as a function of time. Useful in estimating attenuation coefficient as a function of distance and identifying defects and other localized losses. See also: Backscattering; Rayleigh Scattering; Scattering.

Optical Transmitter - Receives electrical signals from the Ethernet controller via the fiber optic transceiver's interface cable and converts electrical signals to optical signals. These optical signals are then transmitted on to the network via the transmit fiber of the optical fiber cable through the SMA connector.

Option Card - A card that contains electronic circuits that implement specialized functions. Fit into the computer's Expansion Slots.

OS/2 - The third-generation operating system developed by IBM and Microsoft for use with Intel's 80286 and 80386 microprocessors. Unlike its predecessor, PC/MS-DOS, OS/2 is a multitasking system.

OSI - Abbreviation for Open Systems Interconnection, a logical structure for network operations standardized within the ISO. See also: OSI Standards.

OSI Standards - The International Standards Organization (ISO) has established the Open System Interconnection (OSI). The idea of OSI is to provide a network design framework to allow equipment from different vendors to be able to communicate. Codex of Mansfield, MA. has published an excellent booklet called "The Basics Booklet of Local Area Networking". Here is a shortened except of what Codex says about standards: Standards allow us to buy items such as batteries and bulbs. Many of us have learned "the hard way" that the lack of computer standards can make it impossible for computers from different vendors to talk to each other. Because a major goal of a LAN is to connect varied systems, standards are being developed to specify the set of rules networks will follow. The OSI Model is a design in which groups of protocols, or rules for communicating, are arranged in layers. Each layer performs a specific data communications function. The concept of layered protocols is analogous to the steps we follow in making a phone call:

Step 1	Listen for dial tone
Step 2	Dial a phone number
Step 3	Wait for ring
Step 4	Exchange greetings
Step 5	Communicate message
Step 6	Say goodbye
Step 7	Hangup

Each of these steps, or OSI "layers" builds upon the one below it. Although each step must be performed in preset order, within each layer there are several options. Within the OSI model, there are seven layers. The first three are the physical, data link and networks layers, all of which are concerned with data transmission and routing. The last three - session, presentation and applications - focus on user applications. The fourth layer transmission provides an interface between the first and last three layers. The X.25 protocol which created a standard for data transmission and routing is equivalent to the first three layers of the OSI Model. Is quickly becoming the standard for how LAN products should be built. See also: X.25.

Out-Of-Hand signaling - A method of signaling which uses a frequency that is within the passband of the transmission facility, but outside of a carrier channel normally used for information transmission... An additional signal is sent "along side" the information-carrying signal. This second signal has a frequency that is usually just outside the frequency of the carrier signal. It is method used to control the parameters of data conversations.

Output - Information that's transferred from a computer's internal storage to any device.

Output Power (LED) - Radiant power expressed in watts.

-P-

PABX (Private Automatic Branch Exchange) - Equipment originally used as a means of switching telephone calls within a business site and from the site to outside lines. Can also be used for low speed transmission of data, in addition to voice.

Packet - A series of bits forming a block into which all, or part of a data message is put to be sent through a network consisting of a Data Frame, preceded by the Preamble and the Start Frame Delimiter. Each packet has a defined format, with some additional bits forming a head preceding the data and a tail following it;

these carry information which the network needs to know about the packet, including its destination and source. The packets are formed by the controller in the sending host system and the data is extracted and reassembled by the controller at the receiving end... A collection of bits that contain both control information and data. The basic unit of transmission in a packet- switched network. Control information is carried in the packet, along with the data, to provide for such functions as addressing, sequencing, flow control, and error control at each of several protocol levels. A packet can be of fixed or variable length, but generally has a specified maximum length... A group of bits, including address, data and control elements, that are switched and transmitted together. Think of a packet as one sentence or one group of numbers being sent at a time. See also: Packet Switching.

Packet Buffer - Memory space set aside for storing a packet awaiting transmission or for storing a received packet. The memory may be located in the network interface controller or in the computer to which the controller is connected. See also: Buffer.

Packet Filtering - The ability of a bridge, router or gateway to limit propagation of packets across two or more interconnected networks by declining to forward a packet whose source and destination address are known to the device to be on the same network. This is typically implemented by the bridge maintaining a table of observed addresses and which of the attached networks they were seen on.

Packet Format - The exact order and size of the various control and information fields of a packet, including header, address, and data fields.

Packet Overhead - A measure of the ratio of the total packet bits occupied by control information to the number of bits of data, usually expressed as a percent.

Packet Switching - A data communications technique in which data is transmitted by means of addressed packets and a transmission channel is occupied for the duration of transmission of the packet only. The channel is then available for use by packets being transferred between different data terminal equipment... A data transmission method, using packets, whereby a channel is occupied only for the duration of transmission of the packet. The packet switch sends the different packets from different data conversations along the best route available in any particular order. At the other end, the packets are reassembled to form the original message which is then sent to the receiving computer. Because packets need not be sent in a particular order, and because they can g any route as long as they reach their destination, packet switching networks can choose the most efficient route and send the most efficient number of packets down that route before switching to another route to send more packets. The other advantage of packet switching is the unified format that every message is molded into.

PAD (Packet Assembler/Disassembler) - An interface device which buffers data sent to/from character mode devices and assembles and disassembles the packets needed for X.25 operation; an extension of CCITT X.25.

Parallel Interface - An interface which permits parallel transmission, or simultaneous transmission of the bits making up a character or byte, either over separate channels or on different carrier frequencies of the same channel.

Parallel Transmission - A technique that sends a number of bits simultaneously over separate cables. Normally used to send a byte (eight bits) at a time to a printer. Compare with serial transmission.

Parity - The method of ensuring each data byte transmitted or received. Each 1 bit is counted in a byte. The number of odd or even 1 bits in the byte is the parity. Parity may be even, odd, or none.

Parity Bit - An additional bit added to a group of bits, typically to a 7-bit character. That additional parity bit means that adding up all the bits in every byte will produce an odd or even number, depending on whether you choose odd or even parity. Parity bits are added for error detection. See also: Parity Checking.

Parity Checking - A method of error checking. The receiving and sending computer decide to send with either even or odd (or no) parity. Let's say even. Every byte, then, will have an even number of "1's" in it. The parity bit is appended or not appended to every byte to makes sure the addition is always the same. The receiving computer can then look at the parity bit along with the rest of the byte, count the number of "1's" and detect an error if that number is not even. Not a foolproof method of error detection. But it's better than the alternative, which is nothing. See parity bit.

Passive Fiber Optic Segment (PFOS) - A mixing segment including one Passive Star and all of the attached fibers. See Segment.

Passive Star - A passive device that is used to link fiber pairs together to form a Passive Fiber Optic Segment (PFOS). Optical signals received at any input port on the Passive Star are distributed to all of its output ports (including the output port of the port pair from which it was received). A Passive Star is typically comprised of a passive star coupler, fiber optic connectors and a suitable mechanical housing.

Passive Star Coupler - A passive fiber optic device providing division of op[tical power received at any of N input ports among all N outport ports. This division of optical power is approximately uniform.

PBX (Private Branch Exchange) - See PABX

PCM - See Pulse Code Modulation.

Peak Wavelength - The wavelength at which the optical power of a source is at maximum.

Peripheral Interface Cable - See SCSI Peripheral Cable.

Peripheral Device - A piece of computer hardware--such as a disk drive, printer, or modem--used in conjunction with a computer and under the computer's control. Peripheral devices are usually physically separate from the computer and connected to it by wires or cables.

Personal Connection - A set of hardware and software products from Ungermann- Bass designed to connect personal computers to Net/One and enable them to transparently share disk and printer resources, emulate 3270 terminals to access IBM hosts via an SNA gateway, and cooperatively exchange data and applications programs among workstations and hosts. Personal Connection is both a PC LAN and an extension of Net/One, allowing integration of PCs into the corporate computing network.

PFOSC (Passive Fiber Optic Star Coupler) - See Star Coupler.

Phase Modulation - Modulation is the process of using a medium to carry information. A flashlight beam could be "modulate" by turning it on and off, thus sending digital information. An electrical sine wave traveling down a twisted wire pair can also be modulated to carry information... A sine wave is defined by its frequency, amplitude and phase. These are the only three parameters of a sine wave that can thus be changed to carry information. Phase modulation is typically used in higher speed modems.

Photon - A quantum of electromagnetic energy. The energy of a photon is hg where h is Planck's constant and g is the optical frequency.

Photoconductivity - The conductivity increase exhibited by some nonmetallic materials, resulting from the free carriers generated when photon energy is absorbed in electronic transitions.

Photocurrent - The current that flows through a photosensitive device (such as a photodiode) as the result of exposure to radiant power. See also: Dark Current; Photodiode; Radiant Power.

Photovoltaic Effect - Production of a voltage difference across a p-n junction resulting from the absorption of photon energy. The voltage difference is caused by internal drift of holes and electrons.

Physical Layer - Layer one of the OSI reference model; encodes, modulates and transmits data across physical links (i.e., the transmission medium, such as coaxial cable) on the network; also defines the network's physical signaling characteristics.

Physical Layer Signalling (PLS) - That portion of the physical layer, contained within the DTE, that provides the logical and functional coupling between the MAU and Data Link Layer.

Physical Media Dependent (PMD) [FDDI] - Details hardware specifications: 62.5/125µm multi-mode fiber optic cable; a wavelength of 1,300nm and a frequency of 125MHz. These specifications allow the use of llight-emitting diodes (LEDs) as light sources instead of more expensive lasers. PMD also specifies a duplex Media Interface Connector (MIC), which utilizes keying and polarization to prevent incorrect physical configurations. Additionally, PMD outllines the peak optical power, optical rise and data dependent jitter so as to guarantee a worst case bit error rate (BER) of 2.5 times 10 to the -10th power and a normal BER of 1 times 10 to the -12th power for an FDDI network. Additionally, PMD specifies a maximum distance between stations of two kilometers, 1,000 phycical connections and a total fiber path length of 200 kilometers when using the prescribed fiber optic cable and default timer values.

Physical Medium Attachment (PMA) - The portion of the MAU that contains the functional circuitry.

Physical Protocol (PHY) [FDDI] - Describes the encoding scheme used by the network and the related method of timing and re-timing transmissions. In FDDI, each node uses its own internal clock for the transmission or re-transmission of data, as opposed to systems that utilize centralized timing. PHY specifies 4B/5B NRZI (non-return to zero invert), data transmission, which refers to the encoding of 4 bits of intormation into a 5 bit pattern, a particularly jefficient encoding scheme. In NRZI, a binary 1 is represented by a transition at the beginning of a bit interval. The PHY standard also specifies both distributed clocking and elastic buffering. Each station on the FDDI ring has its own clocking source and its own elasticity buffer of at least 10 bits. Each station receives data with clocking information from the previous station, but re-transmits the data with clocking information from its own autonomous clock. This effectively limits timing jitter, a common problem with systems with centralized clocking. This is not a problem in an FDDI network however, as stations with adequate priority can transmit numerous frames sequentially.

Physical Signalling (PLS) - That portion of the physical layer, contained within the DTE, that provides the logical and functional coupling between MAU and Data Link Layers.

Pigtail - A short length of optical fiber permanently fixed to a component to couple power to a transmission fiber.

Pin-Diode - A device used to convert optical signals to electrical signals in a receiver. For relatively fast speeds and moderate sensitivity in the 0.75µm to 1.1µm area wavelength, the silicon photodiode is most commonly used. Avalanche photodiodes (APD) combine the detection of optical signals with internal amplification of photocurrent. The internal gain is realized through avalanche multiplication of carriers in the junction region. The advantage in using an APD is its higher signal-to-noise ratio, especially at high bit rates.

Pin Photodiode - A diode with a large intrinsic region sandwiched between P and N doped semiconducting regions. Photons absorbed in this region create electron-hole pairs that are separated by an electric field, thus generating electric current in a load circuit.

Planck's Constant - The number h that relates the energy E of a photon with the frequency g of the associated wave through the relation $E = hg$; $h = 6.626 \times 10^{-34}$ joule second. See also: Photon.

Plastic Clad Silica (PCS) Fiber - A fiber with a glass core and a plastic cladding.

Point-to Point - 1) Point-to-point transmission - Transmission of data between only two stations or nodes, i.e., one sender and one receiver. 2) Point-to-Point link - A circuit which connects two (and only two) nodes without passing through an intermediate node.

Polarity - Electricity works on two pieces of wire. One is positive, i.e. the polarity of that wire is positive. The other wire is negative. Reversing polarity (putting the negative on the positive wire, and vice versa) can convey information. In particular, reversing polarity is used in the nationwide telephone network to indicate the person you are calling has answered the phone.

Polling - An access method used in star networks. The central hub controls the "conversation" by monitoring the signals of each node. The hub asks each node, "Do you want to speak?" If they do, the nodes send their messages. Polling offers all nodes equal access to the network. Another, (preferred) method uses "interrupts." See interrupts... A method of controlling the sequence of transmission by devices on a multipoint line by requiring each device to wait until the controlling processor requests it to transmit... A centrally-controlled method of calling a number of terminals to permit them to transmit information. As an alternative to contention, polling ensures that no single terminal is kept waiting for as long a time as it might under a contention network.

Port - A place which data can enter or leave the network. Examples are the serial and parallel "ports" on the back of most PCs. A port is also a name for a plug or receptacle, like those on the back of most PCs... The entrance or physical access point to a computer, multiplexor, device, or network where signals may be supplied, extracted or observed...A setment or IRL interface of a repeater unit.

POSTSCRIPT - A page-description language in which a computer communicates to a laser writer.

Power - The rate at which energy is absorbed, received, transmitted, transferred, etc. per unit time. Unit: watts.

Power Converter - An "AC-to-DC" converter which converts the +12 to +15 VDC power received from the controller/host (via the fiber optic transceiver interface cable) to the voltages required by the fiber optic transceiver optoelectronic circuitry.

Power Efficiency - The ratio of emitted optical power from a source to the electrical input power.

Power Switch - A switch usually located on the back of the main unit that turns the computer on and off.

Preamble - A sequence of encoded bits which is transmitted before each frame to allow synchronization of clocks and other circuitry at other sites on the channel. In the Ethernet specification, the preamble is 64 bits.

Presentation Layer - The sixth layer of the OSI model of data communications. It controls the formats of screens and files. Control codes, special graphics and character sets work in this layer. See also: OSI Standards.

Primary Buffer - A part of a computer's memory where fast incoming or outgoing data is kept until the computer has the change to process it. See also: Buffer, Secondary Buffer.

Printer - A computer attachment that produces printed copy such as numbers and words on paper.

Printer Driver - A program that lets you print on a corresponding printer. A printer device on an MS-DOS computer fulfills the same function as a **printer resource** on the Macintosh, but you need to install a printer driver for each application program.

Printer Port - A port on the Macintosh used to attach LocalTalk connectors...on other computers used to attach to printer.

Printer Server - A computer and/or program providing LAN users with access to a centralized printer. A person using the LAN will send a message to the printer server computer. This computer will the assign it a piece of memory or disk space to store its file while it waits to be printed. With a printer server, users can send to the printer any time. Their print jobs are usually handled in the order they are received. But priorities can be given and can be bumped to the top of the queue. Print servers allow fewer printers to satisfy more users. Print servers are also especially useful for expensive, laser or high speed printers, because they (the print servers) spread the cost of these expensive machines over many users, thus making expensive printers more affordable.

Printer Software - The software that controls the interaction of the computer and the printer.

Print Spooler - Software which allows print jobs to be taken away from the individual computer or workstation and put into a queue while waiting to print, giving the user full use of the computer during actual printing.

Printing Resource - A system file that lets you print on a corresponding printer attached to the computer. Sometimes called a Printer Driver.

PROFS - Professional Office System. A set of productivity software applications that run under VM/CCMS; most frequently used for electronic mail.

Program - A set of coded instructions that tells a computer how to perform a particular functions...A collection of instructions telling a computer to do a specific job, or set of related jobs. Most programs work some of the time. Few work all the time. The key is to figure which program you have. One way of increasing the chances your favorite program won't work is to run it with other programs, like programs which are RAM-resident.

PROM - Programmable Read Only Memory.

ProNet - A family of token-passing local area networks from Proteon, Natick, MA. The fastest of Proteon's networks is an extremely fast 80 Mbps.

Proprietary LAN - A LAN that runs the equipment of only one vendor. A proprietary LAN cannot join IBM PCs to DEC minicomputers. DEC and Wang both make proprietary LANs. Some people say such LANs are more "bug-free". They also tend to be more expensive. They also tend to tie you to one vendor, although some makers are now coming out with bridges which connect proprietary LANs to non-proprietary LANs.

Protocol - A set of rules for communicating between computers. These govern format, timing, sequencing, and error control. Think of protocol as the rules for communicating. Without these rules, the computer won't make sense of the stream of incoming bits... A set of rules and conventions that govern the orderly and meaningful exchange of information between or among communicating parties. Both hardware protocols and software protocols can be defined... A procedure for ordering the exchange of formatted information packets between correspondents. Protocols are "interpreted" by hardware and software within the adapter. Key elements of a protocol are timing, syntax and semantics. **Timing** includes speed matching, so a computer with a 9,600 bps port can talk to one with a 1,200 bps port, and proper sequencing of data if it arrives out of order. **Syntax** specifies the signal levels to be used and the format in which the data is to be sent. **Semantics** encompasses the information needed for coordination among machines and for data handling. A list of the five most popular communications protocols: **Xmodem** - the common standard among communications programs. Basically, Xmodem relies on checksum error checking to detect transmission errors. Its variants include **Xmodem-CRC** (cyclic redundancy checking) and **Xmodem 1K**. The former sends data in blocks of 128K. The latter in larger blocks, permitting faster transfer over phone lines...**Ymodem Batch** - similar to Xmodem 1K but sends a header containing the file name along with the file itself. In some implementations, the header also includes the file size and date...**Ymodem G** - not an error-correcting protocol. It is a streaming protocol built around the philosophy that no news is good news. Thus it sends the entire file before waiting for an acknowledgement. If the receiving side detects an error in midstream, it aborts the transfer. YmodemG is designed for use with high-speed modems that have built-in error-correcting protocols...**Zmodem** is a streaming protocol that allows for error detection, error correction as well as file management capabilities. The receiver can interrupt to request retransmission of any garbled data. Zmodem is fast and reliable on poor phone lines. If a link is interrupted during a transfer, it can redial and pick up the transfer without having to start over...**Kermit** is a popular protocol developed by Columbia University in New York.

Protocol Converter - A device for translating the data transmission code and/or protocol of one network or device to the corresponding code or protocol of another network or device, enabling equipment with different conventions to communicate with one another... A device that translates one communications protocol to another. Compare with bridge and gateway, which are different animals and may contain protocol converters... and usually much more.

PU - **Physical Unit.** A term used in the SNA environment to identify a printer, terminal or PC address. In particular, PU 2.1 is an IBM protocol that facilitates cooperative processing by supporting multiple conversations during a session.

Public Data Network (PDN) - A packet-switched or circuit-switched network available for use by many customers. PDNs may offer value-added services at a reduced cost because of communications resource sharing, and usually provide increased reliability due to built-in redundancy.

Pulse Code Modulation - PCM - A very common way of converting an analog signal - say from a telephone conversation - to a digital signal. Picture this: a voice analog signal looks like a sine wave. In PCM, you take many "pictures" of the sine wave many times each second. You give each "picture" a number. Then you send the numbers of those pictures in digital form to the other end. If you send sufficient numbered "pictures" you can recreate the analog voice signal at the other end. The more times you sample (the more digital information you send), the closer the end result will be to the original voice. PCM is a very good method of representing voice. PCM samples the voice 8,000 times a second. It measures each sample in 8 bits. This means it encodes one second of voice conversation into 64,000 bits, i.e. 8 by 8,000.

Pulse Dispersion - The widening of a pulse as it traverses the length of a fiber. This property limits the useful bandwidth of the fiber and is usually expressed in terms of nanoseconds of widening per kilometer. The principal mechanisms are material dispersion and multimode distortion effect.

Pulse Spreading - The dispersion of incoming optical signals along the length of an optical fiber.

-Q-

Queue - Just what it sounds like. A line of tasks, such as computer jobs or messages, waiting for service - for processing, printing, storing, etc. Any task waiting in a queue can be assigned a "priority" - so that important tasks may jump ahead.

QPSX - Queued Packet and Synchronous switch, a metropolitan area network implementation utilizing a distributed queuing access control method and dual bus architecture quite similar to that envisioned for the emerging IEEE 802.6 standard.

-R-

Radiant Power - The time rate of flow of radiant energy, expressed in watts.

RAM - Random Access Memory. A chip or collection of chips where data can be entered, read and erased. The basic idea of RAM is to speed your computer up. Your CPU could use your floppy as RAM, accessing your floppy every time it needed information. But this would be excruciatingly slow. RAM is the fastest memory device. The fast speed of RAM is good. However, RAM loses its contents when you lose or turn off power. Compare with read-only memory and EPROM.

RAM Cache - RAM you can designate to store certain information an application uses repeatedly. Using the RAM cache greatly speeds-up work but must be used sparingly or not at all with applications that require huge amount of memory.

RAM-Resident Program - A program loaded into computer memory (RAM) where it stays until the machine is reset, rebooted or turned off - or the program is knocked out of RAM. Sitting in memory, a RAM-resident program is quickly available to the user at the touch of a key. SuperKey, SideKick, Smartkey, and Popword are examples of RAM-resident programs. Some programs allow you to "knock" them out of RAM without rebooting. This may work if the RAM-resident program you're knocking out was the last you loaded. But if it's not, you may find yourself with a "hole" in RAM. This is a real delicate situation and one likely to cause your computer to "lock up" and cause you to lose lots of data you haven't saved.

Ray - A geometric representation of a light path through an optical medium: a line normal to the wave front indicating the direction of radiant energy flow. Compare: mode.

Rayleigh Scattering - Scattering by refractive index fluctuations (inhomogeneities in material density or composition) that are small with respect to wavelength. The scattered field is inversely proportional to the fourth power of the wavelength. See also: Material Scattering; Scattering

Read-Only Memory - (ROM) A chip or collection of chips that cannot be written to by normal computer circuitry. ROM must be programmed by special circuitry, generally using high voltages and high voltages and high electrical currents. It is impossible to change the instructions on ROM. Its contents do not change when the computer is turned off. Compare with random access memory, RAM, and EPROM.

Real Time - When the computer works on data as it is created. For instance, a computer figuring out the price of all the phone calls in an office as these phone calls are made. "Real-time" is also used in manufacturing to control robots. Computers which do real-time processing are usually very, very fast and expensive. Compare with batch processing and interactive... Generally, an operating mode under which receiving the data, processing it, and returning the results takes place so quickly as to actually affect the functioning of the environment, guide the physical processes in question, or interact instantaneously with the human user(s). Examples include a process control system in manufacturing, or a computer-assisted instruction system in an educational institution.

Rebooting - Repeating a boot. Turning on or resetting the computer. You do this when your PC "locks" inexplicably. "Locks" means that no matter which key or combination of keys you touch on your keyboard, you can't get your computer to do anything. In addition to "unlocking" your computer, you also reboot your computer to clear RAM of RAM-resident programs. Rebooting is done by pressing the Control, Alt, and Delete keys at the same time on IBM or IBM-compatible personal computers, or by pressing the reset button, if your computer has one. You can reboot any computer by turning its power off, then turning it back on. This is usually not a good idea, since the surge of power that accompanies a computer being turned on and off will reduce the life of many of its electronic components. Some experts recommend leaving computers running full-time. They also recommend turning your screen off, or at least running a public domain program such as Scrnsave Comm., which turns off your screen after three minutes of no use.

Receiver - A detector and electronic circuitry to change optical signals to electrical signals. See also: Detector.

Reconfiguration - Rearranging the equipment on a LAN...eg: moving computers or cables around.

Redundancy - 1) The part of a message or system that can be thrown away without losing the essential information or service. 2) The part of a system that duplicates the essential tasks to take over should the original fail. Redundancy is built into many systems - or you can build redundancy in, at your option - to insure your system will always work... That portion of the total information contained in a message which can be eliminated without the loss of essential information, such as characters used only for checking. Also used to describe a computer or communications facility in which there is a spare "back-up" device for each important component of the system.

Reference Surface - The surface of an optical fiber which is used to contact transverse alignment elements of a connector or other component.

Reflectance - The ratio of reflected power to incident power. Note: In optics, frequently expressed as optical density or as a percent; in communication applications, generally expressed in dB.

Reflection - The abrupt change in direction of a light beam at an interface between two dissimilar media so that the light beam returns into the medium from which it originated.

Refraction - The bending of a beam of light at an interface between two dissimilar media or in a medium whose refractive index is a continuous function of position (graded-index medium).

Refractive Index - The ratio of light velocity in a vacuum to its velocity in the transmitting medium.

Refractive Index Profile - The description of refractive index as a function of radius in a fiber.

Regenerate - See Repeater.

Relay - A device with an electrically controlled magnet whose magnetic field works a set of contacts that establish and interrupt electrical circuits. Like a gate that opens and closes.

Reliability - A measure of how dependably a system performs once you actually use it. Very different from MTBF. And very differed from availability. Also see MTBF.

Remote File Access - Retrieving files located on a distant computer from a local computer.

Remote Station - Any piece of equipment attached to a LAN by a telephone company supplied link. Technically, that includes all devices that aren't servers. Usually it refers to a workstation at a distant location, linked to the main LAN by a modem. See modems... Data terminal equipment located at a distance from the data processing site and requiring electronic communication for access...Data terminal equipment for communication with a data processing system in a distant location.

Repeater - An optoelectronic device that receives an optical signal, converts it to electrical signals, amplifies, reshapes or reconstructs the electrical signals and transmits it in optical form. In Ethernet, a device for connecting one coaxial segment to another within the same LAN. The repeater transmits signals both ways between the segments, and has appropriate procedures for dealing with possible collisions and any fault conditions which may occur. Codenoll repeaters provide for the interconnection of two fiber optic LAN segments, or connect a fiber optic LAN segment to Ethernet LAN based on coaxial cable as the transmission medium... A receiver and transmitter combination used to regenerate an attenuated signal... For local area networks, a device which increases the signal cover of a single LAN segment by joining it to another, so that packets sent on one segment can be "repeated" (or copied) onto the other, increasing the local environment... A device that amplifies signals from one piece of cable and passes them on to another piece of cable without changing the signals' contents. Microwave repeaters do the same thing, except the signals are transmitted through air or space. Repeaters increase the maximum length of LAN connections. Repeaters may also regenerate the digital signal - "squaring it" and "cleaning" it up - but not changing it. Regenerating the signal removes noise and thus reduces the likelihood of errors. You can only regenerate digital signals. You cannot regenerate analog signals. You can regenerate digital signals because a machine can tell what's a signal and what's noise in a digital signal. But no machine exists to do that with an analog signal.

Repeater Set - A repeater unit plus its associated MAUs and, if present, AU Interfaces (AUIs). A repeater set can receive and decode data from any segment under worst case noise, timing, and signal amplitude conditions. It retransmits the data to all other segments attached to it with timing and amplitude restored. The retransmission of data occurs simultaneously with reception. If a collision occurs, the repeater set propagates the collision evvent throughout the network by transmitting a Jam signal.

Repeater Unit - The portion of a repeater set that is inboard of its PMA/PLS interfaces.

Resident Program - See RAM-Resident Program.

Resolution - The degree of precision with which an object is represented. A printer's resolution is determined by the number of dots per inch (dpi).

Resource - A file in a a system folder that tells the computer how to work with a device. Sometimes called a Driver.

Resource Compiler - A program that allows users to create resources, such as icons, that can be recognized by a PC application.

Response Time - The time a system takes to react to a given act; the interval between completion of an input message and receipt of an output response. In data communications, response time includes transmission times to the computer, processing time at the computer (including access of file records), and transmission time back to the terminal.

Responsivity - The ratio of an optical detector's electrical output to its optical input, the precise definition depending on the detector type; generally expressed in amperes per watt or volts per watt or incident optical power.

Retransmit - To send a packet again if the original packet is not acknowledged, if it is received in error, or if a collision is detected.

Reverse Direction - The direction of signal flow toward the head-end in a broadband LAN. Low frequencies travel in this direction.

Return to Zero Code - See Halfstep Signalings

RF - Radio Frequency - A generic term referring to the technology employed in the CATV industry and broadband local area networks. Uses electromagnetic waveforms for transmission, usually in the megahertz (MHz) range.

RFC - Request For Comments, the official standards documentation vehicle for the DARPA Internet community.

RF Modem - Radio Frequency Modem - Device used to convert digital data signals to analog signals (and from analog to digital), then modulate/ demodulate them to/from their assigned frequencies.

Ring - A network topology in which stations are connected to one another in a closed logical circle. Typically, access to the media passes sequentially from one station to the next by means of polling from a master station, or by passing an access token from one station to another... A LAN topology (organization) where each workstation is connected to two other workstations. This forms a loop (or a ring). Data is sent from workstation to workstation around the loop in the same direction. Each PC acts as a repeater by resending messages to other PCs. Rings have a predictable response time, determined by the number of PCs. The more PCs, the slower the LAN. Network control is distributed in a ring network. Loss of one PC may disable the entire network, however. But many ring LANs have a way of recovering very fast should one PC die. Another PC may jump in and take over the control functions. Or you can simply remove a PC physically from the ring, therefore automatically joining the two on either side together. This is how the IBM Token-Passing Ring works. See also: Topology, Bypass Cable, FDDI LAN.

Ring Management (RMT) [FDDI] - Performs the management of the MAC layer components within a station and the rings to which they are locically attached. RMT includes the detection of faults at the MAC layer, such as stuck beacon identification and duplicate address information.

Ring-Wrap [FDDI] - Reconfiguration of a faulty ring or rings, accomplished by rerouting data from the primary ring onto the secondary ring at two or more locations in the network, thus effectivelyl bypassing cable faults. This redundancy results in a highly reliable network that may be used for demanding applications.

RIP - (RFC-1058) Routing Information Protocol, one of the ways that routing information is promulgated in TCP/IP networks.

Rise Time - The time for the leading edge of a pulse to increase from 10% to 90% of its peak value.

ROM - Abbreviation for read-only memory. This type of microchip cannot be altered by the user, it can only be read out.

Router - An electronic device interconnecting two or more networks that operate at the Network Layer (level 3) of the OSI model.

Routing - The dynamic exchange of network interconnection and topology information among the systems on interconnected networks.

RPC - Remote Procedure Call, a protocol whereby applications running on one host on a network can request services from a different host on the network with a software interface that looks like a local procedure call or subroutine, used as a building block for distributed network applications.

RS-232-C - A technical specification published by the EIA that specifies the mechanical and electrical characteristics of the interface for connecting DTE and DCE. It defines interface circuit functions and their corresponding connector pin assignments. The standard applies to both asynchronous and synchronous serial, binary data transmission at speeds up to 20 Kbps in full- or half-duplex mode. RS-232-C defines 20 specific functions. The physical connection between DTE and DCE is made through plug-in, 25-pin connectors. RS-232-C is functionally compatible with the CCITT Recommendation V.24...Also shortened (incorrectly) to RS-232. An interface that connects DTE and DCE. A technical specification published by the Electronics Industries Association, the EIA. It defines the mechanical and electrical characteristics for connecting DTE and DCE. It defines what the interface does, circuit functions and their corresponding connector pin assignments. The standard applies to both synchronous and asynchronous, binary data transmission. Most personal computers use the RS-232-C interface to attach modems. Some printers also use RS-232-C. You should be aware that despite the fact that RS-232-C is an EIA "standard", you cannot necessarily connect one RS-232-C equipped device to another one (like a printer to a computer) and expect them to work intelligently together. That's because different RS-232-C devices are often wired or pinned differently and may also use the same wires for different functions. The "traditional" RS-232-C plug has 25 pins.

RS-422 - A standard operating in conjunction with RS-449 that specifies electrical characteristics for balanced circuits (circuits with their own ground leads).

RS-423 - A standard operating in conjunction with RS-449 that specifies electrical characteristics for unbalanced circuits (circuits using common or shared grounding techniques). Another EIA standard for DTE/DCE connection which specifies interface requirements for expanded transmission speeds (up to 2 Mbps), longer cable lengths and 10 additional functions.

RS-449 - Applies to binary, serial, synchronous or asynchronous communications. Half- and full- duplex modes are accommodated and transmission can be over 2-or 4-wire facilities such as point-to-point or multipoint lines. The physical connection between DTE and DCE is made through a 37-contact connector; a separate 9-connector is specified to service secondary channel interchange circuits, when used. Used in conjunction with RS-422 and RS-423.

-S-

SAA - Systems Application Architecture. An IBM-developed set of standards that provides identical user interfaces for applications running on PCs, minicomputers and mainframes.

Saturation - When received optical signal is too strong for the maximum power allowed by the receiver, optical saturation prevents regeneration of the input signal, thus resulting in distortion (errors) in the received signal.

Scanner - A device that converts images into computer-readable form.

Scattering - The change in direction of light rays or photons after striking a small particle or discontinuity. It may also be regarded as the diffusion on a light beam caused by the inhomogeneity of the transmitting medium. See also: Backscattering; Material Scattering; Mode; Rayleigh Scattering.

SCSI - Acronym for Small Computer System Interface. Pronounced "SKUH-zee".

SCSI Cable Terminator - A device that reduces interference on the SCSI network.

SCSI ID Number - The identifying number of a SCSI device. All linked SCSI devices must have a unique SCSI number.

SCSI Peripheral Cable - A cable linking two SCSI peripheral devices on a SCSI chain.

SCSI Port - The connection point for SCSI cables on SCSI devices.

SDLC (Synchronous Data Link Control) - An IBM communications line discipline or protocol associated with SNA. SDLC provides for control of a single communications link or line, accommodates a number of network arrangements, and operates in half- or full- duplex over private or switched facilities...An IBM communications protocol associated with Systems Network Architecture (SNA). SDLC provides for control of a single communications link or line, accommodates a number of network arrangements, and operates in half- or full- duplex over private or switched lines.

Segment (802.3) - The continuous media connection comprised of similar medium type between Medium Dependent Interfaces (MDIs) in an 802.3 LAN. There are two types of segments: (1) Link: A point-to-point full duplex media connection. (2) Mixing: Multiple access media that may be connected to more than two MDIs.

Semantics - See Protocol.

Semaphore - A message sent when a file is opened or a disk accessed to prevent other users from opening the file or accessing the disk while it is being used. Its purpose is to preserve the integrity of your data by preventing simultaneous access to a file or disk.

Sequencing - The process of dividing a data message into smaller pieces for transmission, where each piece has its own sequence number for reassembly of the complete message at the destination end.

Serial Interface - The "lowest common denominator" of data communications; the simplest possible mechanism for changing the parallel arrangement of data within computers to the serial (one bit after the other) form used on data transmission lines and vice versa. At least one serial interface is usually provided on all computers for the connection of a terminal, a modem or a printer. Sometimes also called a serial port. See also RS-232-C. See also serial interface card... An interface which requires serial transmission, or the transfer of information in which the bits composing a character are sent sequentially. Implies only a single transmission channel.

Serial Interface Card - A printed circuit card which drops into one of the expansion slots of your computer and changes the parallel internal communications of your computer into the one-bit-at-time serial transmission for sending information to your modem or to a serial printer.

Serial Port - The connection on the back of the main unit for devices that use a Serial Interface.

Serial Transmission - Transmission where one bit of information is sent at a time on a channel. Compare with parallel transmission, in which eight bits - one character - are sent simultaneously. See asynchronous, channel, bit, synchronous...A mode of transmission in which each bit of a character is send sequentially on a single circuit or channel, rather than simultaneously as in parallel transmission.

Series 600 Products - Ungermann-Bass network modem products which provide dedicated point-to-point and multipoint circuits on standard CATV broadband cable systems. The NM-640 enables up to 28 full duplex point-to-point or multipoint circuits on each of up to eight 6-MHz CATV channels, and allows up to 19.2 Kbps synchronous or asynchronous communication. The NM-670 enables up to 14 full duplex point-to-point or multipoint circuits on each of up to three 6-MHz CATV channels, and allows synchronous communication at 56 Kbps.

Server - A computer providing a service to LAN users, such as shared access to a file system, a printer, or an electronic mail system. Usually a combination of hardware and software.

Session - A logical network connection between two workstations typically a user station and a server - for the exchange of data. Also a data conversation between two devices, say, a dumb terminal and a mainframe.

Session Layer - The fifth layer of the OSI model of data communications. It does the log keeping, security and administrative tasks.

SFT (Simple File Transfer) - A Net/One file transfer server offered by Ungermann-Bass which allows a user to transfer files between dissimilar host computers.

Shareware - Software you can try out without purchasing, but which you are honor-bound to pay for if you continue to use.

Shell - A program with which users interact, insulating the users from the complexities of the system.

Shielding - The process of protecting a cable (consisting of one or more plastic-coated conductors) with a grounded metal surrounding so that electrical signals from outside the cable cannot interfere with transmission inside the cable. Shielding will also lessen the chance that the information movement along the cable will interfere with other signals on other adjacent cables. The need for shielding stems from this phenomenon: If you send an electrical signal along one pair of cables, those cables will give off a small amount of electrical energy - called magnetic radiation. That radiation will cause electromagnetic interference with a cable close by. If you "shield" the pair carrying the electrical signal, you will cut down the susceptibility of those cables to interference from other cables. LANs should always be installed with the best quality shielded cable. They will run far better with shielded cable. Never skimp on the quality of the cable you're installing for LAN.

Short-Haul Modem - A data set designed for use in communicating data up to distances of 25 miles. The private line metallic circuits linking such devices permit moderately high speed operation, up to 19,200 bps and faster.

Signal Conditioning - The amplification and/or modification of electrical signals to make them more appropriate for transmission over a certain medium - cable, microwave, etc.

Signal Level - The root-mean square (rms) voltage measured during the peak of the RF signal. It is usually expression microvolts referred to an impedance of 75 ohms, or in dBmV... The strength of a signal, generally expressed in either absolute units of voltage or power, or in units relative to the strength of the signal at its source.

Signal-to-Noise Ratio (SNR) - The ratio of the signal level to the noise level, related to bit error rate expressed in dB. The relative power of the power of the signal to the noise on the cable.

Signal Quality Error Test (SQE/Heartbeat) - At the end of each transmission by a transceiver, it must send a short burst of 10 MHz waveform on the collision lead to permit the controller to check proper operation of the collision signal path. The burst starts .6µs after the end of the packet, must last at least 5 cycles, and end within 2µs after the end of the packet. The times are measured at the transceiver connector. There is no collision test signal when just receiving.

Simplex - Transmission in only one direction. Generally a communications system or device capable of transmission in one direction only. See Duplex.

Single Attach Station (SAS) [FDDI] - Only connects to the primary ring. Connections to the SAS are made in a star-like arrangement, rather than a circular arrangement as used for Dual Attach Stations. And concentrators (see Wiring Concentrator) may be cascaded to form a tree-like architecture.

Single Cable - A one-cable system in broadband LANs in which a portion of the bandwidth is allocated for send signals, and a portion for receive signals, with a guard band in between to provide isolation from interference.

Single-Density - A floppy disk that holds less information than double- density systems.

Single Mode Fiber - A fiber that allows only one path for light because of the small core diameter of the fiber: less than 10 microns: mnemonic. A fiber waveguide on which only one mode will propagate, providing the ultimate in bandwidth. It must be used with laser light sources...Single mode fiber has much less signal dispersion than multimode and hence is capable of much higher data rates over much longer distances. While there are few commercial single mode networking implementations in use today, future higher speed networks are expected to be based on single mode fibers.

Single-Sided - Floppy disks that store information only on one side.

Skew Ray - A ray that does not intersect the optical axis of a fiber (in contrast with a meridional ray). See also: Axial Ray; Meridional Ray.

Slot - A narrow socket inside the computer where you can install peripheral cards.

Slotted Ring - A LAN architecture in which a constant number of fixed-length slots (packets) circulate continuously around the ring. A full/empty indicator within the slot header indicates when a workstation or PC attached to the LAN may place information into the slot. Think of a slotted ring LAN as an empty train that constantly travels in a circle, being filled and emptied at different terminals (workstations).

SLIP - (RFC-1055) Serial IP, a way of supporting TCP/IP on asynchronous serial terminal lines.

Small Computer System Interface (SCSI) - A specification of mechanical, electrical, and functional standards for connecting peripheral devices such as hard disks, printers, and optical disks to small computers.

SMTP - (RFC-1067) Simple Mail transport Protocol, an Internet protocol for the exchange of electronic mail.

SNA (Systems Network Architecture) - The network architecture developed by IBM...Proprietary network protocol suite; specifies how IBM mainframes communicate hierarchically with unintelligent peripherals, such as terminals and printers.

SNADS - SNA Distribution Services. Protocol that allows distributing electronic mail and attached documents between PCs and terminals connected to an IBM mainframe.

SNMP - (RFC-1067) Simple Network Management Protocol.

Software - The programs that make a computer useful.....The set of instructions for carrying out various applications and tasks. Examples are programs like WordStar, Lotus 1-2-3 and dBASE III. Compare with and see also hardware and firmware.

Source - Originator of data... The means (usually LED or laser) used to convert an electrical information-carrying signal into a corresponding optical signal for transmission by an optical waveguide... A device that, when driven with electrical signals will produce optical signals.

Source Address - The part of a message which indicates who sent the message. The top left hand address on the envelope.

SPAN - Space Analysis Network.

Spectral Bandwidth - The difference between wavelengths at which the radiant intensity of illumination is half its peak intensity.

Spectral Width - A measure of the wavelength range of a sources output spectrum: can be specified as the full width at half maximum (FWHM). One method of specifying spectral linewidth is the full width at half maximum, specifically the difference between the wavelengths at which the magnitude is one-half of its maximum value.

Speed Of Light - (c) 2.998×10^8 per second.

Splice - A permanent connection between two fibers. May be thermally fused or mechanically applied.

Splicing - Permanent joining of identical or similar fiber ends.

Splitter - A passive device used in a cable system to divide the power of a single input into two or more outputs of lesser power. Can also be used as a combiner when two or more inputs are combined into a single output. See also: Combiner; Coupler; Star Coupler.

Spontaneous Emission - Radiation emitted when the internal energy of a quantum mechanical system drops from an excited level to a lower level without regard to the simultaneous presence of similar radiation. Examples of spontaneous emission include: 1) radiation from an LED, and 2) radiation from an injection laser below the lasing threshold. See also: Injection Laser Diode; Light Emitting Diode.

Spool - Abbreviation for Simultaneous Peripheral Operation On Line. A program or piece of hardware that controls a buffer of data going to some output device, most commonly a printer or a tape drive. A spool allows users to send data to a device, say a printer, even while that device is busy. The spool will control the feeding of data to the printer by using a buffer or by creating a temporary file in which to store the data to be printed. See buffer, compare with queue.

SQE (Heartbeat) Test - This is a test to insure that the collision presence circuit and path is working. This test is generated by the transceiver after it transmits a packet from the host. Since the collision circuitry is critical to the sound function of all CSMA/CD networks, the inclusion of this check is relevant. The problem with heartbeat is that it is not part of the Ethernet Version 1.0 specification. Therefore, Version 1.0 equipment may not function with transceiver that generates the heartbeat signal... Additionally, IEEE 802.3 specifications state that IEEE 802.3 compliant repeaters must not be attached to transceivers that generate heartbeat. (This has to do with a jam signal that prevents redundant collisions from occurring on the network). It is important that the transceiver or multiport transceiver be correctly configurated to work with the Ethernet or IEEE 802.3 controller. Therefore, any confusion as to the configuration should be answered by the controller board user manual. In many cases, Version 1.0 style equipment manuals will not mention the SQE (heartbeat) test, if the test is not mentioned a safe assumption would be to configure the transceiver such that heartbeat is not generated.

Station Management (SMT) [FDDI] - The part of the FDDI standard that overlaps the PMD, PHY and MAC layers. SMT specifies how the other FDDI layers will work together. It also provides for error detection and fault isolation. SMT 6.0 includes: Connection Management (CMT), Ring Management (RMT), and Frame-Based Services and Functions.

ST Connector - A type of connector used on fiber optic cable utilizing a spring loaded twist and lock coupling similar to the BNC connectors used with coaxial cabling.

Star - A network topology consisting of one central node with point-to-point links to several other nodes. Control of the network is usually located in the central node or switch, with all routing of network message traffic performed by the central node. Each station communicates with all other stations through the central node, potentially creating a central point of failure...A LAN topology in which all workstations are wired directly to a central workstation that establishes, maintains, and breaks connections between the workstations. The center of a star is called the hub. The advantage of a star is that it is easy to isolate a problem node. However, if the central node fails, the entire network fails. The star network we're all most familiar with is our local telephone exchange. At the center (the hub) rests the central office. Spanning out in a star are the lines going to the individual workstations (telephones) in people's houses and offices.

Star Bit - In asynchronous transmission, the first element in each character. The start bit tells the receiving device to recognize the incoming information. See also: Asynchronous, Serial Transmission.

Star Coupler - An optical element that allows connection of many fibers to a single fiber. A passive optical device in which power from one or several input fibers is distributed among a number of output optical fibers. Is powered by one or more input optical fibers and distributes ("broadcasts") light signals to all stations in a Local Area Network (LAN). A star coupler can have from 4 to 31 stations connected in a single LAN. It conforms to data signal broadcast functions required by Ethernet IEEE 802.3 CSMA/CD standards. See also: Combiner; Coupler; Splitter.

Starlan - AT & T Information Systems' LAN. Made of up three components: the Network Access Unit, the Premises Distribution System, and the Network Extension Unit. AT & T Starlan connects to the Information Systems Network, a larger data communications switching system, which has many of the characteristics of a very high speed data PBX. AT & T-IS is based in Morristown, NJ.

Start Up - To get the system running. Starting up is the process of first reading the operating system program from the disk, and then running an application program.

Startup Disk - A disk with all the necessary program files to set the computer into operation.

Station - Equipment, such as a computer, terminal, or file server attached to a LAN. Examples are PCs and workstations. Usually stations have keyboards, screens and some processing power. Sometimes, printers are considered stations. "Station" is an imprecise word. It comes from the time before the turn of the 20th century when the telephone industry was regulated by the Interstate Commerce Commission and all telephones were called "stations", because the ICC also regulated the railroads. That's why many of today's telecommunications terms sound like railroad terms... A network node.

Status Message - A message sent to the computer (usually from the printer) that tells you of the status of your job or to alert you of a problem.

Step Index - A fiber in which the core is of a uniform refractive index.

Steradian (sr) - The unit of solid angular measure, being the subtended surface area of a sphere divided by the square of the sphere's radius. There are $4p$ steradians in a sphere. The solid angle subtended by a cone of half-angle $\emptyset$ is $2p(1-\cos\emptyset)$ steradians.

Stimulated Emission - Radiation emitted when the internal energy of a quantum mechanical system drops from an excited level to a lower level when induced by the presence of radiant energy at the same frequency. An example is the radiation from an injection laser diode above lasing threshold. See also: Spontaneous Emission.

Stop Bit - In asynchronous transmission, the last transmitted element in each character. This bit completes the character and alerts the receiver to get ready to accept another character. See also: Star Bit, Asynchronous.

Stop/Start Transmission - A method of transmission in which a group of bits are preceded by a start bit and followed by a stop bit. Also called asynchronous transmission.

Store and Forward - A method of switching messages in which a message from one workstation is received at a central computer. This central computer acts as a switch. It stores the message. The central computer then figures the destination address and finds an available communication circuit. It then forwards the message to its destination. Many electronic mail systems work like this.

Stringer Cable - A wire rope used in the steam tunnels to support electrical or fiber optic cable.

Subchannel - A frequency subdivision created from the capacity of one physical channel by broadband LAN technology. Bands of frequencies of the same or different sizes are assigned to transmission of voice, data, or video signals. Actual transmission paths are created when each assigned band is divided, using FDM, into a number of subchannels. The bandwidth of the subchannels is determined by the form of information to be transmitted.

Subnet - RFC-950 specifies a way or partitioning a given TCP/IP network into "subnets" by utilizing a portion of the host part of the address to represent a physical subset of the local network, thus creating an additional layer in the address hierarchy.

Subsplit - A method of frequency division that allows two-way traffic on a single cable. Incoming signals go to the frequency translator between 5-30 MHz; outgoing signals go from the frequency translator between 54-400 MHz. No signals occupy 30-54 MHz... The most common form of transmission in the CATV industry. In the sub-split scheme, the bandwidth utilized to send toward the head-end (reverse direction) is much smaller, from approximately 5 MHz to 30 MHz, and the bandwidth utilized to send away from the head-end (forward direction) is very large, from approximately 55 MHz to 300 MHz. The guard band between forward and reverse directions (30 MHz to 55 MHz) provides isolation from interference.

Synchronous - Transmission in which there is a constant time between successive bits, characters, or events. The timing is achieved by the sharing of a single clock. Each end of the transmission synchronizes itself with the use of clocks and information sent along with the transmitted data. Most popular communications method for mainframes. Compare with asynchronous. In synchronous transmission, characters are spaced by time, not by start and stop bits. Because you don't have to add these bits, synchronous transmission of a message will take fewer bits (and therefore less time) than an asynchronous transmission. But because precise clocks and careful timing are needed in synchronous transmission, it's usually more expensive to set up synchronous transmission.

Synchronous Transmission - Transmission in which there is a constant time between successive bits, characters, or events. The timing is achieved by sharing of clocking... A transmission method in which the synchronizing of characters is controlled by timing signals generated at the sending and receiving stations (as opposed to start/stop communications). Both stations operate continuously at the same frequency and are maintained in a desired phase relationship. Any of several data codes may be used for the transmission, so long as the code utilized the required line control characters. (Also called "bi-sync," or "binary synchronous.")

Syntax - See Protocol.

System - A coordinated collection of interrelated and interacting parts organized to perform some function or achieve some purpose--for example, a computer system comprising a processor, a keyboard, a monitor, and a disk drive.

System File - A file that computers use to start up and to provide system-wide information

Switched Telephone Network - A network of telephone lines normally used for dialed telephone calls. Generally synonymous with the Direct Distance Dialing network, or any switching arrangement that does not require operator intervention.

-T-

T-1 - A 1.544-megabit-per-second communications circuit provided by long distance communications carriers for voice or data transmissions; T-1 lines are divided into 24 64-kilobyte channels.

T-1 Carrier - A digital transmission system developed by AT & T which sends information at 1.544 Mbps. With T-1 you can simultaneously transmit 24 voice conversations, each encoded at 64,000 bits per second. You can transmit more voice signals if you encode each conversation with fewer bits. T-1 circuits are becoming the preferred backbone communications channel for large corporation. Many PBXs can connect directly to T-1 lines. Some LANs can also, usually through gateways.

T-Connector - A connector used to connect a computer to thin Ethernet cable.

Tap - A passive 5-300 MHz box-like device, normally installed in-line with a broadband branch cable.

Passive circuits tap off only the information signals to its small Type F outlet ports... An electrical connection permitting signals to be transmitted onto or off a bus. The link between the bus and the drop cable that connects the workstation to the bus... 1) Baseband - The component or connector that attaches a transceiver to a cable. 2) Broadband - (Also called a directional tap or multitap) a passive device used to remove a portion of the signal power from the distribution line and deliver it onto the drop line.

Target Token Rotation Time (TTRT) [FDDI] - A means by which all stations on the ring send claim frames with their bid values for TTRT. Each station stores the value of TTRT contained in the latest received claim frame and absorbes claim frames with values higher than its own. When a station receives its own claim frame (the frame that has gone around the ring and has the lowest bid for TTRT value), that station sends a token to initialize the ring and that TTRT is used to allocate ring capacity.

TCP/IP (Transmission Control Protocol/Internet Protocol) - A protocol specification that conforms to the latest DOD ARPANET standard. The TCP/IP protocol module corresponds to layers three and four of the ISO protocol model...A layered set of protocols that allows sharing of applications among PCs and a high-speed communications environment. Because TCP/IP's protocols are standardized across all its layers, including those that provide terminal emulation and file transfer, different vendor's computing devices that run TCP/IP can exist on the same cable and communicate with each other across that cable. Corresponds to layers four (transport) and three (network) of the OSI reference model.

TDM (Time Division Multiplexing) - A method of utilizing channel capacity efficiently in which each node is allotted a small time interval, in turns, during which it may transmit a message or a portion of a message (for instance, a data packet). Nodes are given unique time slots during which they have exclusive command of the channel. The messages of many nodes are interleaved for transmission and then demultiplexed into their proper order at the receiving end. In time division multiplexing, users of a single channel take turns transmitting over the channel.

TELNET - An Internet protocol for remote terminal emulation, specified in RFC-1043.

Teleprocessing - Information handling in which a data processing system uses communications lines.

10BASE 2 - Subset of the IEEE 802.3 standard specifying characteristics for a 10 Mbit/sec CSMA/CD LAN utilizing RG58 50 ohm baseband coaxial cable in segments of up to 185 meter in length.

10BASE5 - Subset of the IEEE 802.3 standard specifying characteristics for a 10 Mbit/sec CSMA/CD LAN utilizing 1/2 inch thick 50 ohm baseband coaxial cable in segments of up to 500 meters in length.

10BASET - Subset of the IEEE 802.3 standard specifying characteristics for a 10 Mbit/sec CSMA/CD LAN utilizing radial twisted pair cabling.

10Net - The LAN from Fox Research, Dayton, OH. It is a baseband, Ethernet CSMA/CD LAN running one one twisted pair. See also: Ethernet.

Terminal - A work station designed to send or receive data that allows access to a main computer... A point in a network at which data can either enter or leave. Examples are modems, personal computers, dumb terminals, and telephones.

Terminal Emulation - A program which runs in a workstation or terminal that makes the workstation look, to both the user and the software, like a specific type of data terminal.

Terminal Server - A special purpose computer used to enable multiple asynchronous terminals to communicate with host computers by connecting over a network.

Terminator - A 75-ohm resistive connector used to terminate the end of a cable or an unused tap. The device is used to minimize cable reflections... A resistive connector used to terminate the end of a cable or an unused tap into its characteristic impedance. The terminator prevents interference-causing signal reflections.

Text - (1) Information presented in the form of readable characters. (2) The display of characters on a display screen. Compare Graphics.

Text File - A file containing information expressed in text form, with the contents encoded in the ASCII format.

Thick-wire Baseband - Common name for the cable used in IEEE 802.3 10BASE5.

Thin-net - A low cost implementation of Ethernet intended principally for connecting clusters of personal computers. See Thin-wire Baseband.

Thin-wire Baseband - Common name for the cable used in IEEE 802.3 10BASE2.

Throughput - The speed at which work is performed by a computer, or data is passed through a network. The total measure of useful information processed or communicated during a specific time period. Expressed in bits per second or packets per second.

Time Division Multiplexer - A device which permits the simultaneous transmission of many independent channels into a single high-speed data stream by dividing the signal into successive alternate bits.

Time Division Multiplexing - A method of sharing a communication channel among several users by allowing each to use the channel for a given period of time in a defined, repeated sequence... There are two basic methods of multiplexing, i.e. sharing lines among many users. The first is called frequency division multiplexing. Each "conversation" (voice, data or video) is allocated a separate radio frequency band on the cable. This is how CATV works. The second is called time division multiplexing. It works only with digital signals. It simply slices one piece of one conversation after another - in time. Think of time division multiplexing as a fast-moving railroad train. Each car carries a piece of the conversation. Let's say there are five data conversations going on. The first car gets a piece of the first conversation. The second car gets a piece of the second conversation... The sixth car would carry the second piece of the first conversation, and so on. Time division multiplexing works because the train is running fast, while the individual conversations are not that fast. Therefore, they don't notice they're being shunted onto and off a train and they don't notice the other conversations. See also: T-1

Timing Considerations - Whenever a fiber optic network becomes extremely large, and/or the distance between stations is increased to several kilometers, timing becomes critical. Ethernet specifications call for maximum timing constraints on the information presented to the network. The mechanism that detects collisions does so when a packet is transmitted. In a situation in which the round trip delay time between the two farthest stations on a network exceed the time it takes to transmit the minimum packet size, the collision will not be detected by the farthest station which first initiated transmission, and it will have no knowledge that its packet was distorted... As in "flux budget" calculations, there is a method to calculate the round trip delay time of a network. Shown here are the nominal delays associated with each component in the fiber Ethernet system:

Transceiver cable	5 nsec/Meter
Codenet transceiver	25 nsec
Fiber Cable	5 nsec/Meter
Fiber-to-fiber Repeater	800 nsec
(includes F/O transceivers)	

Roundtrip propagation delay should not exceed 50 microseconds. Fiber's low loss characteristics are frequently utilized to connect segments of existing coaxial networks. Because fiber requires fewer repeaters per unit length, it is possible to extend networks beyond the traditional 2 1/2 kilometers without violating the allowable round trip propagation delay time. For campus-type, building interconnects, networks comprised entirely of fiber optics can cover a much larger area than coaxial cable, and stay well within the delay time constraints... In addition, internal timing constraints in Codenet fiber optic Ethernet transceiver limit the maximum star-to-station distance to one kilometer. Also see Protocol.

Timesharing - Describes a computer that may be accessed by many users at once. Access is controlled by allocating certain amounts of time and memory in the computer to each user.

Token - A unique combination of bits. When a LAN workstation receives a token, it has now been given permission to transmit. See also: Token Passing.

Token Bus - A token access procedure used with a broadcast topology or network...A LAN with bus topology that uses token passing as its access method.

Token Passing - An access method. A token is passed from workstation to workstation thereby passing permission to send a message. When you have the token, you can send. You then attach your message to the token which "carries" them around the LAN. Every station in between you and whom you want to send to, "sees" the message, but only the receiving workstation accepts it (because that's whom it's addressed to.) When the receiving station gets the message, it either generates another token. Or the central network controller generates a new token. The entire process of generating and passing tokens around the LAN takes fractions of a second. See Token... A mechanism whereby each device receives and passes the right to use the channel. Tokens are special bit patterns or packets, usually several bits in length, that circulate from node to node when there is no message traffic. Possession of the token gives a node exclusive access to the network for transmitting its message, thus avoiding conflict with other nodes that wish to transmit.

Token Ring - A LAN network architecture and access control scheme where the stations are physically connected in a structure that can be logically viewed as a ring, and access to the medium is synchronized by a special control packet called the "token" that is passed from station to station along the ring. A station can only transmit when it has received the token from an adjacent station.

TOP - Technical and Office Protocol. A set of standards specific to office automation products; some TOP protocols reflect the OSI reference model.

Topology - Description of the physical connections of a network. Or the description of the possible logical connections between nodes, indicating which pairs of nodes are able to communicate. Think of it as a map of the "road" between all the things attached to a LAN. Examples are bus, ring, star and tree...1) Physical Topology - The configuration of network nodes and links. Description of the physical geometric arrangement of the links and nodes that make up a network as determined by their physical connections. 2) Logical Topology - Description of the possible logical connections between network nodes, indicating which pairs of nodes are able to communicate, whether or not they have a direct physical connection. Examples of network topologies are as follows: Bus, Ring, Star, Tree

Total Internal Reflection - The total reflection that occurs when light strikes an interface at angles of incidence (with respect to the normal) greater than the critical angle. See also: Critical Angle; Step Index.

Trailer - Information at the end of a data message indicating the conclusion of that particular message packet.

Transceiver - A combined transmitter and receiver. An essential element of all LANs, its function is required at each node of the network. In Ethernet, the transceiver transmits data packets from the controller onto the bus, receives packets from the bus and passes them on to the controller, and detects collisions. Ethernet transceivers are switched off when their associated host is not in use, without affecting the operation of the network as a whole. .. A device required in baseband networks which takes the digital signal from a computer or terminal and imposes it on the baseband medium.

Transceiver Cable - Cable connecting the transceiver to the network interface controller, allowing nodes to be placed away from the baseband medium.

Transceiver Cable Extender - Equipment to extend the link between an Ethernet or IEEE 802.3 station and its medium access unit (or transceiver) beyond the 50 meter maximum length allowed for transceiver cables. This typically involves a fiber optic link.

Transceiver Cable Interface - The specification for the protocols and rules of signal handshaking across the 15 pin interface specified in the IEEE 802.3 and Ethernet standards for communication between a station and a transceiver. A piece of communication equipment that conforms to the standards for the transceiver cable interface will work with any station that supports Ethernet or IEEE 802.3.

Transceiver Case - The protective metal package that physically and optically connects the fiber optic transceiver electronics to the optical fiber star network.

Transceiver Connector - The connector used to connect a computer to a transceiver on standard Ethernet or other types of cable.

Transfer Menu - Typically a way to navigate between programs without returning directly to the Finder.

Transients - Intermittent, short-duration signal impairments. "Junk" on electrical power lines cause these. "Junk" is interference from electrical motors, etc.

Transmission Loss - Total loss encountered in transmission through a system. See also: Attenuation; Reflection.

Transmission Media - Anything such as wire, coaxial cable, fiber optics, air, or vacuum, that is being used to carry an electrical signal which has information.

Transmitter - A driver and a source used to change electrical signals to optical signals... The electronic package that converts an electrical signal for conversion to an optical signal.

Transparent - Describes the operation of a network such that a user is unaware that a service is being provided by the network; rather, the service appears to be provided by resources local to the workstation.

Transport Layer - The fourth layer of the OSI model of data communications. High level quality control (error checking) and some alternate routing is done at this level.

Tree - A LAN topology in which there is only one route between any two of the nodes on the network. The pattern of connections resembles a tree, or the letter T.

Trunk - A single circuit between two points, both of which are switching centers and/or individual distribution points.

Trunk Cable - Coaxial cable used for distribution of RF signals over long distances throughout a cable system.

Turnaround Time - The actual time required to reverse the direction of transmission from send to receiver (or vice versa) on a half-duplex circuit.

Twisted Pair - Two insulated wires wrapped around (i.e. twisted around) each other. The pair of wires may be surrounded by a shield, a jacket, or additional insulation, or similar pairs of wires. Twisted pair wiring is most often used to connect telephones, terminals and computers to PBXs. Twisted pair wiring has a misleading usefulness - its ubiquity. However it suffers several disadvantages: 1) PBX wiring is often too thin. 2) PBX wiring has been often been spliced many times. Too many times. Some of the splices are usually not very good. 3) PBX wiring goes to the wrong place. Namely, it "homes" back to the central PBX cabinet. Existing PBX is not good for ring LANs, for example. And finally, worst of all, most twisted wire pair is not shielded. This makes it susceptible to electromagnetic interference. However, twisted pair wiring is easier to install, and easier to change than coaxial cable, though its bandwidth (information carrying capacity) is usually a lot smaller.

-U-

UDP - User Datagram Protocol. The TCP/IP transaction protocol used for applications such as remote network management and name service access; this lets users assign a name (such as "VAX2") to a physical or numbered address.

ULSI - Ultra Large-Scale Integration, the technique of putting millions of transistors on a single integrated circuit. Compare with LSI, VLSI.

ULTRIX - Digital Equipment Corporations version of UNIX.

UNIX - Computer operating system from AT & T. Also works on some personal computers. Considered to be very flexible and very powerful. Has enjoyed popularity among engineers and technical professionals. Personal computer versions of Unix are Xenix (for IBM PCs and compatibles) and AU/X for Macintosh. Sadly, there aren't many business programs running on UNIX-based programs can also be moved around between small, medium, and large computers. See also: Multitasking.

UNIX Kernel - The basic part of the UNIX operating system which allows application software to talk to hardware (the disk, the Ethernet driver, the printer, etc.) available to the computer.

User-Friendly - Hardware or software with simple instructions and prompts that help the user become familiar with the computer.

Utility Program - A special-purpose application that alters a system file or lets you perform some useful function related to a system file.

UUCP - UNIX to UNIX copy program.

-V-

Vendor Code - Software written by the same company that manufactured the computer system on which it is running (or not running).

Vendor Independence - The ability to allow devices manufactured by different vendors, often using different protocols, to talk, (to communicate) with each other.

VINES - VIrtual NEtworking Software which is the core of the LAN from Banyan, Westboro, MA.

Virtual Circuit - A communications link that appears to be a dedicated point-to-point circuit. Also, a system that delivers data packets in guaranteed sequential order, just as they would arrive over a real point-to-point circuit. There is no guarantee that the packets were sent in order or over the same route. Yet they are reassembled together correctly at the receiving end. By this reassembling, it appears there is an actual point-to-point connection...Provision of a circuit-like service by the software protocols of a network, enabling two end points to communicate as though via a physical circuit; a logical transmission path. The network nodes provide the addressing information needed in the packets that carry the source data to the destination. Advantages of virtual circuits over physical circuits include the resolution of speed mismatch between the data-generating and data-consuming end points, data retransmission in case of transient communication errors, and transformation of the information that traverses the circuits.

Virtual Disk - A portion of a physical disk drive appearing to a dedicated user as a local disk resource.

Virtual Memory - Feature provided by some operating systems which allows programs to address more memory than is actually installed in the computer.

VLSI - Very Large-Scale Integration. The art of putting hundreds of thousands of transistors onto a single quarter-inch square integrated circuit. Compare with LSI, ULSI.

VM/CCMS - Virtual Machine/Conversational Monitor System. Virtual Machine is an IBM operating system that permits external operating systems (such as DEC's Multiple Virtual System [MVS]) to reside on top of it. CCMS is the user interface for the VMS operating system.

VMS - Virtual Memory System. The operating system developed by Digital Equipment Corporation for its VAX computers. VMS provides some of the features found in UNIX as well as comprehensive file management and database management software, and DECnet networking services.

Volatile - An adjective describing a data storage device that loses its contents when power is lost (turned off). An example is the RAM chip/s inside a computer. The opposite to volatile is nonvolatile, which describes a data storage device, which holds its contents when the power is turned off. Examples of nonvolatile memory storage devices are floppy and hard disks and EPROMs.

VT - Virtual Terminal, OSI version of TELNET.

VTAM - Virtual Terminal Access Method. The host-based software that allows communication between an IBM 3270 mainframe and dumb terminals.

-W-

WangNet - Wang Laboratories' proprietary broadband LANs. Used as the brand name for Wang's LAN products.

Waveguide - A guide (Dielectric or Conducting) capable of conducting electromagnetic radiation at single or multiple modes - see fiber... A conducting or dielectric structure able to support and propagate one or more electromagnetic field patterns (modes). See also: Mode.

Wide Area Network - A data communications network designed to serve an area of hundreds or thousands of miles. Public and private packet switching networks and the nationwide telephone network are good examples of wide area networks.

Wideband - A communications channel offering a transmission bandwidth greater than a voice-grade channel. Synonymous with broadband. Compare with baseband... A channel broader in bandwidth than a voice-grade channel.

Winchester Disk - Also known as a hard disk. The Winchester magnetic storage device was pioneered by IBM for use in its 3030 disk system. The devices were called Winchester because "Winchester" was IBM's code name for the secret research project that led to their invention. A Winchester hard disk drive consists of several "platters" of metal stacked on top of each other. Each of the platter surfaces is coated with magnetic material and is "read" and "written" to by "heads" which float across the surface. The whole system works roughly like the old-style jukebox. There are several advantages to a Winchester disk system: 1) It can store, read and write enormous quantities of information. Some Winchesters have a capacity of over one gigabyte. 2) You can access any information on a Winchester faster than any other computer storage medium. 3) Winchesters are reliable and relatively inexpensive...Winchesters also have disadvantages 1) They are very sensitive to rough handling. 2) They are very sensitive to the organization of their directory track, and 3) When Winchesters "crash" touch the surface and you can lose an enormous amount of precious data - possibly millions of bytes of data...Winchester drives are typically at the center of most LANs, acting as central file servers. Make sure you back up the critical 64 changing information on your Winchester - at least twice a day.

Windows - A technique of displaying information on a screen in which the viewer sees what appears to be several sheets of paper much as they would appear on a desktop. The viewer can shift and shuffle the sheets on the screen. Windowing can show two files simultaneously. For example, in one window you might have a letter you're writing to someone and, in the other window, you might have a boilerplate letter, from which you can take a paragraph or two and drop it in your present letter. Being able to see the two letters on the screen makes writing the new letter easier...Windowing also is a technique for running several programs simultaneously - each one running in a separate window. For example, in one window you might run a word processing program. In another, you might be calculating a spreadsheet. In a third, you might be picking up your electronic mail from one of the services you subscribe to.

Wiring Concentrator (CON) [FDDI] - Used to attach a station to the FDDI ring. The concentrator attaches to the ring and provides additional ports which extend the primary ring to the other stations. The secondary ring is not connected to these additional ports.

Word Length - The number of bits or characters in a "word". It is usually figured as the optimal size for processing, storage or transmission. Word lengths are often based on the internal processing capability of the computer's main CPU - i.e. 16-bit words for the IBM PC-AT and 32-bit words for the IBM System 370.

Word Processing - A text-editing program allowing the writing and correcting of text and typesetting.

Workstation - Input/output device at which an operator works. Usually a personal computer, sometimes a terminal. A device which you can send data to or receive data from a computer.

-X-

X.25 - A CCITT standard which defines the interface between a PDN and a packet-mode user device (DTE); also defines the services that these user devices can expect from the X.25 PDN including the ability to establish virtual circuits through a PDN to another user device, to move date from one user device to another, and to destroy the virtual circuit when through... One of many CCITT standards which defines the standard communication protocol by which mainframes access packet switching networks. The X.25 protocol defines the interface between a public data network and a packet-mode user device. Some LANs, for example, have such a device, often called a PAD, packet Assembler- Disassembler. "X" stands for packet-switched network. See also: Packet Switching; OSI Standards.

X.28 - Defines the interface between PADs and non-packet mode DTEs.

X.29 - Defines the interface between PADs and packet-mode DTEs or other PADs.

X.3 - Describes functions of PAD and various parameters used to specify its mode of operation.

X.400 - OSI Electronic Mail interchange standard.

X.500 - OSI Service Directory standard.

X3T9 - The official ANSI designation for the standard specification of the Fiber Distributed Data Interface.

XMIT - Transmit.

XMODEM - see Protocol.

XMODEM BATCH - see Protocol.

XMODEM CRC - see Protocol.

XMODEM G - see Protocol.

XMODEM 1K - see Protocol.

XNS - Xerox Network Systems, their communications protocol for communication over Ethernet...Peer-to-peer protocol that has been incorporated into several local area networking schemes, including 3Com's 3+ and 3+Open network operating system.

XON/XOFF - Abbreviation for transmitter on/transmitter off. A commonly-used method of adjusting the flow of information between a computer and a slower, device, such as a printer or a modem. The computer sends the information. At one point the printer says "Hold it! Wait a moment. I'm full up. I need a few moments to digest what you sent me." This "Hold It. Wait a Moment" signal is called a XOFF, which is Control S. The computer now stops transmitting. And it sits, waiting an XON, which is a Control Q, which the printer sends, when it's ready to begin receiving more data.

X Window System - A window based user interface developed at MIT's project Athena, defined in RFC-1013.

-Y-

YMODEM BATCH - see Protocol.

YMODEM G - see Protocol.

-Z-

Zephyr - A notice transport and delivery system developed at MIT's Project Athena for use by network-based services and applications with a need for efficient and reliable communication with their clients.

ZMODEM - see Protocol.

Zone - A network in a series of interconnected networks, joined through Bridges.

VI

The Data Sheets which follow were complete and accurate at the time of publication. Codenoll also carries a line of data links and LEDs mounted in device receptacles. For information on these, or for new products, contact

> Codenoll Technology Corporation
> Customer Service
> 914-965-6300

Fiber Optic Ethernet Transceivers and PC Adapter Cards, and Network Management System

CodeMan AllNet Enterprise Network Manager, p. 414
CodeNet 8380/8382, p. 416
CodeNet 8381, p. 418
CodeNet 8330/8331/8332, p. 420
CodeNet 8300/8301, p. 422
CodeNet 8320/8321/8322, p. 424

Data Sheet

Codeman™ AllNet Enterprise Network Manager

Features

- **Integrated management of local and wide-area networks (LAN/WAN).**
- **Standards-based, supports SNMP.**
- **Consistent, logical visual presentation.**
- **Window interface simultaneously monitors network management functions.**
- **On-screen icons and color shading indicate network status and alarms.**
- **Dynamic configuration of network elements from a central point.**
- **Zoom-in on selected network objects to view imbedded levels of network elements.**
- **Network traffic and performance analysis.**
- **Network performance, traffic, events and alarms saved as a history file.**
- **Information collected on a polled basis.**
- **Two-way communication with any managed object can be initiated by the network manager at any time.**
- **Supports serial port for out-of-band login and testing of managed objects.**
- **Database approach provides configuration information, location and contact information, and alternative connections.**
- **Mouse support, pull-down menus and on-line, context-sensitive "Help" facility.**

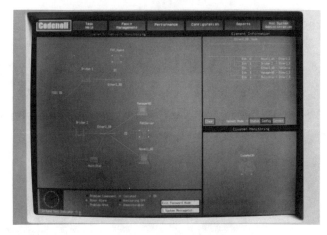

Overview

The Codenoll CodeMan AllNet Enterprise Network Manager is a standards-based Local Area Network (LAN)/Wide Area Network (WAN) management system that includes support for high-speed WAN/MAN links between distant facilities.

CodeMan AllNet provides a uniform, integrated management picture of multi-vendor LAN/WAN networks. The architecture is based on SNMP (Simplified Network Management Protocol) and manages Codenoll's Codenet® FDDI/Ethernet Bridge, FDDI Concentrator, Fiber Optic Ethernet Concentrators, as well as any other SNMP-compatible network device. The graphical on-screen interface, based on X-windows, presents a logical and consistent view of multiple network management functions and makes the CodeMan Enterprise Manager a powerful yet easy to use system.

The CodeMan AllNet Enterprise Network Manager is a software application that runs on a UNIX-based network workstation using the X.11 Windows environment, and that provides integrated monitoring and control of:

- Backbone and premise networks (SMDS, FDDI, Ethernet, Token Ring, T1/T3, 802.6/MAN).
- Ethernet Active Hubs and Passive Fiber Optic Ethernet segments.
- Individual network nodes.
- Codenoll Multi-Protocol Bridges.
- Codenoll FDDI Concentrators and nodes.

Advantages

- Network map quickly pinpoints problems graphically, displaying affected components and scope of impact.
- Using multiple windows, the network manager analyzes and compares problem information from different objects, while maintaining overall network surveillance.
- Provides for central management of enterprise network backbone resources and internetworking connectivity.
- Problem management, configuration, and administration functions available from a single central access point.
- System supports either centralized or departmental control center management configuration.
- Easy to use on-line documentation and context-sensitive help.

Capabilities

The Codenoll Enterprise Network Manager has been designed to control the critical components and points of concentration in a network and maintain performance and status related information from these managed objects. The windowed application allows simultaneous monitoring of multiple network management functions.

- SNMP compatibility allows the integrated management of any SNMP object. Fault diagnostics and other statistics are gathered as appropriate for the managed object.
- Information from managed objects is normally collected on a polled basis, at a frequency appropriate to the object's status.
- Two-way communication with the managed object can be initiated by the network manager at any time.
- Graphics and text representations of logical and physical network configurations are provided.

- On-screen color shading of icons and objects immediately indicates network status and alarms.
- Dynamic configuration of network elements is done from a central point.
- An SQL DBMS is used to create logical groupings of network components into Management Domains for clear network representation. Multiple CodeMan AllNet Managers can be defined, or a single Manager can be defined to manage the entire network.
- Ability to zoom in on selected network objects, and to view imbedded levels of network devices.
- Network performance, traffic, events, and alarms are monitored and saved as a history file for later analysis.
- Database entries for managed objects include configuration information, textual information, location and contact information as well as network address and alternative connection information.
- Future version to support CMIP/CMIS.
- Accumulates and analyzes network statistics.
- Turn nodes/ports/hubs on or off remotely.

CodeMan Architecture

The Codenoll Network Management Architecture consists of a number of software components and levels of network managers that work together. The different CodeMan Network Management products can be combined to satisfy any network management need.

The CodeMan AllNet Enterprise Network Manager will manage a complete network implementation which may include LAN/WAN/MAN components and network equipment from multiple vendors. For larger networks, multiple CodeMan AllNet Managers can be installed, separating the network into multiple "management domains." The CodeMan AllNet Manager works with the CodeMan Premise Manager and the CodeMan Workgroup Manager. The AllNet manager is capable of managing any equipment which uses the SNMP management protocol.

The CodeMan AllNet Manager can also manage multiple Ethernet or FDDI concentrators, as well as individual CodeMan Workgroup Network Managers, in a single facility.

The CodeMan Workgroup Manager takes network management down to the individual PC level. This is a level of network management unavailable with other management systems. The workstations will report their status with network management primitives to the CodeMan Workgroup Manager. The Workgroup Manager will directly monitor and manage individual workstations, and will forward this information in SNMP format to the CodeMan AllNet Manager. Existing PCs can participate in the network management system with minimal impact on memory and CPU usage.

Standard SNMP Support

Following Codenoll's policy of standards-based network solutions, the CodeMan AllNet Enterprise Network Manager is based on Simple Network Management Protocol (SNMP); the UNIX System V operating system; and the X-11 Windows display management system. A future version will support CMIP/CMIS.

System Specifications

System Requirements:
Unix, System V. Tested on SUN OS, and Interactive UNIX
X-11 Windows Runtime
TCP/IP, NFS optional
Informix SQL DBMS
High resolution display driver
Color monitor with a minimum of 1152 × 900 pixel resolution
50MB minimum free disk space, after installation of the operating system
CodeNet or SUN Ethernet interface, capable of IP addressing and operation in promiscuous mode
Supported bus or serial mouse
High density floppy diskette drive

Typical System Configurations:
Sun Sparcstation 1
SUN OS 4.0 or later
200MB minimum disk space (requirement dependent on network size)
SUN Open Windows (X.11 and News) Release 1.0, or later
Informix ISQL runtime
SUN compatible tape drive for software loading and backup
Color monitor (1152 × 900 resolution)

Compaq DeskPro 386/33, or better
8MB of RAM, 3½" or 5¼" high density diskette. 200MB minimum disk space (requirement dependent on network size)
Interactive UNIX, Release 2 or later
Interactive TCP/IP Runtime
Informix 4GL Runtime
Informix SQL Runtime
NEC 4D Monitor (1024 × 768 resolution)
NEC Multisync Graphic Engine
UNIX device driver for MGE
Logitech Serial Mouse or Microsoft Bus Mouse
CodeNet 83XX Ethernet Adapter

386 Options:
Tape backup system, recommended
Serial port for audible alarm
Serial port for out-of-band login and testing
Serial port with Epson compatible printer for reports
IPT Grafsman software package for printing traffic reports

For Information or Assistance

Call the Customer Support Group at Codenoll Technology, (914) 965-6300.

Specifications may change without notice. Codenoll, CodeNet CodeStar, MultiStar and CodeLED are registered trademarks of Codenoll Technology Corporation. Other trademarks mentioned are the property of other companies.

MAKING LIGHT WORK OF NETWORKING

1086 NORTH BROADWAY, YONKERS, NEW YORK 10701 USA • TELEPHONE: (914) 965-6300 • FAX (919) 965-9811

© 1990, Codenoll Technology Corporation
Printed in U.S.A.—October, 1990

Data Sheet

CODENET®-8380/8382
High Performance Fiber Optic Ethernet Transceiver

Features
- Signals comply with IEEE 802.3 standard at AUI (Attachment Unit Interface) via 15-pin AUI connector.
- Fiber optic network operation completely transparent to user's hardware and software.
- Supports distances up to 800 meters to the passive fiber optic Ethernet star.
- Switch selectable operating mode enables unit to function either point-to-point, or with passive star.
- Fiber break detection when used with CodeStar® passive stars.
- Switch selectable "Jabber" function protects system against unnecessary downtime.
- Can be used as light source for network testing of fiber optic cable plant.
- SQE "Heartbeat" test may be defeated via switch selection, allowing unit to be used with IEEE-802.3 electronic repeaters.

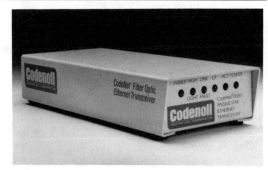

CodeNet-8380/8382

Product Description
The CodeNet 8380/8382 is a fiber optic media access unit (FOMAU) which complies with IEEE 802.3 standard at the AUI connection. All 838X units are equipped with an integral cable which terminates in a male 15 pin AUI connector. The cable provides power and electrical connections to the transceiver and enables connection to Ethernet controllers. Since the 8380/8382 is mechanically and electrically identical to coaxial cable transceivers at the interface cable, Codenet fiber optic Ethernet networks can be added to, and combined with existing co-axial cable based networks, thus allowing Ethernet networks to be expanded and converted to fiber optics at any time.

Functional Description
The 8380/8382 contains several on-board switches which make the unit extremely versatile, allowing it to be used in a variety of configurations. These switches include:
- SQE ("Heartbeat") Test Switch–Enables the user to defeat the SQE or "Heartbeat" test, allowing the unit to be used with IEEE 802.3 electronic repeaters.
- Mode Switch–determines whether the unit will work in passive star (standard) or point-to-point mode. The point-to-point mode enables the 8380/8382 to be used in "remote repeater" and active star applications.
- Test Switch–Allows the unit's LED transmitter to be used as a light source for testing of a network's fiber optic cable plant, greatly simplifying network installation.
- Jabber Switch–In IEEE 802.3 mode, once the unit's "Jabber" circuitry is tripped (indicative of a controller malfunction), a power-up reset is required before the unit will function again. The Jabber switch is factory set to an automatic reset mode, in which the unit will automatically reset itself once the controller clears, eliminating unnecessary down time.
- Break Detection Switch–This switch can defeat the unit's fiber break detection in passive star mode (in point-to-point mode, it is automatically defeated).

The unit also has several indicator LEDs which facilitate troubleshooting of network problems.

Secure Reliable Performance
The high bandwidth of fiber means the highest performance in your CodeNet network now and, since the fiber is the same as that specified for high performance networks like FDDI, your physical plant investment enjoys a long lifetime. Your CodeNet network is completely immune to electrical interference, even in noisy manufacturing or processing facilities. For secure networks, such as government and research installers, fiber optics are the only choice; unauthorized taps are difficult to make and easily detected. With CodeNet you can increase the distance between workstations or servers and still enjoy increased reliability. Compared with CodeNet optical fiber, all other networking media are severely limited, in terms of future upgrades.

Simple Installation and Setup
Onboard standard fiber optic connectors directly attach to standard duplex fiber optic network cable using standard ST-type or optional SMA connectors.

Optimal Network Design
Since CodeNet fiber optic Ethernet networks support both active and passive star hubs to meet user needs, CodeNet transceivers are available for all possible configurations:
- CodeNet-8380 connects to a CodeStar passive star hub.
- CodeNet-8381 connects to a MultiStar active star hub.
- CodeNet-8382 connects to CodeStar passive star hubs with other CodeNet-30XX or 83X2 CodeNet products.

Standard IEEE 802.3 Ethernet

Following Codenoll's policy of standards-based network solutions, each CodeNet-8380/8382 transceiver meets applicable signaling standards of IEEE 802.3 Ethernet for 10 Mbps baseband CSMA/CD LANs.

Specifications — CodeNet-8380/8382 Transceivers

Network Standards:	IEEE 802.3 CSMA/CD Ethernet 10 Mbps baseband
Electrical Interface:	1 meter integrated AUI cable with 15-pin male subminiature D-type Ethernet AUI connection
Optical Interface:	Duplex fiber optic cable with bayonet type ST connectors standard; screw-on SMA connectors optional
Internal Option Switches: (factory setting in bold)	Break Detect (**On**/Off) SQE-Heartbeat (On/**Off**) Point-to-Point Mode (On/**Off**) Jabber Reset (**Auto**/Manual) Test Mode (On/**Off**)
Optics:	Transmitter CodeLED® high radiance edge-emitting Aluminum Galium Arsenide (AlGaAs) Light Emitting Diode (LED) Receiver Silicon PIN-type photodiode Optical Wavelength 830nm ±20nm, peak Cable Sizes 50/125, 62.5/125, 85/125, or 100/140μm duplex, multimode, graded index
Physical:	Transceiver Size (not including interface connections) 3.3 in. W × 8.2 in. L × 1.5 in. H (8.4 cm × 20.8 cm × 3.8 cm)

Operating Environment
0°C to +55°C, 10 to 90% humidity (non-condensing)

Power
500mA @ +12V, maximum

Weight
21 oz. (585 grams)

Optical Specifications[1]:

	CodeNet-8380	CodeNet-8382
Peak Power Output Into:		
62.5 μm core fiber	−9 dBm	−3.5 dBm
100 μm core fiber	−8.5 dBm	−3 dBm
50 μm core fiber	−12.5 dBm	−6.5 dBm
Receiver Sensitivity	−37 dBm	−30 dBm
Optical Flux Budget Into:		
62.5 μm core fiber	28 dB	26.5 dB
100 μm core fiber	28.5 dB	27 dB
50 μm core fiber	24.5 dB	23 dB
Dynamic Range	18 dB	15 dB
Maximum Cable Length:		
Star	800m[2]	800m[2]

[1] Typ. peak power measured at 25°C, 50% duty cycle.
[2] Limited by IEEE 802.3 roundtrip delay time, flux budget permitting.

Diagnostic Indicators:	Power — Green Activity — Green Collision — Yellow Link Fault — Yellow Jabber — Yellow High Light — Yellow

For Information or Assistance

Call the Customer Support Group at Codenoll Technology, (914) 965-6300.

Specifications may change without notice. Codenoll, CodeNet CodeStar, MultiStar and CodeLED are registered trademarks of Codenoll Technology Corporation. Other trademarks mentioned are the property of other companies.

MAKING LIGHT WORK OF NETWORKING

1086 NORTH BROADWAY, YONKERS, NEW YORK 10701 USA • TELEPHONE: (914) 965-6300 • FAX (914) 965-9811

© 1990, Codenoll Technology Corporation
Printed in U.S.A. — October, 1990

Data Sheet

CODENET®-8381
Fiber Optic Inter-Repeater Link (FOIRL) Transceiver for use with MultiStar® & MultiConnect Repeaters

Features
- Fully compatible with IEEE Standard 802.3, Section 9.9.
- Allows for distances up to 4.5 km between repeaters.
- Fiber optic network operation completely transparent to user's hardware and software.
- Low light level monitor detects fiber breakage.
- Optical idle signal provides fiber integrity indication and allows simple measurement of transmitter power.
- Electrical interface accepts Ethernet (Versions 1 & 2) or IEEE 802.3 standard signal formats.
- Switch selectable SQE "Heartbeat" test, automatic "Jabber" reset, and Hi-Lo optical transmit power.

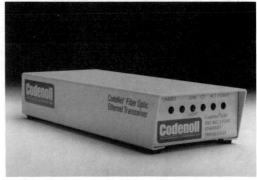

CodeNet-8381

Product Description

The CodeNet-8381 is a fiber optic media access unit (FOMAU) which fully complies with IEEE 802.3 specifications for a vendor independent fiber optic inter-repeater link. The unit may be used to provide a point-to-point Ethernet link between any two IEEE 802.3 compatible repeaters, bridges, servers, or DTEs regardless of vendor. Compatibility is provided at both the electrical interface (AUI connector), and at the optical interface, allowing the CodeNet-8381 to be added to existing networks, using multiple vendor equipment.

The 8381 is equipped with an integral cable which terminates in a male 15 pin AUI connector. Since the unit is mechanically and electrically identical to coaxial cable transceivers at the AUI connector, CodeNet fiber optic Ethernet networks can be added to, and combined with existing coaxial cable based networks, allowing Ethernet networks to be expanded and converted to fiber optics at any time.

Functional Description

The 8381 contains several on-board switches which make the unit extremely versatile, allowing it to be used in a variety of configurations. These switches include:
- **Optical Power Switch**—Selects the optical power (Hi or Lo) of the transmitter LED, allowing the unit to provide an Ethernet link of up to 4.5 km (depending on network propagation delay and flux budget).
- **Jabber Switch**—In IEEE 802.3 mode, once the unit's "Jabber" circuitry is tripped (indicative of a controller malfunction) a power-up reset is required before the unit will function again. The Jabber switch is factory set to an automatic reset mode, in which the unit will automatically reset itself once the controller clears. This eliminates unnecessary down time.
- **SQE ("Heartbeat") Test Switch**—Allows the user to defeat the SQE "Heartbeat" test, enabling the unit to be used with IEEE 802.3 electronic repeaters.

The unit also has several indicator LEDs while facilitate network troubleshooting.

Secure Reliable Performance

The high bandwidth of fiber means the highest performance in your CodeNet network now and, since the fiber is the same as that specified for high performance networks like FDDI, your physical plant investment enjoys a long lifetime. Your CodeNet network is completely immune to electrical interference, even in noisy manufacturing or processing facilities. For secure networks, such as government and research installations, fiber optics are the only choice; unauthorized taps are difficult to make and easily detected. With CodeNet you can increase the distance between workstations or servers and still enjoy increased reliability. Compared with CodeNet optical fiber, all other networking media are severely limited, in terms of future upgrades.

Optimal Network Design

Since CodeNet fiber optic Ethernet networks support both active and passive star hubs to meet user needs, CodeNet transceivers are available for all possible configurations:
- CodeNet-8380 connects to a CodeStar passive star hub.
- CodeNet-8381 connects to a MultiStar active star hub.
- CodeNet-8382 connects to CodeStar passive star hubs supporting CodeNet-30XX or 83X2 CodeNet products.

Simple Installation and Setup

Onboard standard fiber optic connectors directly attach to standard duplex fiber optic network cable using ST-type connectors (SMA connectors optional).

Standard IEEE 802.3 Ethernet

Following Codenoll's policy of standards-based network solutions, each CodeNet-8381 transceiver meets applicable Fiber Optic Inter-Repeater Link (FOIRL) signaling standards of IEEE 802.3 Ethernet for 10 Mbps baseband CSMA/CD LANs.

Specifications — CodeNet-8381 Transceiver

Network Standards:	IEEE 802.3 CSMA/CD Ethernet 10 Mbps baseband, Section 9.9 Fiber Optic Repeater Link (FOIRL)
Electrical Interface:	1 meter integrated AUI cable with 15-pin male subminiature D-type Ethernet AUI connection
Optical Interface:	Duplex fiber optic cable with bayonet type ST connectors standard; screw-on SMA connectors optional. (FOIRL standard is SMA)
Internal Option Switches: (factory settings in bold)	Optical Power (**Hi**/Lo) Jabber Reset (**Auto**/Manual) SQE-Heartbeat (On/**Off**)
Optics:	**Transmitter** CodeLED® high radiance edge-emitting Aluminum Galium Arsenide (AlGaAs) Light Emitting Diode (LED) **Receiver** Silicon PIN-type photodiode **Optical Wavelength** 830nm ±20nm, peak **Cable Sizes** 50/125, 62.5/125, 85/125, or 100/140μ m duplex, multimode, graded index
Physical:	Transceiver Size (not including interface connections) 3.3 in. W × 8.2 in. L × 1.5 in. H (8.4 cm × 20.8 cm × 3.8 cm) Operating Environment 0°C to +55°C, 10 to 90% humidity (non-condensing) Power 500 mA @ +12V, maximum Weight 21 oz. (585 grams)

Optical Specifications[1]:

Peak Power Output Into:	
62.5 μm core fiber	−12 dBm
100 μm core fiber	−11.5 dBm
50 μm core fiber	−15.5 dBm
Receiver Sensitivity	−27 dBm
Optical Flux Budget Into:	
62.5 μm core fiber	15 dB
100 μm core fiber	15.5 dB
50 μm core fiber	11.5 dB
Dynamic Range	18 dB
Maximum Cable Length:	4500m[2]

[1] Typ. peak power measured at 25°C, 50% duty cycle, Hi power [factory default]. Lo Power reduces power by approximately 3 dBm and distance to approximately 3000 m.
[2] Limited by IEEE 802.3 roundtrip delay time, flux budget permitting.

Diagnostic Indicators:	Power — Green Activity — Green Collision — Yellow Link Fault — Yellow Jabber — Yellow

For Information or Assistance

Call the Customer Support Group at Codenoll Technology, (914) 965-6300.

Specifications may change without notice. Codenoll, CodeNet CodeStar, MultiStar and CodeLED are registered trademarks of Codenoll Technology Corporation. Other trademarks mentioned are the property of other companies.

MAKING LIGHT WORK OF NETWORKING

1086 NORTH BROADWAY, YONKERS, NEW YORK 10701 USA • TELEPHONE: (914) 965-6300 • FAX (914) 965-9811

Data Sheet

CODENET®-8330/8331/8332
High Performance Fiber Optic Ethernet PC Network Interface Cards

Features

- Connects an IBM PC XT, AT, PS/2-25, 30 and compatibles to a CodeNet fiber optic Ethernet LAN.
- Driver support for all popular network operating systems, including IBM OS/2 Lan Manager, Novell NetWare, Microsoft Lan Manager, 3Com 3+ and 3+ Open, Digital DECnet, SCO UNIX, Western Digital ViaNet and others.
- Unique PC shared memory interface: dual-ported 8K memory shared with PC for high throughput, fast response. No DMA channels required.
- Meets all applicable Ethernet and IEEE 802.3 signaling standards. 10 Mbps operation, CSMA/CD.
- Easily field upgradable to 16 bit and Micro Channel versions at low cost.
- Computers can be up to 4500 meters apart in point to point and star configurations.
- Wait states are software selectable to optimize network performance.
- Uses just one PC expansion slot.
- The CodeNet 8330/8332 connect to CodeStar® passive star hubs. The CodeNet 8332 can be used on the same star as the CodeNet 30XX Series interfaces. The CodeNet 8331 connects to the CodeNet MultiStar® active star hub.

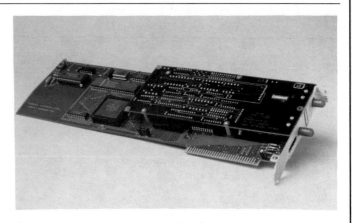

Performance Summary

The CodeNet-8330/8331/8332 cards provide a high-performance link between the user's PCs and the reliability, performance, security and long-distance capability of the CodeNet fiber optic Ethernet network.

These boards are based on the proven speed, performance and throughput of the EtherCard PLUS (WD8003EB) from Western Digital and the Codenoll Fiber Optic Ethernet Transceiver, the most widely used fiber optic Ethernet transceiver in the world. By adding the performance of CodeNet fiber optics to the fastest Ethernet PC network interface, the CodeNet-8330/8331/8332 cards achieve the highest performance in the industry for 8-bit ISA network boards.

Easy Field Upgradability

The CodeNet 8330/8331/8332 Network Boards each consist of two parts: (1) A plug-in board containing the high speed data interface between the computer and the Ethernet controller; and (2) a plug-on daughter board containing a CodeNet fiber optic Transceiver. Either part may be changed in the field, eliminating the need and the money required to purchase a completely new adapter when computers are upgraded. For example, by simply changing the plug-in board the user has a CodeNet 8300 EISA Network Board.

Meets Ethernet and IEEE 802.3 Standards

In line with Codenoll's dedication to standards-based network solutions, The CodeNet-8330/8331/8332 cards meet all applicable signaling standards of IEEE 802.3 Ethernet for 10 Mbps baseband CSMA/CD local area networks.

High Performance, Lower Price

The CodeNet-8330/8331/8332 achieve higher performance because of an up-to-32K buffer memory architecture. Since this memory buffer is dual-ported and directly accessed by the PC, the PC simply uses "move string" instructions to move data to and from the buffer in the same way it moves data around its own memory.

This eliminates the need to set up a hardware-implemented PC Direct Memory Access (DMA) channel data transfer. The result is that the CodeNet cards offer much higher throughput. The CodeNet-8330/8331/8332 are especially suited to servers, gateways, backbone connections and high performance workstations. A software-selectable option allows the user to modify usage wait states to optimize performance.

Application Flexibility

Codenoll provides drivers for IBM OS/2 Lan Manager, Microsoft Lan Manager, Novell NetWare, SCO Unix, Western Digital's ViaNet, Sun's PC NFS, Digital's DECnet, and 3Com's 3+ system. In addition, most of the major suppliers of network operating software support these network adapter cards.

Installation and Setup

Each CodeNet-8330/8331/8332 is supplied with input and output fiber optic connectors for the network interface. Physical connection is direct to industry standard fiber optic cable, so there are no devices outside the computer.

The I/O memory address and interrupt level selection are configured at the factory but can be easily modified during the installation process.

Setup is an easy on-screen operation thanks to the configuration program on the CodeNet Driver diskette, which includes software drivers for major network environments.

The Benefits of Codenoll Fiber Optics

Codenoll networks are completely immune to electromagnetic noise of all types and no form of cable damage can bring down more than one node of the network. Unlike copper wire, Codenoll cable can be run over or around fluorescent lights, in power raceways, near copy machines, x-ray units, large computers, elevator motors, transformers, production machinery, or robots. Because of this dielectric feature, a large number of network errors and resulting retransmissions are eliminated, thus increasing effective network throughput. Bit error rates less than 10^{-15} are common in Codenoll networks.

CodeNet Ethernet is more robust and reliable than coax or twisted pair copper-based systems. No short circuits, ground faults, reflections, or impedance mismatches are possible. It is impossible for any form of cable damage to cause damage to sensitive computer electronics as can happen on copper-based networks. It is also impossible for a CodeNet cable to cause a fire or explosion since there is no electrical current present outside the computer.

It is virtually impossible for an unauthorized agent to tap into the cable without easy detection.

CodeNet permits up to 4.5 kilometers between nodes with or without repeaters in both point-to-point and star configurations.

Specifications—CodeNet-8330 and -8331 Ethernet PC Network Boards

Hardware Compatibility:	IBM PC, PC XT, PC AT, PS/2 (non-Micro Channel bus), EISA bus or their compatibles.
Standards:	IEEE 802.3 Ethernet 10 Mbps baseband, CSMA/CD.
Interfaces:	Duplex fiber optic cable with bayonet type ST connectors as standard, optional screw-on SMA connectors available. Standard AT plug-in board.
LAN Software:	Most major environments including IBM OS/2 Lan Manager, Novell NetWare, Microsoft Lan Manager, 3Com 3+ and 3+ Open, Digital DECnet, SCO UNIX, and others.
Technical:	RAM buffer 8 Kbytes, upgradable to 32 KB I/O Base Address 0200 to 03E0 (Hex) in software selectable 32-byte increments Interrupt Channel IRQ 2, 3, 4, 7. RAM Base Address C0000 through E8000 (hex) in 8-Kbyte increments. ROM Size 16, 32 or 64 Kbytes ROM Base Address C0000 through EC000 Transceiver Transmitter: CodeLed high radiance edge-emitting Aluminum Galium Arsenide Light Emitting Diode (LED); Receiver: Silicon PIN-type Photodiode Optical Wavelength 830nm ±20nm, peak Cable 50/125, 62.5/125, 85/125, or 100/140 duplex, multimode, graded index.
Physical:	Board Size Plugs into an expansion slot in any IBM PC, EISA bus, PC XT, PC AT or non-Micro Channel PS/2, or their compatibles. 10.67 cm (4.2 in.) × 33.35 cm (13.40 in.). IBM full size PC card. Operating Environment 0°C to +55°C, 10-90% humidity, non-condensing Power 1.0A @ +5V, 0.3A @ +12V, typical

Optical Transmitter Power:#

	CodeNet-8330	CodeNet-8331	CodeNet-8332
Peak Power Output Into:			
62.5 μm core fiber	−9 dBm	−12 dBm	−3.5 dBm
100 μm core fiber	−8.5 dBm	−11.5 dBm	−3 dBm
50 μm core fiber	−12.5 dBm	−15.5 dBm	−6.5 dBm
Receiver Sensitivity	−37 dBm	−27 dBm	−30 dBm
Optical Flux Budget Into:			
62.5 μm core fiber	28 dB	15 dB	26.5 dB
100 μm core fiber	28.5 dB	15.5 dB	27 dB
50 μm core fiber	24.5 dB	11.5 dB	23 dB
Dynamic Range	18 dB	18 dB	15 dB
Maximum Cable Length:			
Star	800m		800m
Pt. to Pt.		4500m*	

#Typ. power measured at 25°C, 50% duty cycle.
*Limited by IEEE 802.3 roundtrip delay time, flux budget permitting.

Diagnostic Indicators:	Collision—Yellow Transmit—Green Receive—Green Link Fault—Yellow** Receive Overpower—Yellow*	Jabber—Yellow Collision—Yellow

*Not used on 8331
**Low Light Level on 8331

For Information or Assistance

Call the Technical Support Group of Codenoll Technology at (914) 965-6300.

Specifications may change without notice. Codenoll, CodeNet and CodeLed are registered trademarks of Codenoll Technology Corporation. EtherCard PLUS and SuperDisk are trademarks of Western Digital Corporation, used with permission. Other trademarks are the property of various companies.

MAKING LIGHT WORK OF NETWORKING

1086 NORTH BROADWAY, YONKERS, NEW YORK 10701 USA • TELEPHONE: (914) 965-6300 • FAX (919) 965-9811

PN: 05-0058-02-0001

© 1990, Codenoll Technology Corporation
Printed in U.S.A.—August, 1990

Data Sheet

CODENET-8300/8301
Fiber Optic Ethernet/
EISA Network Boards

Features

- **Connects an EISA computer to a Code-Net® fiber optic Ethernet network.**
- **Bus Master with 32 bit wide data path.**
- **Uses the 33 MB EISA bus transfer rate for maximum throughput.**
- **Driver support for all popular network operating systems including Microsoft OS/2 Lan Manager, SCO Unix and NetWare 286 and 386.**
- **Onboard co-processor offloads functions from CPU, including LAN drivers.**
- **Computers can be up to 4500 meters apart in point to point or star configurations.**
- **Transparent to all operating and applications software.**
- **CodeNet-8300 connects to a passive star hub; CodeNet-8301 to an active star hub.**
- **Meets all applicable Ethernet and IEEE 802.3 signaling standards.**
- **Built-in connectors for standard fiber optic cable.**
- **Operates in all computers using the EISA bus.**
- **Ethernet and fiber optic interfaces in one EISA Board.**
- **Simple installation by user in any single card slot.**
- **Optimizes network performance of 386 and 486 based computers.**

Performance Summary

The CodeNet-8300 and 8301 are easily installed by the user in any full length expansion slot of an EISA computer and connected to the network via standard duplex fiber optic cable.

Both boards are based on the field-proven Codenoll high power fiber optic Ethernet transceiver, the most widely used fiber optic Ethernet transceiver in the world.

Codenoll has developed a unique card architecture that allows existing boards to be upgraded as the user's computers are changed to more powerful models. The 8300 and 8301 Network Boards each consist of two parts: (1) A high speed data interface between the EISA computer and the Ethernet controller; and (2) an intelligent interface to the fiber optic network. Either part may be changed in the field, thus eliminating the need and the inherent cost to purchase a completely new adapter card each time computers are upgraded.

Meets Ethernet and IEEE 802.3 Standards

In line with Codenoll's commitment to standards-based network solutions, the CodeNet-8300 and 8301 boards meet all applicable signaling standards of IEEE 802.3 Ethernet for 10 Mbps baseband CSMA/CD local area networks. Long-term compatibility is a guiding principle of Codenoll product design.

Higher Performance, Lower Price

The CodeNet-8300 and 8301 achieve higher performance because of a design architecture that utilizes on-board Bus Master control of the 32-bit-wide EISA data path. This allows burst mode EISA data transfer rates of 33 MB. An onboard 80186 co-processor with packet buffer and driver memories allows true multiprocessing concurrent with the host CPU(s).

Installation and Setup

Each CodeNet-8300 and 8301 is supplied with standard input and output fiber optic connectors for the network interface. Physical connection is direct to industry standard fiber optic cable, so there are no devices such as stand-alone transceivers outside the computer. The installation is clean and uncluttered, reliable and secure.

The Codenoll Fiber Optic Solution for Networks

EISA computers with a CodeNet-8300 or 8301 Network Board can share resources such as file servers, databases, applications software, and peripherals over fiber optic cable. The network board is plugged into any expansion slot of the EISA computer and a standard fiber optic cable connects the card to a CodeStar passive star hub or a MultiStar active star hub. Up to 32 computers can be connected to one passive star. Many CodeStar passive hubs and MultiStar active hubs can be connected together to create LANS supporting thousands of computers in a single network.

The Benefits of Codenoll Fiber Optic Networks

Codenoll networks are completely immune to electromagnetic noise of all types and no form of cable damage can disable more than one node of the network. So, unlike copper wire, Codenoll cable can be run over or around fluorescent lights, in power raceways, near copy machines, X-ray units, large computers, elevator motors, transformers, production machinery, or robots. The user can safely ignore electromagnetic noise and grounding problems. Because of this feature, a large number of network errors and resulting retransmissions are eliminated, thus increasing effective network throughput. Bit error rates of less than 10^{-15} are common in Codenoll networks.

CodeNet fiber optic Ethernet is more robust and reliable than coax or twisted pair copper-based systems. No short circuits, ground faults, reflections, or impedance mismatches. It is impossible for any form of cable damage to affect sensitive computer electronics as can happen with copper-based networks. It is also impossible for a cable to cause a fire or explosion since there is no electrical current present.

It is extremely difficult for an unauthorized agent to tap into the cable and, should that occur, the unauthorized intrusion can be easily detected during normal network maintenance.

CodeNet permits up to 4.5 kilometers between nodes without requiring repeaters in the point-to-point mode and MultiStar active star hub modes. Workstations can be up to 1600 meters apart using a low cost CodeStar passive hub. CodeNet networks are easily extendable within large buildings or between buildings distributed throughout a campus.

Because of the fiber's large bandwidth and resulting data transmission capacity, the network has long life expectancy. The newer and faster network standards utilize the same fiber cable as existing Ethernet networks. The business disruption and cost of installing new cable is greatly reduced or eliminated when upgrading to 100 Mbps FDDI products. These products include the 9540 series FDDI plug-in boards and 9340 series Low Cost FDDI plug-in boards for ISA computers; the 9500 series FDDI plug-in boards and 9300 series LowCost FDDI plug-in boards for EISA computers; and the 100 Mbps FDDI Concentrator for interconnecting minicomputers, mainframes, personal computers and workstations.

Specifications—CodeNet-8300 and 8301 Fiber Optic Ethernet/EISA Network Boards.

Hardware Compatibility:	Any computer equipped with an Extended Industry Standard Architecture (EISA) bus
Standards:	IEEE 802.3 Ethernet 10 Mbps baseband, CSMA/CD
Optical Interfaces:	Duplex fiber optic cable with ST connectors; optional SMA connectors
Electrical Interface:	Standard EISA slot pin connector
LAN Software:	Supports most popular network environments such as: Novell NetWare, LAN Manager, UNIX, and others
Indicators:	Activity, Collision, Fiber Break (8300), Low Light (8301), Jabber, High Light
RAM buffers:	32 KB packet buffer and 16 KB device buffer
I/O Base Address:	0200 to 03E0 (Hex) in software selectable 32-byte increments
Interrupt Channel:	IRQ 2, 3, 5, 7
RAM Base Address:	C0000 through E8000 (hex) in 8-Kbyte increments
ROM Size:	16, 32 or 64 KB
ROM Base Address:	C0000 through EC000
Transmitter:	CodeLed high radiance edge-emitting Aluminum Galium Arsenide Light Emitting Diode (LED)
Receiver:	Silicon PIN-type Photodiode with 50% duty cycle optical transmit test mode.
Jabber Auto Reset:	Switch selectable
Optical Wavelength:	830nm ± 20nm, peak
Optical Connector Types:	ST, SMA-906 optional
Cable:	50/125, 62.5/125 or 100/140 duplex, multimode, graded index

Passive star Transceiver: CodeNet-8300

Fiber Size	100μm	62.5μm	50μm
Peak Optical Output	−8.5dbm	−9dbm	−12.5dbm
Sensitivity	−37dbm	−37dbm	−37dbm
Flux Budget	28.5db	28db	24.5db
Dynamic Range	18db	18db	18db

Active Star (FOIRL) Transceiver: CodeNet-8301

	100μm	62.5μm	50μm
Peak Optical Output (High setting)	−11.5dbm	−12dbm	−15.5dbm
Sensitivity	−30dbm	−30dbm	−30dbm
Flux Budget	18.5db	18db	14.5db
Dynamic Range	21db	21db	21db

Diagnostic Indicators:

Codenet-8300
- Collision (yellow)
- Activity (green)
- Fiber Fault (yellow)
- Jabber (yellow)
- Receive Overpower (yellow)

Codenet-8301
- Collision (yellow)
- Activity (green)
- Low Light (yellow)
- Jabber (yellow)
- Not used (yellow)

Physical Board Size:	Plugs into an expansion slot in any EISA bus computer. 10.67 cm (4.2in.) × 33.35 cm (13.40 in.).
Operating Environment:	0°C to +50°C, 10-90% humidity, non-condensing
Power:	1.0A @ +5V, 0.3A @ +12V, typical

For Information or Assistance

Call the Technical Support Group of Codenoll Technology at (914) 965-6300.

Specifications may change without notice.

Codenoll and CodeNet are registered trademarks of Codenoll Technology Corporation. Other trademarks are the property of various companies.

MAKING LIGHT WORK OF NETWORKING

1086 NORTH BROADWAY, YONKERS, NEW YORK 10701 USA • TELEPHONE: (914) 965-6300 • FAX: (914) 965-9811

© 1989 Codenoll Technology Corporation
Printed in U.S.A. 10/89

Data Sheet

CODENET®-8320/8321/8322
Micro Channel/Ethernet
Network Interface Cards

Features
- **Connects any Micro Channel IBM PS/2 or compatible computer directly to a CodeNet Ethernet fiber optic network.**
- **Dual-ported 16-Kbyte memory shared with processor for high throughput. No Micro Channel bus DMA channels required.**
- **16-bit data bus interface.**
- **10 Mbps transmission, CSMA/CD, IEEE 802.3 Ethernet.**
- **Modular onboard transceiver attaches directly to fiber optic network.**
- **Adaptable to ISA or EISA computers on premises.**
- **Extends distances between devices on networks.**
- **Common architecture with CodeNet-8330 PC network interface card; one driver supports XT/AT and Micro Channel computers.**
- **High performance, low power CMOS LAN controller.**

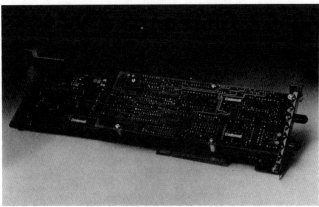

Functional Overview
The CodeNet-8320 Network Interface Card (NIC) combines the performance features of a fiber optic network with a high throughput Micro Channel interface card. Occupying a single expansion slot in any IBM Micro Channel PS/2 or compatible computer, this card provides full Ethernet functionality and the proven performance of Codenoll fiber optic transceivers.

High Performance, Low Price
The high speed buffer memory in the CodeNet-832X NICs means the system uses "move string" instructions to transfer data to and from the buffer in the same way it moves data around its own memory. There is no need to set up a slower, hardware-implemented Direct Memory Access (DMA) channel for data transfer. Thanks to shared memory access and dual porting, a CodeNet-832X card yields higher throughput than most Ethernet adapters, and at an affordable price.

Secure Reliable Performance
The high bandwidth of fiber means the highest performance in your CodeNet network now and, since the fiber is the same as that specified for high performance networks like FDDI, your physical plant investment enjoys a long lifetime. Your CodeNet network is completely immune to electrical interference, even in noisy manufacturing or processing facilities. For secure networks, such as government and research installations, fiber optics are the only choice; unauthorized taps are difficult to make and easily detected. With CodeNet you can increase the distance between workstations or servers and still enjoy increased reliability. Compared with CodeNet optical fiber, all other networking media are severely limited, in terms of future upgrades.

Network Support
CodeNet-832X NICs support most popular network operating systems including: Novell NetWare 286 and 386, 3Com 3+ and 3+Open, LAN Manager, DEC PCSA/DECnet-DOS, Interactive, and SCO UNIX and Banyan Vines. A NetBIOS/OSI Interface Program supports NetBIOS-compatible network applications and operating systems, and an optional TCP/IP protocol package permits communication with many minis and mainframes. Call Codenoll for complete information.

Optimal Network Design
Since Codenet fiber optic Ethernet networks support both active and passive star hubs to meet user needs, CodeNet-832X cards are available for all possible configurations:
- CodeNet-8320 connects to CodeStar\ passive star hub.
- CodeNet-8321 connects to MultiStar\ active star hub.
- CodeNet-8322 connects to CodeStar passive star hubs supporting CodeNet-30XX NICs.

Simple Installation and Setup
Onboard standard fiber optic connectors directly attach to standard duplex fiber optic network cable. The Micro Channel Programmable Option Select program is used to display and select addresses and interrupt channel, eliminating hardware jumper adjustment during installation.

Converts to ISA or EISA Bus
With simple modification, CodeNet-832X cards can be transferred from a Micro Channel computer to either an Industry Standard Architecture (ISA) or Extended ISA computer for a fraction the cost of a new network interface card. Modification is done by the user on site.

Standard IEEE 802.3 Ethernet

Following Codenoll's policy of standards-based network solutions, each CodeNet-832X card meets applicable signaling standards of IEEE 802.3 Ethernet for 10 Mbps baseband CSMA/CD LANs.

Specifications—CodeNet-8320, -8321, -8322 Network Interface Cards

Hardware Compatibility:	IBM Micro Channel bus computers and compatibles
Network Standards:	IEEE 802.3 CSMA/CD Ethernet 10 Mbps baseband
Interfaces:	Duplex fiber optic cable with bayonet type ST connectors standard; screw-on SMA connectors optional
LAN Environments:	DOS-OS/2 Lan Manager (NDIS), Novell NetWare, 3 Plus Open and 3 Plus, DECnet-DOS and PSCA, Unix, Xenix, TCP/IP, NETBIOS/OSI, Interactive and SCO UNIX, Banyan Vines, IBM OS/2 EE. Call regarding others under development.
Electronics:	RAM buffer 16-Kbyte dual port, shared access I/O Base Address 0200 to 03E0 (Hex) in software selectable 32-byte increments Interrupt Channel IRQ 3, 4, 10, 15 RAM Base Address C0000 through DC000 (Hex) in 16-Kbyte increments Boot ROM Size 16, 32 or 64 Kbytes ROM Base Address C0000 through DC000
Optics:	Transmitter CodeLED® high radiance edge-emitting Aluminum Galium Arsenide (AlGaAs) Light Emitting Diode (LED) Receiver Silicon PIN-type photodiode
Physical:	Optical Wavelength 830nm ±20nm, peak Cable Sizes 50/125, 62.5/125, 85/125, or 100/140μm duplex, multimode, graded index Board Size 3.5 in. h × 11.5 in. w (8 cm × 29 cm) Operating Environment 0°C to +55°C, 10 to 90% humidity (non-condensing) Power 1.4A @ +5V; 0.1A @ +12V, typical

Optical Transmitter Power[1]:

	CodeNet-8320	CodeNet-8321	CodeNet-8322
Peak Power Output Into:			
62.5 μm core fiber	−9 dBm	−12 dBm	−3.5 dBm
100 μm core fiber	−8.5 dBm	−11.5 dBm	−3 dBm
50 μm core fiber	−12.5 dBm	−15.5 dBm	−6.5 dBm
Receiver Sensitivity	−37 dBm	−27 dBm	−30 dBm
Optical Flux Budget Into:			
62.5 μm core fiber	28 dB	15 dB	26.5 dB
100 μm core fiber	28.5 dB	15.5 dB	27 dB
50 μm core fiber	24.5 dB	11.5 dB	23 dB
Dynamic Range	18 dB	18 dB	15 dB
Maximum Cable Length:			
Star	800m		800m
Pt. to Pt.		4500m[2]	

[1] Typ. power measured at 25°C, 50% duty cycle.
[2] Limited by IEEE 802.3 roundtrip delay time, flux budget permitting.

Diagnostic Indicators:	Collision—Yellow Transmit—Green Receive—Green Link Fault—Yellow[3] Receive Overpower—Yellow[4]	Jabber—Yellow Collision—Yellow

[3] Low Light Level on 8321
[4] Not used on 8321

For Information or Assistance

Call the Customer Support Group at Codenoll Technology, (914) 965-6300.

Specifications may change without notice. Codenoll, CodeNet CodeStar, MultiStar and CodeLED are registered trademarks of Codenoll Technology Corporation. PS/2, Micro Channel, OS/2 Extended Edition, AT and IBM are trademarks of International Business Machines Corporation. Other trademarks mentioned are the property of other companies.

MAKING LIGHT WORK OF NETWORKING

1086 NORTH BROADWAY, YONKERS, NEW YORK 10701 USA • TELEPHONE: (914) 965-6300 • FAX (919) 965-9811

PN: 05-0058-01-0001

© 1990, Codenoll Technology Corporation
Printed in U.S.A.—August, 1990

VII

Plastic Optical Fiber Ethernet Adapter Cards and Accessories

CodeNet 8681,   p. 428
CodeNet 8601,   p. 430
CodeNet 8621,   p. 432
CodeNet 8631,   p. 434
Plastic Optical Fiber Connector System,   p. 436
Plastic Optical Fiber Tools and Accessories,   p. 438

Data Sheet

CODENET®-8681
Plastic Optical Fiber Ethernet Active Star Transceiver

Features
- **Uses inexpensive plastic optical fiber (POF) to transmit data.**
- **Easier to use and install than twisted pair.**
- **Workstations can be up to 100 meters apart.**
- **Foolproof POF connectors are easier to use and more dependable than any other LAN connector.**
- **Fiber optic network operation completely transparent to user's hardware and software.**
- **Electrical interface accepts Ethernet (Versions 1 & 2) or IEEE 802.3 standard signal formats.**
- **Low light level monitor detects fiber breakage.**
- **Optical idle signal provides fiber integrity indication and allows simple measurement of transmitter power.**
- **Switch selectable SQE "Heartbeat" test, automatic "Jabber" reset.**

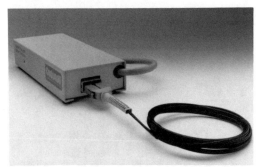

CodeNet-8681: POF Ethernet Tranceiver.

The Codenoll Plastic Optical Fiber Ethernet Solution for Networks

Plastic Optical Fiber (POF) revolutionizes the LAN media choices. It is as inexpensive as twisted pair, and easier to install, to use and to maintain than any other Local Area Network media. No other media has the built-in bandwidth of fiber and corresponding data transmission capacity, resulting in the best return on your network investment. Newer and faster network standards, like FDDI, use exactly the same fiber that is installed today for Codenoll's fiber optic Ethernet networks.

The CodeNet POF Ethernet product line is more robust and reliable than coax or twisted pair solutions. It is completely immune to electromagnetic noise and no form of cable damage, like short circuits, ground faults, reflections, or impedance mismatches, can disable the network. It is impossible for a cable to cause a fire or explosion since there is no electrical current present, and it is virtually impossible for an unauthorized agent to tap into the cable without easy detection. Workstations can be up to 100 meters apart using a MultiStar™ active hub. Any CodeNet 86XX POF network segment can be combined with other CodeNet fiber optic components. CodeNet glass fiber optical networks are easily extendable within large buildings or between buildings throughout a campus. The CodeNet fiber optic products offer maximum flexibility and cost-effectiveness when designing today's Local Area Network. The low cost of plastic fiber, POF connectors, and the unparalleled ease of connectorization make the CodeNet POF products ideal for network workgroups.

Product Description

The CodeNet-8681 is the first media access unit for an active star Ethernet network which uses plastic optical fiber (POF) as its transmission medium. The CodeNet-8681 consists of two boards within the transceiver case: 1) A mother board which contains the interface between the peripheral device and fiber optic transceiver; and 2) A daughter card which contains the fiber optic transceiver. The 8681 has the same mother board as the field proven CodeNet-8381 Fiber Optic Inter-Repeater Link (FOIRL) Transceiver, but has a new daughter card with a new optical transceiver designed for POF.

The CodeNet-8681 is used in Ethernet active star configurations and uses the same signalling conventions as the IEEE FOIRL (Fiber Optic Inter-Repeater Link) Ethernet standard. A POF Ethernet segment can also be connected with any other type of Ethernet media segment... glass optical fiber, 10-Base-T, and thick and thin coax.

On board switches allow the user to select an automatic "Jabber" reset mode and defeat the SQE "Heartbeat" test. Several indicator LEDs on the front of the unit facilitate network troubleshooting.

The 8681 is equipped with an integral cable which terminates in a male 15 pin AUI connector. Since the unit is mechanically and electrically identical to coaxial cable transceivers at the AUI connector, CodeNet fiber optic Ethernet networks can be added to, and combined with existing coaxial cable based networks, allowing Ethernet networks to be expanded and converted to fiber optics at any time.

Optimal Network Design

Since the CodeNet MultiStar active Ethernet hub supports both glass and plastic optical fiber, as well as twisted pair and coax, any Ethernet network may be expanded using fiber optics. POF products are an economical solution for workgroup areas, and glass optical fiber could be used to create links up to 4.5 km long.

Simple Installation and Setup

Codenoll's connector system for POF allows quick and simple attachment to duplex 1000 μm plastic optical fiber. Both the transmit and receive fibers are enclosed in a single, polarized connector. A partial shroud around the connector protects the end faces of the fiber while maintaining easy access for termination.

Codenoll
TECHNOLOGY CORPORATION

Specifications—CodeNet-8681 POF Transceiver

Standards:	Uses IEEE FOIRL (Fiber Optic Inter-Repeater Link) Ethernet signalling conventions
Electrical Interface:	1 meter integrated AUI cable with 15-pin male subminiature D-type Ethernet AUI connection
Optical Interface:	Duplex 1000 μm plastic optical fiber cable with Codenoll POF Connectors
Internal Option Switches: (factory settings in bold)	Jabber Reset (**Auto**/Manual) SQE-Heartbeat (On/**Off**)
Optics:	Transmitter Aluminum Galium Arsenide (AlGaAs) Light Emitting Diode (LED)
	Optical Wavelength 660nm, peak
	Receiver Silicon PIN-type photodiode
	Cable Sizes 1000 μm duplex, high quality PMMA core, step index, Plastic Optical Fiber
Physical:	Transceiver Size (not including interface connections) 3.3 in. W × 8.2 in. L × 1.5 in. H (8.4 cm × 20.8 cm × 3.8 cm)
	Operating Environment 0°C to +55°C, 10 to 90% humidity (non-condensing)
	Power 500 mA @ +12V, maximum
	Weight 21 oz. (585 grams)

Optical Specifications[1]:

Peak Output Power:	−9 dBm
Receiver Sensitivity:	−29 dBm
Optical Flux Budget:	20 dB
Maximum Cable Length:	50 meters

[1] Typ. peak power into 1000 μm PMMA (polymethyl methacrylate) core fiber measured at 25°C, 50% duty cycle.

Diagnostic Indicators:	Power—Green Activity—Green Collision—Yellow Link Fault—Yellow Jabber—Yellow

For Information or Assistance

Call the Customer Support Group at Codenoll Technology, (914) 965-6300.

Specifications may change without notice. Codenoll, CodeNet CodeStar, MultiStar and CodeLED are registered trademarks of Codenoll Technology Corporation. Other trademarks mentioned are the property of other companies.

MAKING LIGHT WORK OF NETWORKING

1086 NORTH BROADWAY, YONKERS, NEW YORK 10701 USA • TELEPHONE: (914) 965-6300 • FAX (914) 965-9811

PN: 05-0058-26-0001

© 1990, Codenoll Technology Corporation
Printed in U.S.A. — November, 1990

Data Sheet

CODENET®-8601
EISA/Plastic Optical Fiber Ethernet Network Interface Card

Features
- Uses inexpensive Plastic Optical Fiber (POF) to transmit data.
- Easier to use and install than twisted pair.
- Workstations can be up to 100 meters apart.
- Foolproof POF connectors are easier to use and more dependable than any other LAN connector.
- Used in Ethernet active star configurations.
- Connects an EISA computer to a CodeNet® fiber optic Ethernet network.
- Bus Master with 32 bit wide data path.
- Uses the 33 MB EISA bus transfer rate for maximum throughput.
- Fileserver support for all popular network operating systems including Microsoft OS/2 Lan Manager and NetWare.
- Onboard co-processor offloads functions from CPU, including LAN drivers.
- Transparent to all operating and applications software.
- Operates in all computers using the EISA bus.
- Ethernet and fiber optic interfaces in one EISA Board.
- Simple installation by user in any single card slot.
- Optimizes network performance of 386 and 486 based computers.
- Low light diagnostic LED detects fiber break.
- Optical idle signal provides fiber integrity indication and allows simple measurement of transmitter power.
- Switch selectable auto-Jabber reset and SQE Heartbeat options.

CodeNet-8601 EISA/POF Ethernet NIC showing modular architecture.

The Codenoll Plastic Optical Fiber Ethernet Solution for Networks

Plastic Optical Fiber (POF) revolutionizes the LAN media choices. It is as inexpensive as twisted pair, and easier to install, to use and to maintain than any other Local Area Network media. No other media has the built-in bandwidth of fiber and corresponding data transmission capacity, resulting in the best return on your network investment. Newer and faster network standards, like FDDI, use exactly the same fiber that is installed today for Codenoll's fiber optic Ethernet networks.

The CodeNet POF Ethernet product line is more robust and reliable than coax or twisted pair solutions. It is completely immune to electromagnetic noise and no form of cable damage, like short circuits, ground faults, reflections, or impedance mismatches, can disable the network. It is impossible for a cable to cause a fire or explosion since there is no electrical current present, and it is virtually impossible for an unauthorized agent to tap into the cable without easy detection. Workstations can be up to 100 meters apart using a MultiStar™ active hub. Any CodeNet 86XX POF network segment can be combined with other CodeNet fiber optic components. CodeNet glass fiber optical networks are easily extendable within large buildings or between buildings throughout a campus. The CodeNet fiber optic products offer maximum flexibility and cost-effectiveness when designing today's Local Area Network. The low cost of plastic fiber, POF connectors, and the unparalleled ease of connectorization make the CodeNet POF products ideal for network workgroups.

Product Description

The CodeNet-8601 is a POF Ethernet Network Interface Card for EISA bus computers. The CodeNet-8601 consists of two boards while still occupying only a single expansion slot; 1) A mother board containing the high speed data interface between the computer and Ethernet controller; and 2) A daughter board which contains the fiber optic transceiver. The daughter board contains a new optical transceiver designed for POF. The modular design of the card enables either board to be changed in the field by the user, allowing the board to be converted to glass fiber or to a different computer bus for a fraction of the cost of a new adapter.

The CodeNet-8601 is used in Ethernet active star configurations and uses the same signalling conventions as the IEEE FOIRL (Fiber Optic Inter-Repeater Link) Ethernet standard. A POF Ethernet segment can be connected with any other type of Ethernet media segment—glass optical fiber, 10-Base-T and thick and thin coax.

The CodeNet-8601 also has front panel status indicator LEDs for quick diagnosis and resolution of network problems.

Higher Performance, Lower Price

The CodeNet-8601 acheives higher performance because of a design architecture that utilizes on-board Bus Master control of the 32-bit-wide EISA data path. This allows burst mode EISA data transfer rates of 33 MB. An onboard 80186 co-processor with packet buffer and driver memories allows true multiprocessing concurrent with the host CPU(s).

Optimal Network Design

Since the CodeNet MultiStar active Ethernet hub supports both glass and plastic fiber optic connections, as well as twisted pair and coax, any existing Ethernet network may be expanded using fiber optics. POF networks are the new standard for workgroup networks, with glass optical fiber being used to create links up to 4.5 km long.

Installation and Setup

Codenoll's connector system for POF allows quick and simple attachment to duplex 1000 μm plastic optical fiber. Both the transmit and receive fibers are enclosed in a single, polarized connector. A partial shroud around the connector protects the end faces of the fiber while maintaining easy access for termination.

Specifications—CodeNet-8601 EISA/POF Ethernet Network Interface Card

Hardware Compatibility:	Any computer equipped with an Extended Industry Standard Architecture (EISA) bus
Standards:	Uses IEEE FOIRL (Fiber Optic Inter-Repeater Link) Ethernet signalling conventions
Optical Interfaces:	Duplex 1000 μm plastic optical fiber cable with Codenoll Plastic Optical Fiber Connectors
Electrical Interface:	Standard EISA slot pin connector
Option Switches: (factory settings in bold)	Jabber Auto Reset (**Auto**/Manual) SQE-Heartbeat (On/**Off**)
LAN Software:	File Server support for: Novell NetWare, LAN Manager, and others
RAM buffers:	32 KB packet buffer and 16 KB device buffer
I/O Base Address:	0200 to 03E0 (Hex) in software selectable 32-byte increments
Interrupt Channel:	IRQ 2, 3, 5, 7
RAM Base Address:	C0000 through E8000 (hex) in 8-Kbyte increments
ROM Size:	16, 32 or 64 KB
ROM Base Address:	C0000 through EC000
Transmitter:	Aluminum Galium Arsenide Light Emitting Diode (LED)
Optical Wavelength:	660 nm, peak
Receiver:	Silicon PIN-type Photodiode
Cable:	1000 μm duplex, high quality PMMA core, step index, Plastic Optical Fiber
Diagnostic Indicators:	Transmit—Green Low Light—Yellow Receive—Green Jabber—Yellow Collision—Yellow
Physical Board Size:	Plugs into an expansion slot in any EISA bus computer. 10.67 cm (4.2in.) × 33.35 cm (13.40 in.).
Operating Environment:	0°C to +50°C, 10-90% humidity, non-condensing
Power:	1.0A @ +5V, 0.3A @ +12V, typical
Weight:	14 oz. (398 grams)

Optical Specifications[1]:

Peak Output Power:	−9 dBm
Receiver Sensitivity	−29 dBm
Optical Flux Budget:	20 dB
Maximum Cable Length:	50 meters

[1] Typ. peak power into 1000 μm PMMA (polymethyl methacrylate) core fiber measured at 25°C, 50% duty cycle.

For Information or Assistance

Call the Technical Support Group of Codenoll Technology at (914) 965-6300.

Specifications may change without notice.

Codenoll, MultiStar and CodeNet are registered trademarks of Codenoll Technology Corporation. Other trademarks are the property of various companies.

MAKING LIGHT WORK OF NETWORKING

1086 NORTH BROADWAY, YONKERS, NEW YORK 10701 USA • TELEPHONE: (914) 965-6300 • FAX (914) 965-9811

PN: 05-0058-25-0001

© 1990, Codenoll Technology Corporation
Printed in U.S.A.   November, 1990

Data Sheet

CODENET®-8621
Micro Channel/Plastic Optical Fiber Ethernet Network Interface Card

Features
- Uses inexpensive plastic optical fiber (POF) to transmit data.
- Easier to use and install than twisted pair.
- Workstations can be up to 100 meters apart.
- Foolproof POF connectors are easier to use and more dependable than any other LAN connector.
- Used in Ethernet active star configurations.
- Connects any Micro Channel IBM PS/2 or compatible computer directly to a CodeNet Ethernet fiber optic network.
- Dual-ported 16-Kbyte memory shared with processor for high throughput. No Micro Channel bus DMA channels required.
- 16-bit data bus interface.
- Modular onboard transceiver attaches directly to fiber optic network.
- Adaptable to ISA or EISA computers on premises.
- Common architecture with CodeNet-8631 PC network interface card; one driver supports XT/AT and Micro Channel computers.
- High performance, low power CMOS LAN controller.
- Low light diagnostic LED detects fiber break.
- Optical idle signal provides fiber integrity indication and allows simple measurement of transmitter power.
- Switch selectable auto-Jabber reset and SQE "Heartbeat" options.

The Codenoll Plastic Optical Fiber Ethernet Solution for Networks

Plastic Optical Fiber (POF) revolutionizes the LAN media choices. It is as inexpensive as twisted pair, and easier to install, to use and to maintain than any other Local Area Network media. No other media has the built-in bandwidth of fiber and corresponding data transmission capacity, resulting in the best return on your network investment. Newer and faster network standards, like FDDI, use exactly the same fiber that is installed today for Codenoll's fiber optic Ethernet networks.

The CodeNet POF Ethernet product line is more robust and reliable than coax or twisted pair solutions. It is completely immune to electromagnetic noise and no form of cable damage, like short circuits, ground faults, reflections, or impedance mismatches, can disable the network. It is impossible for a cable to cause a fire or explosion since there is no electrical current present, and it is virtually impossible for an unauthorized agent to tap into the cable without easy detection. Workstations can be up to 100 meters apart using a MultiStar™ active hub. Any CodeNet 86XX POF network segment can be combined with other CodeNet fiber optic components. CodeNet glass fiber optical networks are easily extendable within large buildings or between buildings throughout a campus. The

CodeNet-8621 Micro Channel/POF Ethernet NIC showing modular architecture.

CodeNet fiber optic products offer maximum flexibility and cost-effectiveness when designing today's Local Area Network. The low cost of plastic fiber, POF connectors, and the unparalleled ease of connectorization make the CodeNet POF products ideal for network workgroups.

Product Description

The CodeNet-8621 is a POF Ethernet Network Interface Card for Micro Channel bus PCs. The CodeNet-8621 consists of two boards while still occupying only a single expansion slot; 1) A mother board containing the high speed interface between the computer and Ethernet controller; and 2) A daughter board which contains the fiber optic transceiver. The daughter board contains a new optical transceiver designed for POF. The modular design of the card enables either board to be changed in the field by the user, allowing the board to be converted to glass fiber or to a different computer bus for the fraction of the cost of a new adapter.

The CodeNet-8621 is used in Ethernet active star configurations and uses the same signalling conventions as the IEEE FOIRL (Fiber Optic Inter-Repeater Link) Ethernet standard. A POF Ethernet segment can also be connected with any other type of Ethernet media segment...glass optical fiber, 10 Base-T, and thick and thin coax.

The CodeNet-8621 also has front panel status indicator LEDs for quick diagnosis and resolution of network problems.

Optimal Network Design

Since the CodeNet MultiStar active Ethernet hub supports both glass and plastic fiber optic connections, as well as twisted pair and coax, any Ethernet network may be expanded using fiber optics. POF networks are the new standard for workgroup networks, with glass optical fiber being used to create links up to 4.5 km long.

Installation and Setup

Codenoll's connector system for POF allows quick and simple attachment to duplex 1000 μm plastic optical fiber. Both the transmit and receive fibers are enclosed in a single, polarized connector. A partial shroud around the connector protects the end faces of the fiber while maintaining easy access for termination.

The Micro Channel Programmable option Select program is used to display and select addresses and interrupt channel, eliminating hardware jumper adjustment during installation.

Specifications—CodeNet-8621 Micro Channel/POF Ethernet Interface Card

Hardware Compatibility:	IBM Micro Channel bus computers and compatibles
Standards:	Uses IEEE FOIRL (Fiber Optic Inter-Repeater Link) Ethernet signalling conventions
Optical Interface:	Duplex 1000 μm plastic fiber optic cable with Codenoll Plastic Optical Fiber connectors
Option Switches: (factory settings in bold)	Jabber Reset (**Auto**/Manual) SQE Heartbeat (On/**Off**)
LAN Environments:	DOS-OS/2 Lan Manager (NDIS), Novell NetWare, 3 Plus Open and 3 Plus, DECnet-DOS and PSCA, Unix, Xenix, TCP/IP, NET-BIOS/OSI, Interactive and SCO UNIX, Banyan Vines, IBM OS/2 EE. Call regarding others under development.
Electronics:	RAM buffer 16-Kbyte dual port, shared access
	I/O Base Address 0200 to 03E0 (Hex) in software selectable 32-byte increments
	Interrupt Channel IRQ 3, 4, 10, 15
	RAM Base Address C0000 through DC000 (Hex) in 16-Kbyte increments
	Boot ROM Size 16, 32, or 64 Kbytes
	ROM Base Address C000 through DC000
Optics:	Transmitter Aluminum Galium Arsenide (AlGaAs) Light Emitting Diode (LED)
	Optical Wavelength 660nm, peak
	Receiver Silicon PIN-type photodiode
	Cable Sizes 1000 μm duplex, high quality PMMA core, step index, Plastic Optical Fiber
Physical:	Board Size 3.5 in. h × 11.5 in. w (8 cm × 29 cm)
	Operating Environment 0°C to +55°C, 10 to 90% humidity (non-condensing)
	Power 1.4A @ +5V; 0.1A @ +12V, typical
Weight:	9 oz. (255 grams)

Optical Specifications[1]:

Peak Output Power:	−9 dBm
Receiver Sensitivity:	−29 dBm
Optical Flux Budget:	20 dB
Maximum Cable Length:	50 meters

[1] Typ. peak power into 1000 μm PMMA (polymethyl/methacrylate) core fiber measured at 25°C, 50% duty cycle.

Diagnostic Indicators:	Collision—Yellow Transmit—Green Receive—Green	Low Light—Yellow Jabber—Yellow

For Information or Assistance

Call the Customer Support Group at Codenoll Technology, (914) 965-6300.

Specifications may change without notice. Codenoll, CodeNet CodeStar, MultiStar and CodeLED are registered trademarks of Codenoll Technology Corporation. Other trademarks mentioned are the property of other companies.

MAKING LIGHT WORK OF NETWORKING

1086 NORTH BROADWAY, YONKERS, NEW YORK 10701 USA • TELEPHONE: (914) 965-6300 • FAX (914) 965-9811

© 1990, Codenoll Technology Corporation

Data Sheet

CODENET®-8631
High Performance Plastic Optical Fiber Ethernet/PC Adapter Board

Features

- Uses inexpensive Plastic Optical Fiber (POF) to transmit data.
- Easier to use and install than twisted pair.
- Workstations can be up to 100 meters apart.
- Foolproof POF connectors are easier to use and more dependable than any other LAN connector.
- Used in Ethernet active star configurations.
- Connects an IBM PC XT, AT, PS/2-25, 30 and compatibles to a CodeNet fiber optic Ethernet LAN.
- Driver support for all popular network operating systems, including IBM OS/2 Lan Manager, Novell NetWare, Microsoft Lan Manager, 3Com 3+ and 3+'Open, Digital DECnet, SCO UNIX, Western Digital ViaNet and others.
- Unique PC shared memory interface: dual-ported 8K memory shared with PC for high throughput, fast response. No DMA channels required.
- Wait states are software selectable to optimize network performance.
- Uses just one PC expansion slot.
- Easily field upgradable to 16 bit and Micro Channel versions at low cost.
- Low light diagnostic LED detects fiber break.
- Optical idle signal provides fiber integrity indication and allows simple measurement of transmitter power.
- Switch selectable auto-Jabber reset and SQE Heartbeat options.

CodeNet-8631 POF Ethernet adapter board showing modular architecture.

The Codenoll Plastic Optical Fiber Ethernet Solution for Networks

Plastic Optical Fiber (POF) revolutionizes the LAN media choices. It is as inexpensive as twisted pair, and easier to install, to use and to maintain than any other Local Area Network media. No other media has the built-in bandwidth of fiber and corresponding data transmission capacity, resulting in the best return on your network investment. Newer and faster network standards, like FDDI, use exactly the same fiber that is installed today for Codenoll's fiber optic Ethernet networks.

The CodeNet POF Ethernet product line is more robust and reliable than coax or twisted pair solutions. It is completely immune to electromagnetic noise and no form of cable damage, like short circuits, ground faults, reflections, or impedance mismatches, can disable the network. It is impossible for a cable to cause a fire or explosion since there is no electrical current present, and it is virtually impossible for an unauthorized agent to tap into the cable without easy detection. Workstations can be up to 100 meters apart using a MultiStar™ active hub. Any CodeNet 86XX POF network segment can be combined with other CodeNet fiber optic components. CodeNet glass fiber optical networks are easily extendable within large buildings or between buildings throughout a campus. The CodeNet fiber optic products offer maximum flexibility and cost-effectiveness when designing today's Local Area Network. The low cost of plastic fiber, POF connectors, and the unparalleled ease of connectorization make the CodeNet POF products ideal for network workgroups.

Product Description

The CodeNet-8631 is a POF Ethernet Network Interface Card for ISA bus PCs. The CodeNet-8631 consists of two boards while still occupying only a single expansion slot: 1) A mother board containing the high speed data interface between the computer and Ethernet controller; and 2) A daughter board which contains the fiber optic transceiver. The 8631 mother board is based upon the proven speed, performance and throughput of the EtherCard PLUS (WD8003EB) from Western Digital. The daughter board contains a new optical transceiver designed for POF.

The modular design of the card enables either board to be changed in the field by the user, eliminating the need and the money required to purchase a completely new adapter when computers are upgraded. For example, by simply changing the mother board, the user can create a CodeNet-8621, a CodeNet Micro Channel/POF Ethernet network board.

The CodeNet-8631 is used in Ethernet active star configurations and uses the same signalling conventions as the IEEE FOIRL (Fiber Optic Inter-Repeater Link) Ethernet standard. A POF Ethernet segment can be connected with any other type of Ethernet media segment…glass optical fiber, 10-Base-T, and thick and thin coax.

The CodeNet-8631 also has front panel status indicator LEDs for quick diagnosis and resolution of network problems.

Optimal Network Design

Since the CodeNet MultiStar active Ethernet hub supports both glass and plastic fiber optic connections, as well as twisted pair and coax, any Ethernet network may be expanded using fiber optics. POF networks are the new standard for workgroup networks, with glass optical fiber being used to create links up to 4.5 km long.

High Performance, Lower Price

The CodeNet-8631 achieves higher performance because of its buffer memory architecture. Since this memory buffer is dual-ported and directly accessed by the PC, the PC simply uses "move string" instructions to move data to and from the buffer in the same way it moves data around its own memory.

This eliminates the need to set up a hardware-implemented PC Direct Memory Access (DMA) channel data transfer. The result is that the CodeNet cards offer much higher throughput. The CodeNet-8631 is especially suited to servers, gateways, backbone connections and high performance workstations. A software-selectable option allows the user to modify usage wait states to optimize performance.

Application Flexibility

Codenoll provides drivers for IBM OS/2 Lan Manager, Microsoft Lan Manager, Novell NetWare, SCO Unix, Western Digital's ViaNet, Sun's PC NFS, Digital's DECnet, and 3Com's 3+ system. In addition, most of the major suppliers of network operating software support these network adapter cards.

Installation and Setup

Codenoll's connector system for POF allows quick and simple attachment to duplex 1000 μm plastic optical fiber. Both the transmit and receive fibers are enclosed in a single, polarized connector. A partial shroud around the connector protects the end faces of the fiber while maintaining easy access for termination.

The 8631's I/O memory address and interrupt level selection are configured at the factory but can be easily modified during the installation process.

Specifications—CodeNet-8631 POF Ethernet PC Adapter Board

Hardware Compatibility:	IBM PC, PC XT, PC AT, PC/2 (non-Micro Channel bus), or their compatibles
	EISA bus or their compatibles
Standards:	Uses IEEE FOIRL (Fiber Optic Inter-Repeater Link) Ethernet signalling conventions.
Interfaces:	Duplex 1000 μm plastic fiber optic cable with Codenoll Plastic Optical Fiber Connectors, Plug-in board, standard AT slot pin connector.
Option Switches: (factory settings in bold)	Jabber Reset (**Auto**/Manual) SQE-Heartbeat (On/**Off**)
LAN Software:	Most major environments including IBM OS/2 LAN Manager, Novell NetWare, Microsoft LAN Manager, 3Com 3+ and 3+ Open, Digital DECnet, SCO UNIX, and others
Technical:	RAM buffer 8Kbytes, upgradable to 32 KB
	I/O Base Address 0200 to 03E0 (Hex) in software selectable 32-byte increments
	Interrupt Channel IRQ 2, 3, 4, 7
	RAM Base Address C0000 through E8000 (hex) in 8-Kbyte increments
	ROM Size 16, 32 or 64 Kbytes
	ROM Base Address C0000 through EC000
	Transmitter: Aluminum Galium Arsenide Light Emitting Diode (LED)
	Optical Wavelength 660nm, peak
	Receiver: PIN-type photodiode
	Cable: High Quality PMMA core, step index, duplex 1000 μm Plastic Optical Fiber
Diagnostic Indicators:	Transmit—Green Receive—Green Low Light—Yellow Jabber—Yellow Collision—Yellow
Physical:	Board Size Plugs into an expansion slot in any IBM PC, PC XT, PC AT or non-Micro Channel PS/2, or their compatibles. 10.67 cm (4.2 in.) × 33.35 cm (13.40in.), IBM full size PC card
	Operating Environment 0°C to +50°C, 10-90% humidity, non-condensing
	Power 1.0A @ +5V, 0.3A @ +12V, typical
Weight:	12 oz. (341 grams)

Optical Specifications[1]:

Peak Output Power:	−9 dBm
Receiver Sensitivity:	−29 dBm
Optical Flux Budget:	20 dB
Maximum Cable Length:	50 meters

[1] Typ. peak power into 1000 μm PMMA (poly methyl methacrylate) core fiber measured at 25°C, 50% duty cycle.

For Information or Assistance

Call the Technical Support Group of Codenoll Technology at (914) 965-6300.

Specifications may change without notice. Codenoll, CodeNet and CodeLed are registered trademarks of Codenoll Technology Corporation. EtherCard PLUS and SuperDisk are trademarks of Western Digital Corporation, used with permission. Other trademarks are the property of various companies.

MAKING LIGHT WORK OF NETWORKING

1086 NORTH BROADWAY, YONKERS, NEW YORK 10701 USA • TELEPHONE: (914) 965-6300 • FAX (914) 965-9811

PN: 05-0058-27-0002

© 1990, Codenoll Technology Corporation
Printed in U.S.A.—December, 1990

Data Sheet

CODENOLL Plastic Optical Fiber (POF) Connector System

Features
- Duplex Fiber Connector
- Low cost
- Less than 1 minute to terminate
- No polishing required—hot plate finish is used
- Durable polymer construction
- Polarized connector prevents improper connection
- Partial shroud protects fiber endfaces
- Specifically designed for 1000 μm POF
- Snap lock prevents accidental uncoupling

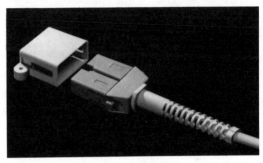

Plastic Optical Fiber (POF) Connector System

Product Description
Codenoll's line of connectors for plastic optical fiber (POF) take advantage of POF's unique properties to create a durable, high performance connection which is low cost and extremely easy to terminate. The Codenoll connector system also includes accessories needed for in building wiring, such as patch panels, in-line connectors, and wall plates.

Functional Description
The connector consists of two parts before finishing: a tail piece containing cable strain relief, and the connector which contains ferrules for the fiber. To terminate a POF cable, its jacket is first stripped back to expose the fiber. The cable is then slipped through the strain relief, and into the ferrules in the connector. The two pieces are then snapped together, locking the fiber in place and held by permanent locking tabs. Excess fiber is trimmed from the front of the connector and the fiber is "polished" by placing the connector ferrules onto a POF Termination Hot Plate for 5–10 seconds. The total time to install a POF connector is typically less than one minute.

Side mounted locking tabs provide a secure connection between the connector and any electro-optic device header, in-line connector, wall plate, or patch panel. The connector is locked in place and cannot be accidentally dislodged.

Codenoll also supplies all tools needed (cable strippers, termination hot plate, trimming tool) for terminating POF cable.

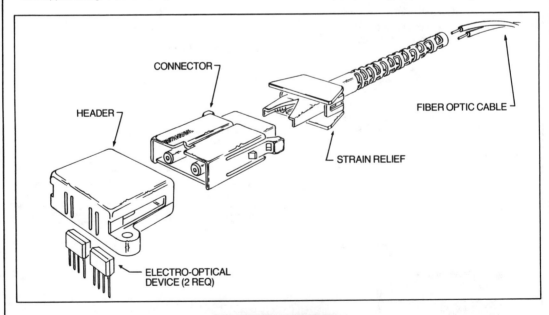

Specifications

Insertion Loss:	Less than 2dBm
Operating Temperature:	−40 to +85°C
Cable:	1000 μm plastic optical fiber
Material:	Thermoplastic
Dimensions:	See drawings

For Information or Assistance

Call the Technical Support Group of Codenoll Technology at (914) 965-6300.

Specifications may change without notice. Codenoll, CodeNet and CodeLed are registered trademarks of Codenoll Technology Corporation. Other trademarks are the property of various companies.

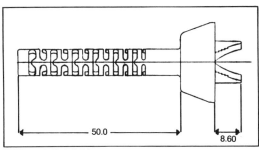

POF Connector Strain Relief

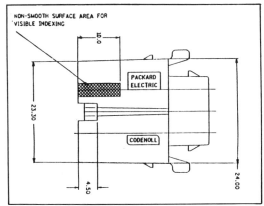

POF Connector

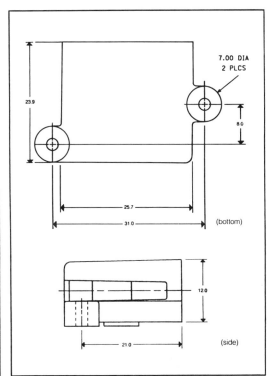

POF Device Header

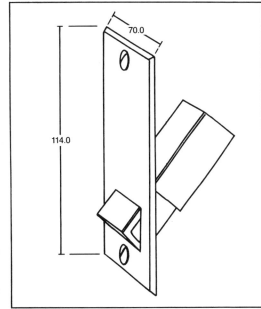

POF Wall Plate

MAKING LIGHT WORK OF NETWORKING

1086 NORTH BROADWAY, YONKERS, NEW YORK 10701 USA • TELEPHONE: (914) 965-6300 • FAX (914) 965-9811

PN: 05-0058-31-0002

© 1990, Codenoll Technology Corporation
Printed in U.S.A.—December, 1990

VIII

Data Sheet

Plastic Optical Fiber (POF) Tools and Accessories

Codenoll Technology provides several products to assist in the testing of plastic optical fiber cable assemblies and the termination of Codenoll's POF connector. These include:

- **POF Cable Test Set**
- **POF Cable Strippers**
- **POF Fiber Trimming Tool**
- **POF Termination Hot Plate**

POF Cable Test Set

Codenoll's POF cable test set allows testing of duplex plastic optical fiber cable assemblies which are terminated with Codenoll's POF connector. The set consists of two portable boxes. One box contains two LED light sources which are drvien at two different frequencies, and the other box contains two PIN diode photodetectors. To test a cable, one end of a terminated cable is plugged into the LED source box and the other end of the cable is connected to the photodetector box. After turning on the source box, you can tell if a cable is wired properly, if it is cross-wired, if it is broken, or if it has high attenuation, simply by pressing a button on the photodetector box.

The units are powered by two 9V batteries and have low battery indicators. The batteries on the LED source box can be recharged with an optional AC adapter.

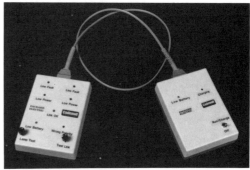

POF Cable Test Set

POF Cable Strippers

Codenoll's cable strippers are used to easily strip the jacket from 1000 μm POF cable without harming the fiber's important cladding.

POF Trimming Tool

Codenoll's fiber trimming tool removes excess POF from the ferrules of a Codenoll POF connector down to the optimum length needed to provide the highest quality end finish using a POF Termination Hot Plate.

POF Termination Hot Plate

After a Codenoll POF connector has been snapped onto a duplex plastic optical fiber cable, a high quality finish is applied to the fiber ends by placing the connector ferrules onto the Termination Hot Plate for 5-10 seconds. This Hot Plate is specially designed to provide a low-loss end finish to POF cable terminated with Codenoll's POF connector.

For high volume production applications, an automated finishing machine is also available. After the excess fiber has been trimmed from the connector, the connector is inserted into a slot in the machine and, by pressing a button, the connector is automatically drawn into a heating element. After a specific amount of time, the connector is automatically withdrawn from the heating element and cooled, reducing cycle time.

For Additional Information
Call the Technical Support Group of Codenoll Technology at (914) 965-6300.

Specifications may change without notice. Codenoll, CodeNet and CodeLED are registered trademarks of Codenoll Technology Corporation.

MAKING LIGHT WORK OF NETWORKING

1086 NORTH BROADWAY, YONKERS, NEW YORK 10701 USA • TELEPHONE: (914) 965-6300 • FAX (914) 965-9811

PN: 05-0058-32-0002

© 1990, Codenoll Technology Corporation
Printed in U.S.A.—December, 1990

Fiber Optic Ethernet Hubs

CodeStar Optical Star Coupler, p. 442
CodeNet 4300 MultiStar Multiport Repeater, p. 444
CodeNet 3311 FOIRL Module, p. 446
CodeNet 8310/8312 Passive Module, p. 448
CodeNet 8611 POF Active Star Module, p. 450
CodeNet 4350 Thin Coax Module, p. 452
CodeNet 4351 Thick Coax Module, p. 454

Data Sheet

CODESTAR™ Optical Star Couplers

Features

- Compatible with all broadcast networks including Codenet™/Ethernet and Serial Bus Coaxial systems
- Completely passive, transmissive design
- High uniformity and reliability
- Bi-directional
- Fully connectorized
- Low coupling (insertion) loss
- Six configurations from 4 ports to 32
- 50, 62.5, 85 or 100 micron graded index fiber
- Protective metal enclosure
- Suitable for rack-mounting or as standalone modules

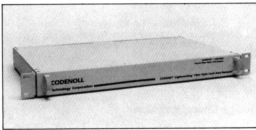
Front View

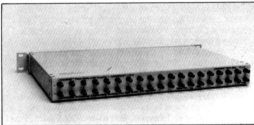
Rear View

Introduction

A Codestar-implemented fiber optic network is a broadcast network, functionally and logically identical to any other broadcast medium, including coaxial cable. Consequently, any broadcast network may be implemented with Codestar fiber optic couplers.

Codestar™ Optical Star Couplers provide a simple means of implementing a fiber optic network for data communication through the interconnection of multiple nodes without the use of repeaters. Codestar couplers are entirely passive, transmissive devices, in which fibers are fused together to form optical mixers through the fused biconical taper technique.

Any light launched into one of the fibers on one side of the optical mixer will be equally divided among, and output through, all of the fibers on the opposite side of the mixer. This characteristic ideally suits Codestar couplers for use in a fiber optic implementation of any broadcast system (such as Ethernet), from the so-called serial bus coaxial cable systems to fiber optic Ethernet systems.

Networks larger than 32 nodes, to several hundred nodes, may be constructed with multiple cluster Codestar couplers interconnected to other stars in up to 3 layers. When going from one star to another, it is necessary to use a repeater, such as the Codenet 3037A.

General Description

Codestar fiber optic couplers are housed in rugged metal enclosures suitable for either standalone or rack-mounting applications (see photo above). Units measure 19″ × 10¾″ × 1⅞″ or 19″ × 10¾″ × 3¾″. These couplers can be supplied with a choice of 50, 62.5, 85 or 100 micron core graded index fiber. All Codestars are sold fully connectorized with SMA-type connectors as standard.

One side of the Codestar is called the "input only" side, and the other side of the Codestar the "output only" side. The network is interconnected by extending one of the input side fibers and one of the output side fibers to each node. Each node then has the capability of transmitting into the Codestar and of having its transmitted data received by all other nodes in the system. Similarly, each node has the ability to receive all the data that is transmitted by all other nodes on the network.

(continued on overleaf)

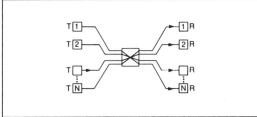

Figure 1. Codestar Transmissive Star Topology

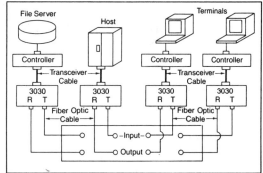

Figure 2. Codestar Star Coupler
(Showing Fiber Optic Cable Connections to Codenet-3030)

Any light coupled into any of the N input ports designated T (see Figure 1 above) will be approximately equally divided among, and exit through each of the N output ports designated R. The relationship between the input and output powers through the star coupler is expressed in terms of an effective coupler "loss" given by:

$$\text{"loss"} = 2C + E + 10 \log N$$

where C is the connector loss (typically 1.5 dB)
E is the star coupler "excess loss"
(typically 1 to 4 dB)
N is the number of optical ports

Cross coupling between input (or output) ports is typically less than 30dB.

Design Characteristics

Although a Codestar looks schematically like a conventional electronic ("active") star, it differs in that it is completely transmissive and passive. This characteristic of the Codestar contributes to high system reliability and virtually eliminates the single point of failure inherent in active ("electronic") star networks. A Codestar-implemented fiber optic network is a broadcast network functionally and logically identical to any other broadcast medium including coaxial cable. Consequently any broadcast network may be implemented with Codestar fiber optic couplers.

Codestar couplers are an integral part of the Codenet fiber optic Ethernet physical layer, when implemented with the Codenet-3030 series Ethernet transceivers (See Figure 2) and Codenet-3035A or 3037A Repeaters. Such a system conforms to all aspects of IEEE standard 802.3 CSMA/CD and Xerox, Digital Equipment Corporation and Intel Ethernet Version 2.0 networks.

Performance Characteristics

Performance characteristics of the Codestar star couplers are as follows:

© Codenoll Technology Corporation, 1988
Printed in U.S.A.—February, 1988
Data Subject to change—(2/88)

Table 1. Codestar Coupler Performance Characteristics

Model Number	Number of Ports	Coupling (Min.)	Loss (dB) (Max.)	Standard Fiber Sizes
2004	4	7	10	50/125, 62.5/125, 85/125, 100/140
2008	8	10	13	50/125, 62.5/125, 85/125, 100/140
2016	16	13	17	50/125, 62.5/125, 85/125, 100/140
2032	32	17	21	50/125, 62.5/125, 85/125, 100/140
2007*	7	15	18	100/140
2019*	19	19	22	100/140

*Lightbus® series

Codestar Couplers Available

Codenoll offers two series of passive fiber optic star couplers. The standard Codestar series of 4, 8, 16 and 32 port couplers are low loss port couplers which allow implementation of networks having either a large number of nodes or covering large areas. The Lightbus® series of couplers provide 7 and 19 port solutions to less demanding network applications at a very low cost per port.

Note

Codestar couplers having non-standard dimensions, fiber or connectors are available on special order. Please contact Codenoll for full details.

Ordering Information

This product is available from your system integrator, your dealer or authorized Codenoll reseller.

MAKING LIGHT WORK OF NETWORKING

1086 NORTH BROADWAY, YONKERS, NEW YORK 10701 • TELEPHONE: (914) 965-6300 • TELEX: 646-159

Data Sheet

Codenet-4300 MultiStar Modular Multiport Repeater

Combining Full 802.3 Repeater Functionality With a Multi Media (Fiber Optic, Thick Ethernet, Thin Ethernet, Twisted Pair Wiring Segments) Multi Segment Capability

Product Family Features

MultiStar Base Unit (Codenet-4300)
- Front loading card cage design allows easy installation of up to 15 modules
- Includes connector for MultiStar Expansion Unit
- Tabletop, wall, or rack mountable

MultiStar Expansion Unit (Codenet-4301)
- Connects to the repeater control card on the Base Unit to extend module capacity by 15

Fiber Optic Modules
- **Codenet-3311** fiber optic inter repeater link (FOIRL) for connecting two MultiStars
- **Codenet-3331** for connecting a MultiStar to the fiber optic corporate backbone via a passive fiber optic star coupler

Thin Ethernet Module (Codenet-4350)
- Eliminates the need for an external transceiver and transceiver cable when connecting to a thin Ethernet segmen
- Provides direct connection to a thin Ethernet coaxial segment (RG-58 C/U)
- Will support thin Ethernet segment lengths of up to 1,000 ft. (305 meters)

Transceiver Interface Module (Codenet-4351)
- Provides connection for Ethernet/IEEE 802.3 transceiver cable (DIX/AUI interface)
- Allows local coaxial segments to attach to the corporate backbone via a standard transceiver
- Onboard jumper allows the module to operate with Signal Quality Error (SQE) or "heartbeat" type transceivers

Codenet-4300 MultiStar Modular Multiport Repeater

Product Family Benefits

Partitioning
- Manual Partitioning - Each segment may be individually isolated from MultiStar via a switch on the module's front panel
- Auto-partitioning - Individual modules will automatically isolate faulty segments from the network and will auto-reconnect when the fault is corrected
- Partition Status Indicator - A LED on each module provides partition status

Repeater Modules
- Activity LED for monitoring segment traffic
- Regulates Ethernet preamble and can restore up to 16 lost bits
- Packets are retimed to restore signal integrity to original levels
- Extends packet fragments to 96 bits to assure 100% collision detection

Product Overview

MultiStar is a modular, multiport repeater that provides a flexible, central platform for multi segment, multimedia Ethernet networks. Its modular design allows networks--composed of different types of media--to be designed, installed, and expanded to meet unique site requirements.

MultiStar provides full Ethernet/IEEE 802.3 compatible repeater functionality allowing single Ethernet segments to be combined to form larger company wide networks.

(Continued on overleaf)

Product Overview (Continued)

MultiStar can be configured with up to 15 modules of any type and in any combination. Up to two additional MultiStar Expansion Units may be added to allow direct connection of up to 45 segments.

Expanding or modifying the network is simplified by the modular architecture of the repeater, which allows networks to be configured to meet the needs of most facilities. By specifying the desired modules, media types may be combined to provide the best possible network configuration.

MultiStar provides a centralized management point which greatly simplifies problem isolation and network troubleshooting. Auto-partitioning and auto-reconnection of faulty segments from the network, improves network uptime by isolating network problems to the offending segment and leaving unaffected network segments operational. Status indicators and manual segment partitioning allow quick diagnosis and resolution of network problems.

Technical Specifications

Base (Codenet-4300) and Expansion Unit (Codenet-4301)

Mechanical
　Material: Fabricated steel chassis
　Size: 17" W x 19.5" D x 5.5" H
　Rack Mount: Shelf mountable in standard 19" rack
　Wall Mount: Mounting keyholes on base for wiring closet wall

Electrical
　Power: 90-270V, 50-60Hz, self adjusting
　Consumption: 200 watts maximum
　　　　　　　Worst case power factor = 0.6
　Protection: Backplane circuit breaker switch

Environmental
　Temperature: 0° to 50° C (operating)
　Humidity: 5% to 95% (non-condensing)

Standards
　UL　　　　　　　　Ethernet, Version 2
　CSA　　　　　　　IEEE 802.3
　FCC (Class B)　　ISO DIS 8802/3
　VDE (Class B)　　TUV

Transfer Interface Module (Codenet-4351)
　Connector Type: Standard 15-pin subminiature D

Thin Ethernet Module (Codenet-4350)
　Connector Type: Standard BNC connector
　Isolation Voltage: 500V RMS

Ordering Information

Products described in this Data Sheet are available from your system integrator, your dealer, or your authorized Codenoll reseller. When ordering these products refer to these Model numbers:

MultiStar Base Unit	Codenet-4300
MultiStar Expansion Unit	Codenet-4301
Point-to-Point FOIRL Card	Codenet-3311
Fiber Optic Star Connect Card	Codenet-3331
Thin Ethernet Module	Codenet-4350
Transceiver Interface Module	Codenet-4351

Related Products
　Pair Tamer™ Set for Telephone Wiring (Codenet-4352)
　(See Codenoll Data Sheet, issued separately)

Note: When ordering products for countries out of the USA, append appropriate country code to product number as described below:
　　　　Canada - CA
　　　　United Kingdom - UK
　　　　Australia - AA
　　　　Europe - ME

Benefits of a Fiber Optic Network

The use of fiber optic cable in a network gives the highest reliability known. Fiber optic cable is totally immune to electromagnetic interference of all types. Unlike coaxial cables, fiber optic cables may be routed over fluorescent lights, in power raceways, near copy machines, x-ray machines, mainframe computers, elevator motors, transformers, factory machines, robots, in fact, anywhere, without regard to electromagnetic noise, grounding problems, or differences in ground potential (ground loops). Fiber Optic Ethernet, therefore, eliminates the cause of a large number of data errors and the consequent retransmissions. This can greatly increase the throughput of any network.

About Codenoll Technology Corporation

Founded in 1980, Codenoll is a supplier of fiber optic Local Area Network (LAN) hardware. The compatible fiber optic LAN products (tradenamed Codenet and Codelink) are interchangeable with copper wire, allowing computers to communicate and share data, files and resources. By using fiber optics, lower cost, higher reliability, longer distances, immunity to environmental disturbances and greater security may be obtained without any modifications whatsoever.

Pair Tamer is a registered trademark of 3Com Corp.

MAKING LIGHT WORK OF NETWORKING

1086 NORTH BROADWAY, YONKERS, NEW YORK 10701 • TELEPHONE: (914) 965-6300 • TELEX: 646-159

Data Sheet

CODENET®-3311
Fiber Optic Inter-Repeater Link (FOIRL) Module for use in MultiStar® & MultiConnect Repeater Chassis

Features
- Fully compatible with IEEE Standard 802.3, Section 9.9.
- May be used to link one MultiStar/MultiConnect to another on a fiber optic link at distances up to 4.5 km.
- Low Light Level monitor detects fiber breakage; enhances data integrity.
- Optical "Idle" Signal allows simple measurement of transmitter power...provides a built-in optical source for testing fiber cables and connectors.
- Electrical interface to Codenoll MultiStar/3Com Multi-Connect system(s).
- User selectable Hi-Lo optical transmitter power allows use of non-standard optical fiber sizes up to 100μm core diameter.
- On-board "bleep" function facilitates network problem diagnosis.
- Automatic/manual partitioning (when faulty) with automatic/manual reconnection.

CodeNet-3311 Fiber Optic Inter-Repeater Link (FOIRL) Module

Product Description

The CodeNet-3311 is a fiber optic plug-in module for the Codenoll MultiStar/3Com MultiConnect active star Ethernet system. The unit provides an optical interface which fully complies with the IEEE 802.3 (Section 9.9) specifications for a vendor independent fiber optic inter-repeater link (FOIRL).

The CodeNet-3311 may be used to provide a point-to-point optical link between two MultiStar/MultiConnect stars, or between a MultiStar/MultiConnect and any fiber optic media access units (FOMAUs) which comply with IEEE 802.3 (Section 9.9) specifications...such as Codenoll's CodeNet-8381 FOMAU. Full compatibility of the optical interface allows the user to add MultiStar/MultiConnect repeaters to existing optical networks using equipment from multiple vendors.

The CodeNet-3311 has front panel status indicators and manual reset/partition controls for quick diagnosis and resolution of network problems. An on-board jumper allows selection of either high or low optical transmit power allowing the unit to provide Ethernet links up to 4.5 km (depending on total network propagation delay and flux budget).

The CodeNet-3311 automatically partitions itself in the event of a fault from the error-free segments, and automatically reconnects when the fault condition no longer exists.

When a segment is autopartitioned, its partition LED glows continuously. Once a segment is autopartitioned, the segment is sampled every 400 msec by transmitting a 523-bit packet called a "bleep." If the "bleep" packet is transmitted successfully without a collision, the segment is reconnected. The partition LED then begins flashing, indicating that the problem has been resolved. The reset switch on the front panel clears the LED and does not affect the segment's operation.

Optimal Network Design

Since CodeNet fiber optic Ethernet networks support both active and passive star hubs to meet user needs, CodeNet MultiStar Modules are available for all possible configurations:
- CodeNet-8310 connects to a CodeStar passive star hub.
- CodeNet-3311 connects to a MultiStar active star hub.
- CodeNet-8312 connects to CodeStar passive star hubs supporting CodeNet-30XX or 83X2 CodeNet products.

Simple Installation and Setup

Onboard standard fiber optic connectors directly attach to standard duplex fiber optic network cable using standard ST connectors or optional SMA connectors.

Secure Reliable Performance

The high bandwidth of fiber means the highest performance in your CodeNet network now, and since the fiber is the same as that specified for high performance networks like FDDI, your physical plant investment enjoys a long lifetime. Your CodeNet network is completely immune to electrical interference, even in noisy manufacturing or processing facilities. For secure networks, such as government and research installations, fiber optics are the only choice; unauthorized taps are difficult to make and easily detected. With CodeNet you can increase the distance between workstations or servers and still enjoy increased reliability. Compared with CodeNet optical fiber, all other networking media are severely limited, especially in terms of future upgrades.

Standard IEEE 802.3 Ethernet

Following Codenoll's policy of standards-based network solutions, each CodeNet-3311 transceiver meets applicable signaling standards of IEEE 802.3 Ethernet for 10 Mbps baseband CSMA/CD LANs.

Specifications—CodeNet-3311 MultiStar Module

Network Standards:	IEEE 802.3 CSMA/CD Ethernet 10 Mbps baseband, Section 9.9 Fiber Optic Repeater Link (FOIRL)
Electrical Interface:	Module for Codenoll's MultiStar or 3Com MultiConnect Ethernet active star
Optical Interface:	Duplex fiber optic cable with bayonet type ST connectors standard; screw-on SMA connectors optional (FOIRL standard is SMA).
Partition Switch:	Manual Partition Reset Manual Partition Clear Autopartition Indicator
Internal Option Jumpers: (factory settings in bold)	Optical Power (**Hi**/Lo)
Optics:	Transmitter CodeLED® high radiance edge-emitting Aluminum Galium Arsenide (AlGaAs) Light Emitting Diode (LED) Receiver Silicon PIN-type photodiode Optical Wavelength 830nm ± 20nm, peak Cable Sizes 50/125, 62.5/125, 85/125, or 100/140μ m duplex, multimode, graded index

Physical: Board Size
4.7 in. W × 12.0 in. L (12 cm × 30 cm)

Operating Environment
0°C to +55°C, 10 to 90% humidity (non-condensing)

Power
500mA @ +5V; 40mA @ +12V, typical

Weight
8 oz. (226 grams)

Optical Specifications[1]:

Peak Power Output Into:
62.5 μm core fiber	−12 dBm
100 μm core fiber	−11.5 dBm
50 μm core fiber	−15.5 dBm
Receiver Sensitivity	−27 dBm

Optical Flux Budget Into:
62.5 μm core fiber	15 dB
100 μm core fiber	15.5 dB
50 μm core fiber	11.5 dB
Dynamic Range	18 dB
Maximum Cable Length:	4500m[2]

[1] Typ. peak power measured at 25°C, 50% duty cycle, Hi power [factory default] Lo Power reduces power by approximately 3 dBm and distance to approximately 3000 m.
[2] Limited by IEEE 802.3 roundtrip delay time, flux budget permitting.

Diagnostic Indicators: Activity—Yellow
Link Fault—Yellow
Partition—Red

For Information or Assistance

Call the Customer Support Group at Codenoll Technology, (914) 965-6300.

Specifications may change without notice. Codenoll, CodeNet CodeStar, MultiStar and CodeLED are registered trademarks of Codenoll Technology Corporation. MultiConnect is a trademark of 3Com Corporation. Other trademarks mentioned are the property of other companies.

MAKING LIGHT WORK OF NETWORKING

1086 NORTH BROADWAY, YONKERS, NEW YORK 10701 USA • TELEPHONE: (914) 965-6300 • FAX (914) 965-9811

PN: 05-0058-16-0001

© 1990, Codenoll Technology Corporation
Printed in U.S.A.—December, 1990

Data Sheet

CODENET®-8310/8312 Passive Fiber Optic Ethernet Module for use in MultiStar® & MultiConnect Repeater Chassis

Features

- Links a Codenoll passive star network segment to other Ethernet segments.
- Automatic fiber break detection (when used with passive star).
- Onboard "Jabber" function indicates faulty controller.
- Electrical interface to Codenoll MultiStar/3Com MultiConnect system.
- Onboard transmitter test function (switch selectable) may be used as light source for network testing of fiber optic plant.
- Automatic/manual partitioning (when faulty) with automatic/manual reconnection.
- Onboard "bleep" function facilitates network problem diagnosis.
- Full optical collision detection.

CodeNet Passive Ethernet Segment Module

Product Description

The CodeNet-8310/8312 MultiStar Module allows the user to directly link a passive fiber optic network segment to a Codenoll MultiStar/3Com MultiConnect Ethernet active star. The Codenet-8310/8312 is completely compatible with IEEE 802.3 hardware and software and is plug-in-compatible with the MultiStar Repeater. The Codenet-8312 is also backward-compatible with earlier Codenoll fiber optic products.

Functional Description

The CodeNet-8310/8312 MultiStar Module provides for direct interfacing between an individual fiber optic network segment and a MultiStar Repeater. The 8310/8312 Module contains an integrated fiber optic transceiver which eliminates the need for separate fiber optic transceivers.

The CodeNet-8310/8312 Fiber Optic Repeater Module also contains front panel status indicators and manual reset/partition controls for quick diagnosis and resolution of network problems.

The CodeNet-8310/8312 is used with a CodeStar® passive fiber optic star (which constitutes an optical segment). The CodeNet-8310/8312 includes built-in collision detection, and a fiber break feature for improved network reliability and performance. Once a fiber break has been detected, the CodeNet-8310/8312 automatically partitions itself from the rest of the segments and when the problem has been resolved it will automatically reconnect the fiber segment.

The CodeNet-8310/8312 also contains a switch selectable test circuit, which allows it to be used as a light source for network testing of the fiber optic cabling. The CodeNet-8310/8312 Module is also capable of asserting the "Jabber" function when a faulty controller is detected. The unit will automatically partition itself from the error-free segments and will automatically reconnect when a fault condition no longer exists.

When a segment is auto-partitioned, its partition LED glows continuously. Once a segment is autopartitioned, the segment is sampled every 400 msec by transmitting a 523 bit packet called a "bleep." If the "bleep" packet is transmitted successfully without a collision, the segment is reconnected. The partition LED then begins flashing, indicating that the problem has been resolved. The reset switch on the front panel clears the LED and does not affect the segment's operation.

Optimal Network Design

Since CodeNet fiber optic Ethernet networks support both active and passive star hubs to meet user needs, CodeNet-831X cards are available for all possible configurations:
- CodeNet-8310 connects to a CodeStar passive star hub.
- CodeNet-8311 connects to a MultiStar active star hub.
- CodeNet-8312 connects to CodeStar passive star hubs supporting CodeNet-30XX or 83X2 CodeNet products.

Simple Installation and Setup

Onboard standard fiber optic connectors directly attach to standard duplex fiber optic network cable using standard ST connectors or optional SMA connectors.

Secure Reliable Performance

The high bandwidth of fiber means the highest performance in your CodeNet network now, and since the fiber is the same as that specified for high performance networks like FDDI, your physical plant investment enjoys a long lifetime. Your CodeNet network is completely immune to electrical interference, even in noisy manufacturing or processing facilities. For secure networks, such as government and research installations, fiber optics are the only choice; unauthorized taps are difficult to make and easily detected. With CodeNet you can increase the distance between workstations or servers and still enjoy increased reliability. Compared with CodeNet optical fiber, all other networking media are severely limited, especially in terms of future upgrades.

Standard IEEE 802.3 Ethernet

Following Codenoll's policy of standards-based network solutions, each CodeNet-8310/8312 card meets applicable signaling standards of IEEE 802.3 Ethernet for 10 Mbps baseband CSMA/CD LANs.

Specifications — CodeNet-8310/8312 MultiStar Module

Network Standards:	IEEE 802.3 CSMA/CD Ethernet 10 Mbps baseband
Electrical Interface:	Module for Codenoll's MultiStar or 3COM MultiConnect Ethernet active star
Optical Interface:	Duplex fiber optic cable with bayonet type ST connectors standard; screw-on SMA connectors optional
Partition Switch:	Manual Partition Reset Manual Partition Clear Auto Partition Indicator
Internal Option Switches: (factory settings in bold)	Break Detect (**On**/Off) SQE-Heartbeat (On/**Off**) Point-to-Point Mode (On/**Off**) Jabber Reset (**Auto**/Manual) Test Mode (On/**Off**)
Optics:	Transmitter: CodeLED® high radiance edge-emitting Aluminum Galium Arsenide (AlGaAs) Light Emitting Diode (LED) Receiver: Silicon PIN-type photodiode Optical Wavelength 830nm ±20nm, peak Cable Sizes 50/125, 62.5/125, 85/125, or 100/140µ m duplex, multimode, graded index

Physical: Board Size
4.7 in. W × 12.0 in. L (12 cm × 30 cm)

Operating Environment
0°C to +55°C, 10 to 90% humidity (non-condensing)

Power
700mA @ +5V; 40mA @ +12V, typical

Weight
8 oz. (226 grams)

Optical Specifications[1]:

Peak Power Output Into:	CodeNet-8310	CodeNet-8312
62.5 µm core fiber	−9 dBm	−3.5 dBm
100 µm core fiber	−8.5 dBm	−3 dBm
50 µm core fiber	−12.5 dBm	−6.5 dBm
Receiver Sensitivity	−37 dBm	−30 dBm
Optical Flux Budget Into:		
62.5 µm core fiber	28 dB	26.5 dB
100 µm core fiber	28.5 dB	27 dB
50 µm core fiber	24.5 dB	23 dB
Dynamic Range	18 dB	15 dB
Maximum Cable Length:		
Star	800m[2]	800m[2]

[1] Typ. peak power measured at 25°C, 50% duty cycle.
[2] Limited by IEEE 802.3 roundtrip delay time, flux budget permitting.

Diagnostic Indicators:	Activity — Yellow Low Light — Yellow Partition — Red

For Information or Assistance

Call the Customer Support Group at Codenoll Technology, (914) 965-6300.

Specifications may change without notice. Codenoll, CodeNet CodeStar, MultiStar and CodeLED are registered trademarks of Codenoll Technology Corporation. MultiConnect is a trademark of 3COM Corporation. Other trademarks mentioned are the property of other companies.

MAKING LIGHT WORK OF NETWORKING

1086 NORTH BROADWAY, YONKERS, NEW YORK 10701 USA • TELEPHONE: (914) 965-6300 • FAX (914) 965-9811

© 1990, Codenoll Technology Corporation

Data Sheet

CODENET®-8611
Plastic Optical Fiber Ethernet Active Star Module for use in MultiStar® & MultiConnect Repeater Chassis

Features

- Uses inexpensive plastic optical fiber (POF) to transmit data.
- Easier to use and install than twisted pair.
- Workstations can be up to 100 meters apart.
- Foolproof POF connectors are easier to use and more dependable than any other LAN connector.
- Low Light Level monitor detects fiber breakage; enhances data integrity.
- Optical "Idle" Signal allows simple measurement of transmitter power...provides a built-in optical source for testing fiber cables and connectors.
- Electrical interface to Codenoll MultiStar/3Com Multi-Connect system(s).
- On-board "bleep" function facilitates network problem diagnosis.
- Automatic/manual partitioning (when faulty) with automatic/manual reconnection.

CodeNet-8611 Plastic Optical Fiber Ethernet Active Star Module in Multistar Hub.

The Codenoll Plastic Optical Fiber Ethernet Solution for Networks

Plastic Optical Fiber (POF) revolutionizes the LAN media choices. It is as inexpensive as twisted pair, and easier to install, to use and to maintain than any other Local Area Network media. No other media has the built-in bandwidth of fiber and corresponding data transmission capacity, resulting in the best return on your network investment. Newer and faster network standards, like FDDI, use exactly the same fiber that is installed today for Codenoll's fiber optic Ethernet networks.

The CodeNet POF Ethernet product line is more robust and reliable than coax or twisted pair solutions. It is completely immune to electromagnetic noise and no form of cable damage, like short circuits, ground faults, reflections, or impedance mismatches, can disable the network. It is impossible for a cable to cause a fire or explosion since there is no electrical current present, and it is virtually impossible for an unauthorized agent to tap into the cable without easy detection. Workstations can be up to 100 meters apart using a MultiStar™ active hub. Any CodeNet 86XX POF network segment can be combined with other CodeNet fiber optic components. CodeNet glass fiber optical networks are easily extendable within large buildings or between buildings throughout a campus. The CodeNet fiber optic products offer maximum flexibility and cost-effectiveness when designing today's Local Area Network. The low cost of plastic fiber, POF connectors, and the unparalleled ease of connectorization make the CodeNet POF products ideal for network workgroups.

Product Description

The CodeNet-8611 is a plastic optical fiber plug-in module for the Codenoll MultiStar/3Com MultiConnect active star Ethernet system. The CodeNet-8611 consists of two boards: 1) A mother board which contains the interface between the Multistar and optical transceiver; and 2) A daughter card which contains the fiber optic transceiver. The 8611 has the same mother board as the CodeNet-8311 Fiber Optic Inter-Repeater Link (FOIRL) MultiStar Module, but has a new daughter card with an optical transceiver designed for POF.

The CodeNet-8611 is used in Ethernet active star configurations and uses the same signalling conventions as the IEEE FOIRL (Fiber Optic Inter-Repeater Link) Ethernet standard. A POF Ethernet segment can also be connected with any other type of Ethernet media segment...glass optical fiber, 10-Base-T, and thick and thin coax.

The CodeNet-8611 also has front panel status indicators and manual reset/partition controls for quick diagnosis and resolution of network problems.

The CodeNet-8611 automatically partitions itself in the event of a fault from the error-free segments, and automatically reconnects when the fault condition no longer exists.

When a segment is autopartitioned, its partition LED glows continuously. Once a segment is autopartitioned, the segment is sampled every 400 msec by transmitting a 523-bit packet called a "bleep." If the "bleep" packet is transmitted successfully without a collision, the segment is reconnected. The partition LED then begins flashing, indicating that the problem has been resolved. The reset switch on the front panel clears the LED and does not affect the segment's operation.

Optimal Network Design

Since the CodeNet MultiStar active Ethernet hub supports both glass and plastic optical fiber, as well as twisted pair and coax, any Ethernet network may be expanded using fiber optics. POF products are an economical solution for workgroup areas, and glass optical fiber could be used to create links up to 4.5 km long.

Simple Installation and Setup

Codenoll's connector system for POF allows quick and simple attachment to duplex 1000 μm plastic optical fiber. Both the transmit and receive fibers are enclosed in a single, polarized connector. A partial shroud around the connector protects the end faces of the fiber while maintaining easy access for termination.

Specifications—CodeNet-8611 MultiStar POF Module

Standards:	Uses IEEE FOIRL (Fiber Optic Inter-Repeater Link) Ethernet signalling conventions
Electrical Interface:	Module for Codenoll's MultiStar or 3Com MultiConnect Ethernet active star
Optical Interface:	Duplex 1000mm plastic optical fiber cable with Codenoll POF Connectors
Partition Switch:	Clear Autopartition indicator Reset Manual Partition Manual Partition
Internal Option Switches: (factory settings in bold)	Jabber Reset (**Auto**/Manual) SQE-Heartbeat (On/**Off**)
Optics:	Transmitter Aluminum Galium Arsenide (AlGaAs) Light Emitting Diode (LED) 660nm Optical Wavelength 660nm, peak Receiver Silicon PIN-type photodiode Cable Sizes 1000 μm duplex, high quality PMMA core, step index, plastic optical fiber

Physical:	Board Size 4.7 in. W × 12.0 in. L (12 cm × 30 cm) Operating Environment 0°C to +55°C, 10 to 90% humidity (non-condensing) Power 500mA @ +5V; 40mA @ +12V, typical Weight 8 oz. (226 grams)

Optical Specifications[1]:

Peak Output Power:	−9 dBm
Receiver Sensitivity:	−29 dBm
Optical Flux Budget:	20 dB
Maximum Cable Length:	50 meters

[1]Typ. peak power into 1000 μm PMMA (polymethyl methecrylate) core fiber measured at 25°C, 50% duty cycle.

Diagnostic Indicators:	Activity—Yellow Link Fault—Yellow Partition—Red

For Information or Assistance

Call the Customer Support Group at Codenoll Technology, (914) 965-6300.

Specifications may change without notice. Codenoll, CodeNet CodeStar, MultiStar and CodeLED are registered trademarks of Codenoll Technology Corporation. MultiConnect is a trademark of 3Com Corporation. Other trademarks mentioned are the property of other companies.

MAKING LIGHT WORK OF NETWORKING

1086 NORTH BROADWAY, YONKERS, NEW YORK 10701 USA • TELEPHONE: (914) 965-6300 • FAX (914) 965-9811

PN: 05-0058-29-0001

© 1990, Codenoll Technology Corporation
Printed in U.S.A. — November, 1990

Data Sheet

Codenet-4350
Thin Ethernet Coax Transceiver Interface Module for use in MultiStar & MultiConnect Repeater Chassis

Features

- Connects thin Ethernet coax to Codenet-4300 MultiStar and 3Com's MultiConnect, and twisted pair to both repeaters via Codenet-4352 PairTamers

- Automatic/manual partitioning (when faulty) with auto/manual reconnection

- Signal input/output levels meet IEEE 802.3 specifications

- Onboard jumper for network segment termination

- Onboard "bleep" function facilitates network problem diagnosis

- Complies with FCC Class A, Subpart J, Part 15 and VDE Class B Specifications

- Cable lengths to 1,000 ft.

- Easy thumbscrew installation

Codenet-4350 Transceiver Interface Module

General Description

The Codenet-4350 Transceiver Interface Module connects thin coaxial cable (thin Ethernet) to the Codenet MultiStar or 3Com's MultiConnect Repeater. When used with Codenoll-4352 PairTamers, it connects twisted pair segments to these repeater chassis.

The Codenet-4350 contains a 50 Ohm BNC connector on its face plate and incorcorporates a transceiver that allows direct connection to thin coaxial cable, which can be up to 1,000 ft. in length. The module provides selectable internal grounding and jumper termination of the network segment.

The Codenet-4350, a 3.5 inch by 12 inch pc board, easily inserts into a CodeStar or MultiConnect Repeater cabinet and is secured via a pair of thumbscrews. It is not necessary to depower the repeater during installation.

The Codenet-4350 Transceiver Interface Module contains an activity indicator and a partition indicator on its front panel. The unit may be partitioned manually via a three position toggle switch on its front panel.

Paritioning may be accomplished either manually, using the three position partition switch on the unit's front panel, or automatically. Autopartitioning occurs: after 64 consecutive collisions; if the receive packet length is greater than 6.4 msec, or if the transmit collision is greater than 6.4 msec.

When autopartitioning a segment, the unit's front panel partition LED glows continuously. Once a segment is autopartitioned, the segment is sampled every 400 msec by transmitting a 512 bit "bleep" packet. If the packet is transmitted successfully without a collision, the segment is reconnected and autopartitioning is terminated. The partition LED then begins flashing, indicating that the problem has been resolved. The reset switch only clears the LED and does not affect the segment's operation.

Manual partitioning all unused segments (as well as faulty segments), prevents data errors caused by accidental reconnection of a segment that has not been grounded or terminated correctly.

Technical Specifications

Partitioning
- Manually using the partion switch (down position)
- Autopartioned after 64 consecutive collisions
- Autopartioned if receive packet length greater than 6.4 milliseconds
- Autopartioned if transmit collision greater than 6.4 milliseconds

Partitioning Sample Packet (Bleep)
- 512 bits. A "bleep" is sent approximately every 400 milliseconds when a segment is partitioned Autopartitioning ends when a "bleep" is transmitted without a collision

External Connector
- 50 ohm BNC thin Ethernet connector

Signal input and output level meet IEEE 802.3 specifications

Distance to XCVR
- N/A

Power Consumption
- 5.0 watts maximum
  Internal amp resettable fuse

Host Interface
- Codenoll MultiStar or MultiConnect Repeater bus interface standard

Installation Attachment
- Two (2) thumbscrews for fastening the mounting bracket to the MultiStar or MultiConnect Repeater chassis

Network Ground
- Jumper selectable on board with a screw for network grounding of the network segments through the chassis to the AC ground of the power connector

Network Termination
- Jumper selectable on board for terminating one end of the network segment

Compliances
- FCC Class A, Subpart J, Part 15
- VDE Class B

Propagation Delays
- 550nS receive per pair
- 120nS transmit per pair

Jitter Output
- 3.5nS maximum

Cable Length
- 1,000 ft. thin coax, RG-58 or equivalent

Signal-to-noise
- 5:1 Receive

Indicators

NAME	COLOR	DESCRIPTION
ACTIVITY	Yellow	Flashes as packets are received by the module from the segment it's attached to
		Not lit if there is no activity on the segment
PARTITION	Red	Flashes if the segment has been autopartitioned and then reset
		Glows continuously if the module is partitioned

Weight & Dimensions
- 1 lb. (0.2Kg); 3.5 in. x 12 in.

MAKING LIGHT WORK OF NETWORKING

1086 NORTH BROADWAY, YONKERS, NEW YORK 10701 • TELEPHONE: (914) 965-6300 • TELEX: 646-159

Data Sheet

Codenet-4351
Thick Ethernet Coax Transceiver Interface Module for use in MultiStar & MultiConnect Repeater Chassis

Features

- Connects thick Ethernet coax via an external transceiver to Codenet-4300 MultiStar and/or 3Com's MultiConnect Repeater(s)

- Automatic/manual partitioning (when faulty) with auto/manual reconnection

- Signal input/output levels meet IEEE 802.3 specifications

- Onboard "bleep" function facilitates network problem diagnosis

- SQE may be enabled or disabled without affecting network performance

- Complies with FCC Class A, Subpart J, Part 15 and VDE Class B Specifications

- Transceiver cable lengths to 50 meters

- Easy thumbscrew installation

Codenet-4351 Transceiver Interface Module

General Description

The Codenet-4351 Transceiver Interface Module connects standard coaxial (thick Ethernet) cable via an external transceiver to the Codenet MultiStar or 3Com's MultiConnect Repeater. The module has a 15-pin female DIX connector on its face plate and connects to an external transceiver through an Ethernet IEEE 802.3 transceiver cable. Maximum distance to the transceiver is 50 meters.

The Codenet-4351, a 3.5 inch by 12 inch pc board, easily inserts into a CodeStar or MultiConnect Repeater cabinet and is secured via a pair of thumbscrews.

It is not necessary to depower the repeater during installation.

The Codenet-4351 Transceiver Interface Module contains an activity indicator and a partition indicator on its front panel. The unit may be partitioned manually via a three position toggle switch on its front panel.

Partitioning may be accomplished either manually, using the three position partition switch on the unit's front panel, or automatically. Autopartitioning occurs: after 64 consecutive collisions; if the receive packet length is greater than 6.4 msec, or if the transmit collision is greater than 6.4 msec.

When autopartitioning a segment, the unit's front panel partition LED glows continuously. Once a segment is autopartitioned, the segment is sampled every 400 msec by transmitting a 512 bit "bleep" packet. If the packet is transmitted successfully without a collision, the segment is reconnected and autopartitioning is terminated. The partition LED then begins flashing, indicating that the problem has been resolved. The reset switch only clears the LED and does not affect the segment's operation.

Manual partitioning all unused segments (as well as faulty segments), prevents data errors caused by accidental reconnection of a segment that has not been grounded or terminated correctly.

Technical Specifications

Partitioning
- Manually using the partion switch (down position)
- Autopartioned after 64 consecutive collisions
- Autopartioned if receive packet length greater than 6.4 milliseconds
- Autopartioned if transmit collision greater than 6.4 milliseconds

Partitioning Sample Packet (Bleep)
- 512 bits. A "bleep" is sent approximately 400 milliseconds when a segment partitioned Autopartitioning ends when a "bleep" is transmitted without a collision

External Connector
- 15-pin DIX connector with slide lock

PIN	DEFINITION
1	COMMON
2	COLL
3	TRANSMIT
4	COMMON
5	+REC
6	COMMON
7	COMMON
8	COMMON
9	COLL
10	TRANSMIT
11	COMMON
12	-REC
13	+11 - +13VOLTS @ .5 AMPS
14	COMMON
15	SHIELD COMMON TO CHASSIS

Signal input and output level meet IEEE 802.3 specifications

Distance to XCVR
- 50 meters

Power Consumption
- 2.5 watts maximum
  Power consumption is reflected in the power supply efficiency of 5.0 watts. External transceivers will add up to 3 watts of dissipation to the power supply. 1 amp internal resettable fuse.

Host Interface
- Codenoll MultiStar/3Com MultiConnect Repeater bus interface standard

Installation Attachment
- Two (2) thumbscrews for fastening the mounting bracket to the MultiStar and/or MultiConnect Repeater chassis

Shield Ground
- Grounding through the front bracket to chassis

SQE Requirements
- SQE may be enabled or disabled without affecting network performance

Transceiver Cable
- 50 meters maximum, IEEE 802.3 compatible

Propagation Delays
- 550nS receive per pair
- 120nS transmit per pair

Jitter Output
- 3.5nS maximum

Indicators

NAME	COLOR	DESCRIPTION
ACTIVITY	Yellow	Flashes as packets are received by the module from the segment it's attached to
		Not lit if there is no activity on the segment
PARTITION	Red	Flashes if the segment has been autopartitioned and then reset
		Glows continuously if the module is partitioned

Weight & Dimensions
- 1 lb. (0.2Kg); 3.5 in. x 12 in.

MAKING LIGHT WORK OF NETWORKING

1086 NORTH BROADWAY, YONKERS, NEW YORK 10701 • TELEPHONE: (914) 965-6300 • TELEX: 646-159

IX

Fiber Optic FDDI PC Adapter Cards and Concentrators

CodeNet 9540/9543, 9340/9343,   p. 458
CodeNet 9041,   p. 460

Data Sheet

CODENET-9540 and 9543
ISA/FDDI Network Boards
CODENET-9340 and 9343
ISA/FDDI LowCost Network Boards

Features and Advantages

- Connects an ISA or Extended-ISA bus computer to an ANSI standard 100 megabit per second (Mbps) Fiber Distributed Data Interface (FDDI) token ring network.
- Installs in the standard card slot in any ISA or Extended-ISA bus computer including Compaq, IBM AT and non-Micro Channel PS/2, Zenith, Olivetti and their compatibles.
- Full 100 Mbps performance for file servers, backbones, bridges, routers, gateways, workgroups and workstations.
- Software drivers for most popular network software systems including Microsoft LAN Manager, Novell Netware 2.1x, Netware 386 and SCO UNIX.
- The CodeNet FDDI family of products covers single attachment, dual attachment, and redundant or independent rings.
- Network Boards can be used to bridge to 10 Mbps IEEE 802.3 Ethernet, IEEE 802.5 Token Ring and IEEE 802.4 MAP.
- Network Boards work with CodeNet-9041 Concentrator System.
- Conforms to ISO-9384 and ANSI-X3T9.5 published standards.
- Allows up to two kilometers between stations (954X Series).
- Simply installed standard size plug-in Board and standard fiber optic connectors.
- CodeNet-934X Adapters cut cost per station substantially over standard FDDI connection, and save more when used with CodeNet-9041 Concentrator System.
- Modular design makes the Network Board field-upgradable from single ring to redundant dual ring or from redundant dual ring to independent dual counter-rotating ring.
- Fiber optic cable provides the highest possible network reliability and security.
- Physical network system completely unaffected by electrical problems including EMI, RFI, short circuits, and ground loops.

FDDI Advantages

The American National Standards Institute (ANSI) developed the standard for very high speed token passing ring networks: ANSI X3T9.5, Fiber Distributed Data Interface (FDDI), which has also been sanctioned by the International Standards Organization, ISO-9384. The standard specifies substantial performance and configuration improvements over current LANs: 100 Mpbs data rate, up to two kilometers between stations and up to 1000 physical connections per ring.

Full ANSI 100 Mbps Speed

All CodeNet FDDI Network Boards operate at the full ANSI FDDI transmission rating of 100 Mbps. They allow a standard ISA bus personal computer (IBM PC/AT or compatible) to function on a backbone, in a workgroup or as a high performance workstation.

Low Cost FDDI Connection

Only Codenoll provides a full set of completely ANSI standard FDDI Network Boards (954X series) and a low cost series of fully FDDI token ring compatible Network Boards (934X series). Both operate at full ANSI 100 Mbps. The 954X series supports the ANSI standard two kilometers between stations; the 934X LowCost FDDI Network Boards allow up to one kilometer between stations.

Standard ISA Bus

The CodeNet FDDI Network Boards are designed to function on the industry standard architecture (ISA) bus, converting the computer to an ANSI standard FDDI 100 Mbps Class A or Class B station. They are standard size ISA, PC/AT Boards that are easily installed. Connecting the fiber optic FDDI cables and loading the appropriate network software completes the installation.

Product Summary

The CodeNet series of ISA/FDDI Network Boards is a complete family of Network Boards for computers such as Compaq, IBM AT and non-Micro Channel PS/2, Zenith, Olivetti and their compatibles. They provide direct connection as Class A or Class B active stations to ANSI standard FDDI 100 Mbps networks. The CodeNet FDDI Network Boards can be configured for single ring (9540 and 9340), or dual ring redundant (9543 and 9343).

Codenoll
TECHNOLOGY CORPORATION

Software Driven

The CodeNet FDDI boards adapt to different networks with software drivers. Drivers for most popular network software, including Microsoft LAN Manager, Novell Netware 2.1x, Netware 386 and SCO UNIX are available. To change network types or if new network protocols become available, the user need only change the software to be compatible.

Configuration Flexibility

CodeNet FDDI Network Boards are modular so they may be reconfigured in the field. When a computer is upgraded to a new type, the fiber optic (PHY) and controller (MAC) modules are simply transferred to a new base board at considerable dollar savings compared to a new Network Board purchase. Nor is there the inconvenience or business disruption of sending Boards back to the factory for upgrading.

Codenoll Fiber Optic Experience

Codenoll Technology Corporation is a pioneer and leading supplier of high performance fiber optic equipment for computer networks. We have furnished high speed, noise immune, highly reliable networks to thousands of installations worldwide.

For More Information or Assistance

Call the Technical Support Group of Codenoll Technology at (914) 965-6300.

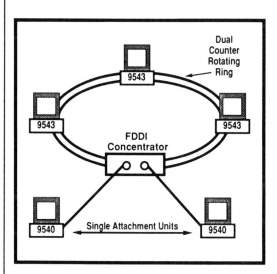

Specifications—CodeNet-954X and 934X FDDI Network Adapter Cards

Standards Supported:	FDDI ANSI-X3T9.5, ISO 9384 SMT V6.1	
Bus Interface:	ISA bus, Compaq, Zenith, Olivetti, IBM AT and non-Micro Channel PS/2 and compatibles	
Network Interface:	FDDI (MIC) duplex plug, standard. FDDI Optical Bypass: Modular connector for FDDI Dual Bypass Switch	
Fiber Sizes:	62.5μm, 50μm, 85μm and 100μm	
Optical Specifications: (62.5μm fiber)	830nm ±30nm	1300nm ±30nm
Output power (avg)	−18dbm	−17dbm
Receiver sensitivity (avg)	−31dbm	−32dbm
Flux budget	13db	15db
Dynamic range	15db	18db
Fiber Data Rate:	125 Megabaud	
Line Code:	4B/5B	
Usable Data Rate:	100 Mbps + clock	
Storage Temperature:	−55°C-80°C	
Physical:	Board size 10.67 cm (4.2 in) × 33.35 cm (13.13 in). ISA full size PC card	
	Operating Environment Temperature: 0°C-50°C Humidity: 10%-90% non-condensing Altitude: 0-10,000 ft.	
	Power 2.0 A @ +5V, typical	

Specifications may change without notice. Codenoll and CodeNet are registered trademarks of Codenoll Technology Corporation. Other trademarks mentioned are the property of other companies.

MAKING LIGHT WORK OF NETWORKING

1086 NORTH BROADWAY, YONKERS, NEW YORK 10701 • TELEPHONE: (914) 965-6300 • TELEX: 646-159

Data Sheet

CODENET-9041 ISA/FDDI Concentrator Hardware/Software System

Features and Advantages

- The system is a combination of ISA network boards and software that make an ISA bus computer, such as Compaq, IBM PC/AT or non-Micro Channel PS/2, Zenith, Olivetti and their compatibles into an FDDI Concentrator.
- Reduces the connection cost per FDDI station by attaching single stations to the ring network via low cost single-PHY boards in the concentrator.
- Complete FDDI device support capability. Either single ring (1 PHY) or dual counter-rotating ring (2 PHY) or combinations may be connected to a CodeNet-9041.
- The CodeNet-9041 can be single attachment, dual attachment, and redundant or independent rings.
- Flexible configuring. The CodeNet-9041 supports both the standard CodeNet ANSI Standard FDDI Adapter Cards (CodeNet-954X series) and the Low Cost FDDI Adapter Cards (CodeNet-934X series) in any combination.
- Full 100 Mbps FDDI performance for connection to backbones, mainframes, minicomputers, servers, gateways, bridges and routers.
- The CodeNet-9041 can be the concentrator for a workgroup using structured star fiber cabling.
- Installs in the standard card slots in any ISA bus computer including Compaq, IBM PC/AT or non-Micro Channel PS/2, Zenith, Olivetti and their compatibles with one slot for each physical connection (PHY).
- The Concentrator is transparent to operating systems. Drivers are available for most popular network software systems including Microsoft LAN Manager, Novell Netware 2.1X, Netware 386 and SCO Unix.
- Conforms to ISO-9384 and ANSI X3T9.5 published standards.
- Allows up to two kilometers between stations (954X series).
- Standard FDDI, ST or SMA connectors.
- Modular design makes the CodeNet-9041 field upgradable from single ring to redundant dual ring or from redundant dual ring to independent dual counter-rotating ring. Additional physical connections (PHYs) may be added in the field into available card slots.
- Fiber optic cable provides the highest possible network security and reliability.
- Physical network system completely unaffected by electrical problems including EMI, RFI, short circuits, and ground loops.

Product Summary

The CodeNet-9041 combines CodeNet-9000 Series Industry Standard Architecture (ISA) cards and CodeNet software to make an ISA bus computer into an FDDI concentrator. A concentrator is a specialized type of FDDI station, a central hub. It attaches to a single or dual counter-rotating FDDI ring. Multiple ports allow attachment of other devices as either single (Class A) or double (Class B) stations in a physical star arrangement. Multiple concentrators can be cascaded in structured star fiber cabling systems.

FDDI Concentrator Advantages

The American National Standards Institute (ANSI) has developed a standard for very high speed token passing ring networks (ANSI X3T9.5), Fiber Distributed Data Interface (FDDI), which has also been approved by the International Standards Organization (ISO-9384). This interface specifies such substantial performance and configuration improvements over current LANs as: 100 Mbps speed, up to two kilometers between stations, and up to 1000 physical connections per ring.

One type of physical connection is defined as a concentrator. The concentrator allows lower priced physical connections to share one main dual ring connection.

The result is a significantly reduced per station cost but each station retains the 100 Mbps speed of the FDDI network. In addition, the concentrator provides much greater network management, expandability and maintenance capabilities.

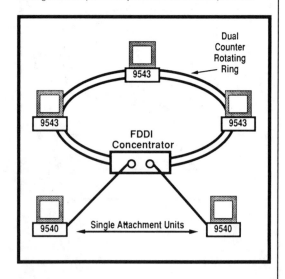

Low Cost FDDI Connection

Codenoll's unique product offerings provide a full set of completely ANSI standard FDDI adapter cards (954X series) and a low cost series of fully FDDI token ring compatible cards (934X series). Both operate at full ANSI 100 Mbps. The 954X series supports the ANSI standard two kilometers between stations; the 934X Series is significantly lower in cost and allows up to one kilometer between stations.

Configuration Flexibility

The CodeNet-9041 FDDI Concentrator is built up using adapter cards which function on the Industry Standard Architecture bus. They convert any ISA bus computer to a standard FDDI 100 Mbps Class A station in a hub configuration.

The user may choose to install single physical connectors (1 PHY) or may install double physical connectors (2 PHY) for full FDDI dual ring stations. They may be "mixed and matched" to the users unique station requirements. In any selection, the user may choose the CodeNet 954X Series ANSI Standard FDDI Adapter Cards or the CodeNet 934X Series LowCost FDDI Adapter Cards for additional cost savings.

Concentrator Software

The ISA bus computer configured with the CodeNet-9041 System is managed by a VRTX software package supplied by Codenoll. This software consists of two entities: The first manages normal FDDI data transfer (MAC), and the second performs station management (SMT). It provides complete management of the Concentrator's outbound, inbound and internal data packet traffic. All packet transfers between the concentrator, the workstations and the main ring are handled in real time to maintain the network's high performance. The CodeNet-9041 Concentrator is completely transparent to network operating systems and the attached hardware.

Individual Board Flexibility

CodeNet FDDI boards are modular so they may be reconfigured in the field. When a computer is upgraded, the fiber optic (PHY) and controller (MAC) modules are transferred to a new base board at considerable dollar savings compared to a new adapter purchase. Nor is there the inconvenience or business disruption of sending cards back to the factory for upgrading.

Codenoll Fiber Optic Experience

Codenoll Technology Corporation is a pioneer and leading supplier of high performance fiber optic equipment for computer networks. We have furnished high speed, noise immune, highly reliable networks to thousands of installations worldwide.

For More Information or Assistance

Call the Technical Support Group of Codenoll Technology at (914) 965-6300.

Specifications—CodeNet-9041 FDDI Concentrator System

Standards Supported:	FDDI ANSI-X3T9.5, ISO 9384
Bus Interface:	ISA bus, including Compaq, IBM PC/AT and non-Micro Channel PS/2, Zenith, Olivetti and compatibles
Network Interface:	FDDI (MCI) duplex plug, standard. ST or SMA connectors, optional.
Fiber Sizes:	50μm, 62.5μm, 85μm and 100μm

Optical Specifications: (62.5μm fiber)

	830μm ± 30μm	1300μm ± 30μm
Output power (avg)	−18dbm	−17dbm
Receiver sensitivity (avg)	−31dbm	−32dbm
Flux budget	13db	15db
Dynamic range	15db	18db

Data Rate:	125 Megabaud
Line Code:	4B/5B
Storage Temperature:	−55°C-80°C
Physical:	Board size 10.67 cm (4.2 in) × 33.35 cm (13.13 in). ISA full size card

Operating Environment
Temperature: 0°C-50°C
Humidity: 10%-90% non-condensing
Altitude: 0-10,000 ft.

Power
2.0 A @ +5V, typical

MAKING LIGHT WORK OF NETWORKING

1086 NORTH BROADWAY, YONKERS, NEW YORK 10701 • TELEPHONE: (914) 965-6300 • FAX: (914) 965-9811 • TELEX: 646-159

Specifications may change without notice. Codenoll and CodeNet are registered trademarks of Codenoll Technology Corporation. Other trademarks mentioned are the property of other companies.

© 1989 Codenoll Technology Corporation
Printed in U.S.A. 10/89

Fiber Optic Cable Assemblies

CFOC 50/125, 62.5/125, 85/125, 100/140,   p. 464
Codenoll POF–Plastic Optical Fiber Duplex Cable Assemblies,   p. 466

Data Sheet

Codenoll Multimode Duplex Fiber Optic Cable Assemblies

Features
- High reliability
- Thinner, lighter and more flexible than coax
- Low cost
- Low loss
- Easy to install
- Color coded connectors for ease of identification and trouble-free installation.
- High bandwidth
- High tensile strength
- Small sizes
- Small bend radius
- Flame retardant type available
- Custom lengths
- 905 and 906 type SMA connectors standard— other types available
- Wide variety of options available

General Description

Codenoll's wide ranging line of highly reliable, duplex multi-mode fiber optic cable assemblies are ideal for local area networks, premises networks, data processing networks and systems, machine control applications, remote sensing applications, aircraft, marine and telemetry applications. These interconnection cables are designed for numerous connections where rearrangement of facilities is often desired. The selection and mating of highest quality cable and connectors result in the lowest connection loss. Interconnection cables are available in standard lengths, or may be ordered to special lengths. All interconnection jumpers undergo rigorous factory testing and inspection prior to shipping in order to ensure high performance and reliability, plus utmost quality.

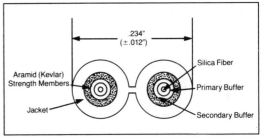

Cutaway view of duplex fiber optic cable showing construction.

Codenoll Fiber Optic Cable Specifications*

	CFOC-50/125	CFOC-62.5/125	CFOC-85/125	CFOC-100/140
Cord/Clad Diam ($\pm 3\mu m$)	50/125μm	62.5/125μm	85/125μm	100/140μm
Maximum Attenuation at 830nm* (db/km)	5.0	5.0	4.5	5.0
Numerical Aperture ($\pm .02$)	.20	.28	.26	.29
Bandwidth (850nm)	400	200	200	100**

*Lower attenuation available on special order.
**Higher bandwidth available on special order.

Specifications

Construction

Fiber Core:	Graded Index Silica
Fiber Clad:	Silica
Fiber Buffer:	Polyester—900 micron ±25
Strength Member:	Kevlar 49* (Aramid) Strands Concentric with Fiber Optic Components
Flame Retardant Jacket (.030" Wall):	PVC* Printed identification on one (1) channel
Connectors:	SMA 905 and 906 styles (please specify)

*Plenum available on special order.

Physical Properties:

Cable Dimensions:	3.0 mm (.118") × 6.0 mm (.234") .130" max × .246" max
Cable Weight:	15 Kg/Km (10.1 #/M')
Maximum Load (Tensile):	600 N (134.8#)—Installation
Maximum Load (Tensile):	100 N (22.4#)—Operating
Minimum Bend Radius:	5 cm (1.97")—Installation Load
Minimum Bend Radius:	3 cm (1.18")—Operating (Unloaded)
Flammability:	VW-1 Rated
Number of Fibers:	2
Lengths Available (Duplex/ Connectorized):	10, 20, 50, 100, 200, 500 and 1000 meters

Also Available On Special Order

Multi-packs, heavy duty breakouts, high density fiber, pre-terminated pulling eyes, armored cable, interior/exterior cables, 6-fiber connectorized breakout cable plus a wide range of fiber diameters, jacket materials and connectors, are available on special order. Consult Codenoll, your dealer or authorized Codenoll reseller for details.

NOTE: A connectorization kit (Codenet-CFTK), containing a complete tool set and supplies—housed in a heavy duty carrying case—is also available. This kit does not contain connectors, which are supplied separately.

Ordering Information

This product is available from your system integrator, your dealer or authorized Codenoll reseller.

©Codenoll Technology Corporation, 1988
Printed in U.S.A.---September, 1988
Data subject to change---9/88

MAKING LIGHT WORK OF NETWORKING

1086 NORTH BROADWAY, YONKERS, NEW YORK 10701 • TELEPHONE: (914) 965-6300 • TELEX: 646-159

Data Sheet

CODENOLL POF
Plastic Optical Fiber
Duplex Cable and Assemblies

Features

- Easier to use and install than twisted pair.
- Rugged connectors specifically designed for POF.
- Higher reliability and security.
- Low cost.
- Polarized connectors are small and easy to use.
- Small bend radius.
- Flame retardant jacket available.
- Custom or standard lengths.
- Supports IEEE Ethernet Standard Networks.

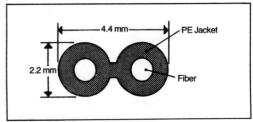
Patch Cable

General Description

Codenoll's line of Plastic Optical Fiber (POF) cable assemblies provide the performance benefits of fiber optics at a price competitive with unshielded twisted pair copper phone wire. Like glass fiber, POF is immune to electrical noise and difficult to tap, but costs much less. Each assembly is terminated with the Codenoll/Packard POF connector, specifically designed to take advantage of POF's unique characteristics to lower connector cost while retaining high performance. Each connector is polarized so the connector can only be inserted the correct way into a patch panel or header. These low cost POF cable assemblies are the perfect media choice for LAN workgroups.

Standard patch cables come with a black polyethylene jacket. Drop cables have a gray PVC jacket around the standard patch cable jacket. Plenum cable is available. POF assemblies come in three standard lengths and undergo rigorous factory testing and inspection prior to shipping to ensure high performance, reliability, and quality.

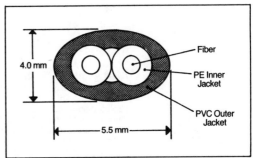
Drop Cable

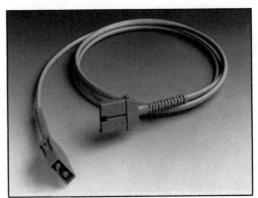
POF Drop Cable Assembly

Codenoll Plastic Optical Fiber Cable Specifications

Optical

Maximum Attenuation (650 nm)	< 150* dB/km	Minimum Bend Radius	2.5 cm	2.5 cm
Numerical Aperture (± .03)	.47	Tensile Strength (5% elongation)	14 kg	20 kg
		Operating Temperature	−40°– +85°C	−40°– +85°C

*Measured with monochromatic light source at 25°C.
**Flame retardant and plenum available on special order.

Construction

Fiber Core:	High Purity Polymethyl methacrylate (PMMA)
Fiber Cladding:	Fluorinated Polymer
Jacket:	Patch Cable: Polyethylene
	Drop Cable: Patch cable sheathed in gray PVC jacket**
Connectors:	Codenoll POF connectors

Physical Properties

	Patch Cable	Drop Cable
Core/Clad Diameter (±6%)	980/1000 μm	980/1000 μm
Cable Dimensions	2.2 × 4.4 mm	4.0 × 5.5 mm
Weight	8 kg/km	23 kg/km

Cable Assembly Information

Type	Length	Part No.
Duplex Patch	1 meter	115-13-1001
Duplex Drop	2 meters	115-13-0002
Duplex Drop	5 meters	115-13-0005
Duplex Bulk	—	000-D-9999-BC

NOTE: A connectorization kit containing a complete tool set and supplies—housed in a heavy duty carrying case—is also available. This kit does not contain connectors, which are supplied separately.

For Information or Assistance

Call the Customer Support Group at Codenoll Technology, (914) 965-6300.

Specifications may change without notice. Codenoll, CodeNet CodeStar, MultiStar and CodeLED are registered trademarks of Codenoll Technology Corporation. Other trademarks mentioned are the property of other companies.

MAKING LIGHT WORK OF NETWORKING

1086 NORTH BROADWAY, YONKERS, NEW YORK 10701 USA • TELEPHONE: (914) 965-6300 • FAX (914) 965-9811

© 1990, Codenoll Technology Corporation
Printed in U.S.A.—December, 1990

Pioneer's Product Categories

BATTERIES
Tadiran

CABINETS & RACKS
Bud

CAPACITORS
Cornell-Dubilier
Kemet
Mallory
Nichicon

CONNECTORS
Ansley/T&B
Augat
Dale
Holmberg/T&B
Packard Electric
Panduit

CRYSTALS
M-Tron

DISPLAYS
Dale
Futaba

FANS & BLOWERS
Nidec-Torin

FILTERS
Corcom

FUSES & BREAKERS
Littelfuse
Potter & Brumfield

HARDWARE
Panduit
Thermalloy
Vector

INSTRUMENTATION
A & M Instruments
B&K-Precision/Maxtec
Beckman
Fluke
Hitachi
Kessler-Ellis
Leader
LFE (API)
Newport
Simpson

INSTRUMENTATION cont.
Triplett
Westcon (Sycon)
Yokogawa/GE

MOTORS
Oriental Motor
Superior Electric

OPTOELECTRONICS
Motorola Semiconductor
Optek Technology
Quality Technologies
Three-Five Systems (III-V)

LAMPS
Sylvania

POTENTIOMETERS & CONTROLS
Bourns
Clarostat
Dale
Techno

POWER SUPPLIES & TRANSFORMERS
Dale
Power-One
Sola
Superior Electric
Triad

RELAYS
Clare Relays
Opto 22
Potter & Brumfield
Teledyne

RESISTORS & NETWORKS
Bourns
Clarostat
Dale
Techno

SEMICONDUCTORS
Actel
Altera
Cirrus Logic
General Instrument
General Semiconductor
Intel
International Rectifier